Studies in Systems, Decision and Control

Volume 380

The series "Studies in Systems, Decision and Control" (SSDC) covers both new developments and advances, as well as the state of the art, in the various areas of broadly perceived systems, decision making and control–quickly, up to date and with a high quality. The intent is to cover the theory, applications, and perspectives on the state of the art and future developments relevant to systems, decision making, control, complex processes and related areas, as embedded in the fields of engineering, computer science, physics, economics, social and life sciences, as well as the paradigms and methodologies behind them. The series contains monographs, textbooks, lecture notes and edited volumes in systems, decision making and control spanning the areas of Cyber-Physical Systems, Autonomous Systems, Sensor Networks, Control Systems, Energy Systems, Automotive Systems, Biological Systems, Vehicular Networking and Connected Vehicles, Aerospace Systems, Automation, Manufacturing, Smart Grids, Nonlinear Systems, Power Systems, Robotics, Social Systems, Economic Systems and other. Of particular value to both the contributors and the readership are the short publication timeframe and the world-wide distribution and exposure which enable both a wide and rapid dissemination of research output.

Indexed by SCOPUS, DBLP, WTI Frankfurt eG, zbMATH, SCImago.

All books published in the series are submitted for consideration in Web of Science.

More information about this series at http://www.springer.com/series/13304

Nardênio Almeida Martins ·
Douglas Wildgrube Bertol

Wheeled Mobile Robot Control

Theory, Simulation, and Experimentation

Springer

Nardênio Almeida Martins
Department of Informatics
State University of Maringá
Maringá, Paraná, Brazil

Douglas Wildgrube Bertol
Department of Electrical Engineering
Universidade do Estado de Santa Catarina
Joinville, Santa Catarina, Brazil

ISSN 2198-4182 ISSN 2198-4190 (electronic)
Studies in Systems, Decision and Control
ISBN 978-3-030-77911-5 ISBN 978-3-030-77912-2 (eBook)
https://doi.org/10.1007/978-3-030-77912-2

This Springer imprint is published by the registered company Springer Nature Switzerland AG
The registered company address is: Gewerbestrasse 11, 6330 Cham, Switzerland

Theory without practice is useless, but practice without theory is naive.

When you are learning, the teacher acts just like a needle; the student is the thread. As your mentor, I can help you by pointing you in the right direction. But, like the needle in the thread, I must separate myself from you in the end, because the strength, the fiber, and the ability to bring all the parts together must be yours.

—Lance H. K. Secretan

To our families and all who believe, persist, and seek the fulfillment of their dreams.

To the treasures of my life, Maria Madalena, Natasha, Nícolas, and my second mother, Augusta de Oliveira Dias (⋆ 08/28/1928 – † 11/18/2020), a simple, wise, patient, tolerant and lovely person that I knew, lived and learned to admire, respect and love.

—Nardênio Almeida Martins

To Jaqueline, Regina, and Nubio.

—Douglas Wildgrube Bertol

Preface

In this book, differential-drive wheeled nonholonomic mobile robots (DWMRs) are considered to be the object of study because they are easy to implement physically (simple mechanical structure and low manufacturing cost), as well as serving several applications, as they adapt relatively well to operating conditions in indoor and outdoor paved environments, in which the soil irregularities are not very severe. Besides, DWMRs have awakened the scientific and technological interests of the international scientific community in the control area due to their simple kinematic model and, mainly, to their nonholonomic restrictions (bidirectional movement). Since DWMRs belong to a class of nonlinear, multivariable, underactuated, nonholonomic systems and, in practice, have uncertainties and/or disturbances, these constitute quite challenging control problems. Uncertainties and/or disturbances can be parametric or structured and nonparametric or unstructured. Among the control problems, the trajectory tracking control problem for DWMRs is treated as an object of study in this book, as it is particularly relevant in practical applications, considering that the control modules of the DWMR must, in general, follow a path previously planned and collision-free. However, as the control problems of the DWMRs are closely related to the kinematic and dynamic modeling and, consequently, with the difficulty in obtaining adequate models for the control purposes, it is considered necessary to elaborate, develop, and implement control projects based on Lyapunov's stability theory that are robust to compensate for the damaging effects of the incidence/influence of uncertainties and/or disturbances. Thus, the control system designs covered in this book use both the kinematic and dynamic models.

For the design of robust kinematic controllers, the first-order sliding control technique is used as a basis to deal with the problems of uncertainties and/or disturbances in the kinematic model in rectangular coordinates. The choice of this control technique is due to its fast response, its good performance in a transient regime, and its robustness concerning the uncertainties and/or disturbances that can be reached by the control inputs when in the sliding mode. The drawbacks of this control technique are the vibration phenomenon due to the discontinuous control portion and the need to know the limits of the disturbances to apply the adequate gain to compensate for them, thus saving the actuators from unnecessary efforts. To circumvent these drawbacks, continuous approximation methods are used, including soft computing techniques, specifically fuzzy logic and artificial neural networks, culminating in techniques of first-order quasi-sliding mode control, adaptive fuzzy sliding mode control, and adaptive neural sliding mode control. In addition to these drawbacks, sliding mode control projects have difficulty selecting sliding surfaces due to the lack of methodologies. In this book, an adequate choice of sliding surfaces is made based on the behavior of the reduced dynamics of the system using the equivalent control method to meet the desired performance and robustness requirements.

Most dynamic controllers of DWMRs generate torques as a control signal; however, commercial DWMRs usually receive velocity signals as control and not torques, as is the case of the following DWMRs: the PowerBot and Pioneer from Mobile Robots Inc., the Khepera from the company K-Team Corporation, the Merlin Miabot Pro Autonomous Mobile Robot from Merlin Systems Corporation, and the Soccer Robot from Microrobot. Therefore, in this book the dynamic model of the PowerBot DWMR is considered and the dynamic control is a PD control provided by the manufacturer, one of the reasons why this book deals with kinematic controller designs that are robust to the incidence of uncertainties and/or disturbances. Some of these robust kinematic controller designs are extended to introduce formation control of DWMRs using the leader–follower control strategy with the separation-bearing technique. Finally, to prove the effectiveness of the robust kinematic controllers contained in this book, simulations are carried out with MATLAB/Simulink software and/or with the MobileSim simulator, as well as experiments in the PowerBot DWMR to illustrate the successful practical application of the theory.

In short, this book focuses on the development and methodologies of trajectory control of DWMRs. The methodologies are based on kinematic models (posture and configuration) and dynamic models, both subject to uncertainties and/or disturbances. The control designs are developed in rectangular coordinates obtained from the first-order sliding mode control in combination with the use of soft computing techniques, such as fuzzy logic and artificial neural networks. Control laws, as well as online learning and adaptation laws, are obtained using the stability analysis for both the developed kinematic and dynamic controllers, based on Lyapunov's stability theory. An extension to the formation control with multiple DWMRs in trajectory tracking tasks is also provided. Results of simulations and experiments are presented to verify the effectiveness of the proposed control strategies for trajectory tracking situations, considering the parameters of an industrial and a research DWMR, the PowerBot.

The main objective of this book is the development of projects and the implementation of control systems for DWMRs in the performance of tasks of tracking trajectories with or without incidence of uncertainties and/or disturbances.

In addition to the main objective, other objectives are also established, such as:

- Address theoretical–scientific and practical aspects of DWMRs;
- Dealing with the mathematical modeling of the kinematics and dynamics of DWMRs;
- Present the design, implementation, and performance of kinematic and dynamic control systems and the stability analysis in trajectory tracking for engineering applications;

- Provide some methods of chattering phenomenon mitigation;
- Provide source codes of the control system designs that enable the developer to use creativity and imagination to modify, alter, or redo control system designs in both kinematic and dynamic scopes as well as proposing new control designs;
- Provide simulations in MATLAB/Simulink software, MobileSim simulator, and practical applications in DWMR PowerBot;
- Promote a tool for pedagogical applications to related areas.

This book is structured as follows. Chapter 1 contains an introduction to mobile robotics with an emphasis on the description and mathematical modeling of DWMRs, such as kinematic model, dynamic model, the inclusion of actuator dynamics, formulation of the dynamic state-space model, and controllability of the dynamic model. Chapter 2 provides the theoretical foundation on control problems and systems, introducing the trajectory tracking problem, PD dynamic control, posture error dynamics, robustness considerations, and generic model for nonlinear systems. Chapter 3 describes the simulation and experimentation environments, dealing with the implementation design considering data and information from the PowerBot DWMR, trajectory adopted, as well as the ideal, realistic, and experimental scenarios. It is emphasized that the contents of Chaps. 1–3 are deemed necessary for the development and understanding of the control projects covered in Chaps. 4–9. Chapter 4 deals with a control based on backstepping methodology very widespread in the technical–scientific literature that does not present aspects of robustness. Chapter 5 reports on the concept, characteristics, robustness, performance, and chattering phenomenon of the first-order sliding mode control and introduces four variants of this control technique. Chapter 6 considers four control variants by first-order quasi-sliding mode control, in which the chattering mitigation is treated using a continuous approximation method, i.e., fractional continuous approximation or proper continuous function, resulting in the loss of invariance property, but guaranteeing robustness. Chapter 7 discusses two variants of the adaptive fuzzy sliding mode control to mitigate the chattering phenomenon with the guarantee of robustness, whose difference between them is in the way in which the adaptation law of the consequences is elaborated by using a fuzzy system as a continuous approximation method. Chapter 8 presents an adaptive neural network sliding mode control, in which is used radial basis function neural network as continuous approximation method, thus mitigating the chattering phenomenon and ensuring robustness. Moreover, Chaps. 4–8 address the control design with the Lyapunov method for stability analysis, simulations using MATLAB/Simulink software and/or MobileSim simulator, experimental results using PowerBot DWMR, analysis and discussion of results, and general considerations. Chapter 9 extends one of the variants of the first-order sliding mode control, first-order quasi-sliding mode control, and adaptive fuzzy sliding mode control to the formation control of DWMRs by using the leader–follower strategy with the separation-bearing technique and considering the decentralized structure and homogeneous architecture. This chapter addresses the problem formulation, control design with Lyapunov method for stability analysis, simulations using MATLAB/Simulink software and/or MobileSim simulator, and general considerations.

The electronic supplementary material contains the applications made in each chapter of the book, such as:

- Simulation codes implemented in MATLAB/Simulink software;
- Scripts for generating graphics and visualizing the results;
- Source codes of the implemented control techniques.

These extra materials of the book can be found at http://extras.springer.com/, which are available to authorized users. The extra material is only distributed to Springer customers. We do not have the mentioned github repo.

Thus, this book is mainly intended for researchers, and undergraduate and graduate students in the areas of robotics with control applications, but it is also recommended for researchers, teachers, students, and professionals in the areas of computer science, informatics, mathematics, physics, electrical engineering, electronic engineering, mechanical engineering, computer engineering, control and automation engineering, mechatronics engineering, who are interested in the knowledge of the state-of-the-art theory and practice, in the control field applied in robotics, specifically the sliding mode control technique.

Maringá, Brazil
April 2021

Nardênio Almeida Martins
Douglas Wildgrube Bertol

Acknowledgements

We always thank God, especially for the health given to us. Our sincere wishes of gratitude to all those people who somehow contributed directly or indirectly to the development and conclusion of this book.

We, the authors, are not geniuses of humanity, but we are grateful for the phrases below from some of them, in this case, those are from Albert Einstein, which makes us have more motivation, dedication, discipline, perseverance, persistence, resilience, pride, and love in the exercise of our profession as a teacher to be even better people, as well as educating better citizens for humanity.

"Certainly my career was not determined by my own will, but by countless factors over which I have no control."

"When we accept our limits, we can go beyond them."

"We were unable to solve a problem based on the same reasoning used to create it."

"A person who has never made a mistake has never experienced anything new."

"Imagination is more important than knowledge. Knowledge is limited. Imagination surrounds the world."

"I have no special talent. I'm just passionately curious."

"Don't try to be a successful person. Try to be a person of value."

Maringá, Brazil
April 2021

Nardênio Almeida Martins
Douglas Wildgrube Bertol

Contents

About the Authors

Nardênio Almeida Martins obtained his M.Sc. in electrical engineering from the Federal University of Santa Catarina (1997) and Ph.D. in automation and systems engineering from the Federal University of Santa Catarina (2010). He is currently Associate Professor in the Department of Informatics and the Graduate Program in Computer Science at the State University of Maringá and Member of the research groups "Robotics" of the Department of Automation and Systems of the Federal University of Santa Catarina—Florianópolis Campus and the "Automation of Systems and Robotics Group" at the State University of Santa Catarina—Joinville Campus, working mainly on the following research topics in robotics: robot manipulators, joint space, operational space, wheeled mobile robots, trajectory tracking, adaptive control, robust control theory, formation control, neural networks, fuzzy logic, and Lyapunov stability theory.

Douglas Wildgrube Bertol obtained his M.Sc. in electrical engineering from the Federal University of Santa Catarina (2009) and Ph.D. in automation and systems engineering from the Federal University of Santa Catarina (2015). He is currently Associate Professor in the Department of Electrical Engineering and the Graduate Program in Electrical Engineering at the Universidade do Estado de Santa Catarina and Member of the Systems Automation and

Robotics Research Group (GASR) at the same university, working mainly in subjects of applied robotics, mobile robots, trajectory tracking, sliding mode control theory, formation control, neural networks, fuzzy logic, and Lyapunov stability theory.

Chapter 1
Background on DWMR System Modeling

1.1 Introduction

In mobile robot applications, three mechanisms of locomotion are well widespread: wheels, track plates, and legs. In this book, only the wheel locomotion mechanism will be considered, since it is the most widely used, easy to implement physically, and relatively well adaptive to the operating conditions in paved internal and external environments where ground irregularities are not very severe. For environments with marked irregularities and tasks that need obstacles transposition, the last two mechanisms of locomotion are more appropriate.

Mobile robots equipped with wheels that theoretically do not suffer deformations constitute a class of mechanical systems that are characterized by nonintegrable kinematic constraints (nonholonomic constraints) and therefore can not be eliminated from the equations of the model.

The derivation of the kinematic and/or dynamic models for differential wheeled mobile robots (DWMRs) are available in the literature [8, 14, 17] of mobile robots and for generic mobile robots equipped with wheels of various types. A systematic procedure for model derivation can be found in [1, 16]. Generally, the dynamic model of a DWMR is performed using the Newton-Euler method or the Lagrangian approach [9]. However, in [19, 20] the modeling of a DWMR is performed through the Kane approach, which emphasizes the advantages of this method for nonholonomic systems and presents the resulting dynamic model.

In [3, 4, 10], the DWMRs can be divided into four different models, namely: posture kinematic model, configuration kinematic model, configuration dynamic model, and the posture dynamic model. The kinematic models describe the DWMR by a function of the velocity and the orientation of the wheels, while the dynamic models describe the DWMR by a function of generalized forces applied by the actuators. The posture models consider as state variables only the position and orientation of the robot, unlike the configuration models which consider, besides the posture variables, other internal variables, such as angular displacement of the wheels. Moreover, an analysis of the structure of the kinematic and dynamic models of DWMRs are

N. A. Martins and D. W. Bertol, *Wheeled Mobile Robot Control*, Studies in Systems, Decision and Control 380, https://doi.org/10.1007/978-3-030-77912-2_1

performed, which are divided into five classes, characterized by the general structure of the equations of the models. Thus, for each class, structural properties of the kinematic and dynamic models are derived, considering mobility constraints.

In this chapter, the kinematic and dynamic DWMR models used in the course of the book are presented and for ease of understanding see [3, 4, 10], which describes a systematic procedure for derivation of kinematic and dynamic models (using the Lagrangian approach) for DWMRs, taking into account mobility constraints and only conventional fixed wheels. Still, it is important to emphasize that the posture models of the DWMR are used as the basis for the synthesis of the controllers only, while configuration models of the DWMR are considered in the simulations realized in Matlab/Simulink software only.

1.2 DWMR Modeling and Description

The DWMR, shown in Fig. 1.1, is a typical example of a nonholonomic mechanical system. This system consists of a rigid body (base) having two conventional fixed wheels driven by independent actuators (for example, direct current motors or DC motors) to perform the movement and orientation, and a third wheel that rotates freely (passive wheel) whose function is only to support the DWMR, and their effects are negligible in the DWMR dynamics.

The posture vector is characterized by the triple $\boldsymbol{\xi} = [x\ y\ \theta]^T$ where x and y are the coordinates of the point C (which is also the mass center or guidance point) in the inertial coordinate system OX_OY_O and θ is the orientation angle of the mass center coordinate system of the DWMR CX_CY_C concerning to the inertial coordinate system OX_OY_O.

The following parameter notation is used in the DWMR study presented in Fig. 1.1 and listed in Table 1.1.

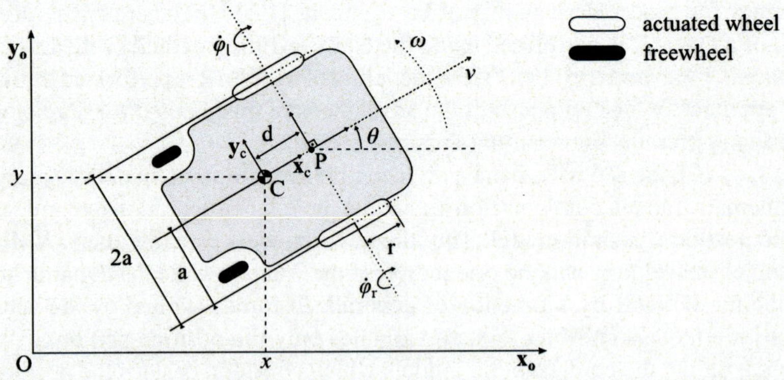

Fig. 1.1 DWMR and coordinate systems

Table 1.1 DWMR parameters

Parameter	Description
P	Intersection of the symmetry axis with the axis of the wheels
C	Mass center or guidance point
d	Distance between C and P
r	Right and left wheel radius
2a	Distance between the actuated wheels and the symmetry axis
m_c	Mass of the DWMR without wheels and motors
m_w	Mass of each wheel and motor assembly
m_t	Total mass of the DWMR
I_c	Moment of inertia of the DWMR without wheels and motors about the vertical axis through P
I_w	Moment of inertia of each wheel and motor about the wheel axis
I_m	Moment of inertia of each wheel and motor about the vertical axis parallel the wheel plane
I	Total inertia moment of the DWMR
$\dot{\varphi}_r, \dot{\varphi}_l$	Angular velocity of the right and left wheels
υ, ω	Linear and angular velocities of DWMR
$\boldsymbol{q}$	Generalized coordinate vector

1.2.1 Kinematic Model

The local coordinates of mechanical systems can be described in terms of the generalized coordinate vector $\boldsymbol{q}$, being $\boldsymbol{q} = [q_1\ q_2\ \ldots\ q_n]^T \in \Re^n$. In many situations, the movement of mechanical systems is subject to various constraints that are permanently satisfied during movement and that take the form of algebraic relations between positions and velocities of points of the system [5].

The DWMR shown in Fig. 1.1 presents three kinematic constraints [3, 4, 7, 10, 21]. The first constraint is that the DWMR can not slide sideways (non-slipping constraint), i.e., only move in the normal direction to the symmetry axis of the actuated wheels. This constraint can be written as:

$$\dot{y}\cos(\theta) - \dot{x}\sin(\theta) = 0, \qquad \text{for C} = \text{P}, \tag{1.1}$$

$$\dot{y}\cos(\theta) - \dot{x}\sin(\theta) - \text{d}\dot{\theta} = 0, \qquad \text{for C} \neq \text{P}. \tag{1.2}$$

The other two constraints are related to the rotation of the wheels (pure rolling constraints), i.e., the actuated wheels can not rotate in false, and are given by:

$$\dot{x}\cos(\theta) + \dot{y}\sin(\theta) + \text{a}\dot{\theta} - \text{r}\dot{\varphi}_r = 0, \qquad \text{for C} = \text{P and C} \neq \text{P}, \tag{1.3}$$

$$\dot{x}\cos(\theta) + \dot{y}\sin(\theta) - \mathrm{a}\dot{\theta} - \mathrm{r}\dot{\varphi}_{\mathrm{l}} = 0, \qquad \text{for C} = \text{P and C} \neq \text{P}, \tag{1.4}$$

where φ_{r} and φ_{l} are angular displacements of the right and left wheels, respectively.

Equations (1.3) and (1.4) can be rewritten as follows:

$$\upsilon + \mathrm{a}\omega = \mathrm{r}\dot{\varphi}_{\mathrm{r}}, \qquad \text{for C} = \text{P and C} \neq \text{P}, \tag{1.5}$$

$$\upsilon - \mathrm{a}\omega = \mathrm{r}\dot{\varphi}_{\mathrm{l}}, \qquad \text{for C} = \text{P and C} \neq \text{P}, \tag{1.6}$$

since

$$\upsilon = \dot{x}\cos(\theta) + \dot{y}\sin(\theta), \tag{1.7}$$

$$\omega = \dot{\theta}. \tag{1.8}$$

Using Eqs. (1.5) and (1.6), one can then correlate the right and left angular velocities of the DWMR ($\dot{\varphi}_{\mathrm{r}}$ and $\dot{\varphi}_{\mathrm{l}}$) with the linear and angular velocities of the mass center of the DWMR (υ and ω), resulting in:

$$\begin{bmatrix} \dot{\varphi}_{\mathrm{r}} \\ \dot{\varphi}_{\mathrm{l}} \end{bmatrix} = \boldsymbol{\Omega} \begin{bmatrix} \upsilon \\ \omega \end{bmatrix} = \begin{bmatrix} \frac{1}{\mathrm{r}} & \frac{\mathrm{a}}{\mathrm{r}} \\ \frac{1}{\mathrm{r}} & -\frac{\mathrm{a}}{\mathrm{r}} \end{bmatrix} \begin{bmatrix} \upsilon \\ \omega \end{bmatrix}, \tag{1.9}$$

and vice versa:

$$\begin{bmatrix} \upsilon \\ \omega \end{bmatrix} = \boldsymbol{\Omega}^{-1} \begin{bmatrix} \dot{\varphi}_{\mathrm{r}} \\ \dot{\varphi}_{\mathrm{l}} \end{bmatrix} = \begin{bmatrix} \frac{\mathrm{r}}{2} & \frac{\mathrm{r}}{2} \\ \mathrm{b} & -\mathrm{b} \end{bmatrix} \begin{bmatrix} \dot{\varphi}_{\mathrm{r}} \\ \dot{\varphi}_{\mathrm{l}} \end{bmatrix}, \tag{1.10}$$

where $\mathrm{b} = \frac{\mathrm{r}}{2\mathrm{a}}$.

In most applications, kinematic constraints Eqs. (1.1)–(1.4) are linear relations with the generalized coordinate vector. Such relations can be described in matrix form as:

$$\boldsymbol{A}(\boldsymbol{q})\dot{\boldsymbol{q}} = 0. \tag{1.11}$$

Since the state vector is represented by five generalized coordinates,

$$\boldsymbol{q} = \left[\boldsymbol{\xi}^{\mathrm{T}}\ \boldsymbol{\varphi}^{\mathrm{T}}\right]^{\mathrm{T}} = \left[x\ y\ \theta\ \varphi_{\mathrm{r}}\ \varphi_{\mathrm{l}}\right]^{\mathrm{T}}, \tag{1.12}$$

the three constraints can be rewritten in the form of Eq. (1.11), i.e.,

$$\boldsymbol{A}(\boldsymbol{q})\dot{\boldsymbol{q}} = \begin{bmatrix} -\sin(\theta) & \cos(\theta) & 0 & 0 & 0 \\ -\cos(\theta) & -\sin(\theta) & -\mathrm{a} & \mathrm{r} & 0 \\ -\cos(\theta) & -\sin(\theta) & \mathrm{a} & 0 & \mathrm{r} \end{bmatrix} \begin{bmatrix} \dot{x} \\ \dot{y} \\ \dot{\theta} \\ \dot{\varphi}_{\mathrm{r}} \\ \dot{\varphi}_{\mathrm{l}} \end{bmatrix}, \qquad \text{for C} = \text{P}, \tag{1.13}$$

$$A(q)\dot{q} = \begin{bmatrix} -\sin(\theta) & \cos(\theta) & -\mathrm{d} & 0 & 0 \\ -\cos(\theta) & -\sin(\theta) & -\mathrm{a} & \mathrm{r} & 0 \\ -\cos(\theta) & -\sin(\theta) & \mathrm{a} & 0 & \mathrm{r} \end{bmatrix} \begin{bmatrix} \dot{x} \\ \dot{y} \\ \dot{\theta} \\ \dot{\varphi}_\mathrm{r} \\ \dot{\varphi}_\mathrm{l} \end{bmatrix}, \quad \text{for C} \neq \text{P}. \tag{1.14}$$

It is emphasized that the referential of DWMR velocity is given by the angular velocity of the right and left wheels ($\dot{\varphi}_\mathrm{r}$ and $\dot{\varphi}_\mathrm{l}$) respectively, i.e.,

$$v = \begin{bmatrix} \dot{\varphi}_\mathrm{r} \\ \dot{\varphi}_\mathrm{l} \end{bmatrix}. \tag{1.15}$$

If the mass and inertia of the wheels and motors are disregarded, it is assumed that the DWMR satisfies the pure rolling and non-slipping conditions [11]. Thus, the matrix $A(q)$ containing the nonholonomic constraints reduces to:

$$A(q) = \begin{bmatrix} -\sin(\theta) & \cos(\theta) & 0 \end{bmatrix}, \quad \text{for C} = \text{P}, \tag{1.16}$$

$$A(q) = \begin{bmatrix} -\sin(\theta) & \cos(\theta) & -\mathrm{d} \end{bmatrix}, \quad \text{for C} \neq \text{P}, \tag{1.17}$$

so that the displacements occur only in the direction of the symmetry axis of the actuated wheels and

$$q = \xi = \begin{bmatrix} x & y & \theta \end{bmatrix}^\mathrm{T}. \tag{1.18}$$

Also, it is pointed out that the referential velocity is supplied by the linear and angular velocities of the DWMR (υ and ω), i.e.,

$$v = \begin{bmatrix} \upsilon \\ \omega \end{bmatrix}. \tag{1.19}$$

It is important to emphasize that the system has, as a configuration space n, the generalized coordinate vector q and the number of constraints given by p, so that the velocity vector is of dimension $\mathrm{m} = \mathrm{n} - \mathrm{p}$, in this case, $\mathrm{m} = 2$ (corresponds to the degrees of freedom of the system).

Be the annihilator [1] of these constraints [5] the Jacobian matrix $S(q)$ of full rank $(\mathrm{n} - \mathrm{p})$ formed by a set of linearly independent and smooth vector fields distributed in the null space of $A(q)$, i.e.,

$$A(q)S(q) = 0. \tag{1.20}$$

According to Eqs. (1.11) and (1.20), it is possible to find an auxiliary velocity vector as a function of time $v \in \Re^{\mathrm{p}\times 1}$ such that for all t:

---

[1] Note that q is in the null space of $A(q)$. In the same way, all columns of the matrix $S(q)$ of Eq. (1.20) are also in the null space of $A(q)$, which justifies the term *annihilator*.

$$\dot{\boldsymbol{q}} = \boldsymbol{S}(\boldsymbol{q})\boldsymbol{v}. \tag{1.21}$$

The configuration kinematic model, represented by the matrix $\boldsymbol{S}(\boldsymbol{q})$, is given by:

- From Eqs. (1.13) and (1.14):

$$\boldsymbol{S}(\boldsymbol{q}) = \begin{bmatrix} \mathrm{b\,a}\cos(\theta) & \mathrm{b\,a}\cos(\theta) \\ \mathrm{b\,a}\sin(\theta) & \mathrm{b\,a}\sin(\theta) \\ \mathrm{b} & -\mathrm{b} \\ 1 & 0 \\ 0 & 1 \end{bmatrix}, \quad \text{for C} = \text{P}, \tag{1.22}$$

$$\boldsymbol{S}(\boldsymbol{q}) = \begin{bmatrix} \mathrm{b}(\mathrm{a}\cos(\theta) - \mathrm{d}\sin(\theta)) & \mathrm{b}(\mathrm{a}\cos(\theta) + \mathrm{d}\sin(\theta)) \\ \mathrm{b}(\mathrm{a}\sin(\theta) + \mathrm{d}\cos(\theta)) & \mathrm{b}(\mathrm{a}\sin(\theta) - \mathrm{d}\cos(\theta)) \\ \mathrm{b} & -\mathrm{b} \\ 1 & 0 \\ 0 & 1 \end{bmatrix}, \quad \text{for C} \neq \text{P}, \tag{1.23}$$

where $\mathrm{b} = \frac{\mathrm{r}}{2\mathrm{a}}$.

Similarly, the posture kinematic model (matrix $\boldsymbol{S}(\boldsymbol{q})$) results in:

- From Eqs. (1.16) and (1.17):

$$\boldsymbol{S}(\boldsymbol{q}) = \begin{bmatrix} \cos(\theta) & 0 \\ \sin(\theta) & 0 \\ 0 & 1 \end{bmatrix}, \quad \text{for C} = \text{P}, \tag{1.24}$$

$$\boldsymbol{S}(\boldsymbol{q}) = \begin{bmatrix} \cos(\theta) & -\mathrm{d}\sin(\theta) \\ \sin(\theta) & \mathrm{d}\cos(\theta) \\ 0 & 1 \end{bmatrix}, \quad \text{for C} \neq \text{P}. \tag{1.25}$$

Another way of describing the posture kinematic model Eqs. (1.24) and (1.25) is utilizing the DWMR velocity in terms of $\dot{x}$, $\dot{y}$, and $\dot{\theta}$ (see Fig. 1.2), i.e.,

$$\begin{cases} \dot{x} = \upsilon\cos(\theta) \\ \dot{y} = \upsilon\sin(\theta) \\ \dot{\theta} = \omega \end{cases}, \quad \text{for C} = \text{P}, \tag{1.26}$$

$$\begin{cases} \dot{x} = \upsilon\cos(\theta) - \omega\mathrm{d}\sin(\theta) \\ \dot{y} = \upsilon\sin(\theta) + \omega\mathrm{d}\cos(\theta) \\ \dot{\theta} = \omega \end{cases}, \quad \text{for C} \neq \text{P}. \tag{1.27}$$

It is important to emphasize that to determine the matrices $\boldsymbol{A}(\boldsymbol{q})$ and $\boldsymbol{S}(\boldsymbol{q})$, it is necessary to carry out the study of the relation between the wheels and the DWMR. An analysis of the deduction of these matrices can be found in [3, 4, 10].

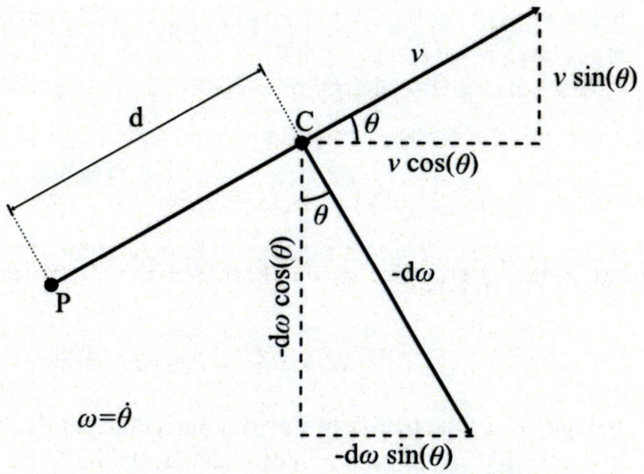

Fig. 1.2 Representation of the DWMR velocities of Fig. 1.1

1.2.2 Dynamic Model

The dynamics, following the Lagrange formalism, for the case of DWMRs with conventional fixed wheels, is given by:

$$\frac{d}{dt}\left(\frac{\partial T}{\partial \dot{\boldsymbol{\xi}}}\right) - \frac{\partial T}{\partial \boldsymbol{\xi}} = \boldsymbol{R}^{\mathrm{T}}(\theta)\boldsymbol{J}_{1\mathrm{f}}^{\mathrm{T}}\boldsymbol{\lambda} + \boldsymbol{R}^{\mathrm{T}}(\theta)\boldsymbol{C}_{1\mathrm{f}}^{\mathrm{T}}\boldsymbol{\mu}, \tag{1.28}$$

$$\frac{d}{dt}\left(\frac{\partial T}{\partial \dot{\boldsymbol{\varphi}}}\right) - \frac{\partial T}{\partial \boldsymbol{\varphi}} = \boldsymbol{J}_2^{\mathrm{T}}\boldsymbol{\lambda} + \boldsymbol{\tau}_{\varphi}, \tag{1.29}$$

where $\boldsymbol{R}(\theta)$ is the rotation matrix between the inertial coordinate system $\mathrm{OX_OY_O}$ and mass center coordinate system of the DWMR $\mathrm{CX_CY_C}$ (Fig. 1.1), which is given by:

$$\boldsymbol{R}(\theta) = \begin{bmatrix} \cos(\theta) & \sin(\theta) & 0 \\ -\sin(\theta) & \cos(\theta) & 0 \\ 0 & 0 & 1 \end{bmatrix}, \tag{1.30}$$

and $L = T - W$ is the Lagrangian of the system with kinetic energy T and potential energy W ($W = 0$, since the DWMR is always moving in a horizontal plane), $\boldsymbol{\lambda}$ and $\boldsymbol{\mu}$ are vectors with Lagrange multipliers associated with the kinematic constraints Eqs. (1.1)–(1.4), and $\boldsymbol{\tau}_{\varphi} = \boldsymbol{\tau}$ are the applied torques by the motors for wheel rotation. Matrices $\boldsymbol{J}_{1\mathrm{f}}, \boldsymbol{C}_{1\mathrm{f}}, \boldsymbol{J}_2$ are defined as in [3, 4, 10].

Equations (1.28) and (1.29) can be grouped, resulting in:

$$\frac{d}{dt}\left(\frac{\partial T}{\partial \dot{\boldsymbol{q}}}\right) - \frac{\partial T}{\partial \boldsymbol{q}} = -\boldsymbol{A}^{\mathrm{T}}(\boldsymbol{q})\boldsymbol{\rho} + \boldsymbol{E}(\boldsymbol{q})\boldsymbol{\tau}, \tag{1.31}$$

being $\boldsymbol{q}$ given by Eq. (1.12), $\boldsymbol{E}(\boldsymbol{q})$ is the n × m transformation matrix of the control inputs and $\boldsymbol{\rho} = [\boldsymbol{\lambda}\ \boldsymbol{\mu}]^{\mathrm{T}}$.

The total kinetic energy of DWMR [21] is expressed by:

$$T = \frac{1}{2}\dot{\boldsymbol{q}}^{\mathrm{T}}\boldsymbol{H}(\boldsymbol{q})\dot{\boldsymbol{q}}, \tag{1.32}$$

where $\boldsymbol{H}(\boldsymbol{q})$ is the n × n positive definite symmetric inertia matrix. Equation (1.31) can be rewritten as:

$$\boldsymbol{H}(\boldsymbol{q})\ddot{\boldsymbol{q}} + \boldsymbol{C}(\boldsymbol{q}, \dot{\boldsymbol{q}})\dot{\boldsymbol{q}} = -\boldsymbol{A}^{\mathrm{T}}(\boldsymbol{q})\boldsymbol{\rho} + \boldsymbol{E}(\boldsymbol{q})\boldsymbol{\tau}, \tag{1.33}$$

being $\boldsymbol{C}(\boldsymbol{q}, \dot{\boldsymbol{q}})$ the matrix of Coriolis and centripetal torques, dependent on the posture and velocity, and with n × n dimension, defined by:

$$\boldsymbol{C}(\boldsymbol{q}, \dot{\boldsymbol{q}})\dot{\boldsymbol{q}} = \frac{d}{dt}\left(\boldsymbol{H}(\boldsymbol{q})\right)\dot{\boldsymbol{q}} - \frac{1}{2}\frac{\partial}{\partial \boldsymbol{q}}\left(\dot{\boldsymbol{q}}^{\mathrm{T}}\boldsymbol{H}(\boldsymbol{q})\dot{\boldsymbol{q}}\right). \tag{1.34}$$

Therefore, knowing the total kinetic energy of the system Eq. (1.32), one can obtain the configuration dynamic model of the DWMR. The differences between configuration and posture models, for example, the models described by [11, 21], are interrelated with the terms considered in the kinetic energy expression T, as well as the number of constraints used.

Considering a dynamic system with uncertainties and disturbances, Eq. (1.33) takes the form:

$$\boldsymbol{H}(\boldsymbol{q})\ddot{\boldsymbol{q}} + \boldsymbol{C}(\boldsymbol{q}, \dot{\boldsymbol{q}})\dot{\boldsymbol{q}} + \boldsymbol{\tau}_{\mathrm{d}} = -\boldsymbol{A}^{\mathrm{T}}(\boldsymbol{q})\boldsymbol{\rho} + \boldsymbol{E}(\boldsymbol{q})\boldsymbol{\tau}, \tag{1.35}$$

with their matrices and vectors having the following denominations:

- $\boldsymbol{\tau}_{\mathrm{d}}$ – n × 1 vector denoting the uncertainties and disturbances, including modeling imprecisions; surface friction; external disturbances; bounded unknown disturbances; unmodeled and unstructured dynamics of actuators, sensors, and other electronic devices; integration errors; noises; operational, sampling, and physical limitations;
- $\boldsymbol{A}(\boldsymbol{q})$ – p × n matrix associated with nonholonomic constraints;
- $\boldsymbol{\rho}$ – p × 1 vector of constraint forces (Lagrange multipliers);
- $\boldsymbol{\tau}$ – p × 1 vector of control or input torques.

The DWMR dynamics of Eq. (1.35) have the following standard properties:

- *Property 1.1: Limitation*—The matrix $\boldsymbol{H}(\boldsymbol{q})$, the norm of matrix $\boldsymbol{C}(\boldsymbol{q}, \dot{\boldsymbol{q}})$ and the vector $\boldsymbol{\tau}_{\mathrm{d}}$ are bounded;
- *Property 1.2: Skew-symmetry*—The matrix $\dot{\boldsymbol{H}}(\boldsymbol{q}) - 2\boldsymbol{C}(\boldsymbol{q}, \dot{\boldsymbol{q}})$ is skew-symmetric. This property is particularly important in the stability analysis of the control system.

In the sequence, the description of the dynamic models of DWMRs obtained is carried out.

1.2.2.1 Dynamic Model Using the Movement Constraints of Each Wheel

Equations (1.28), (1.29), and (1.32) represent a dynamic model, considering the two conventional fixed wheels of the DWMR.

The total kinetic energy T is described as:

$$T = T_1 + T_2 + T_3, \tag{1.36}$$

where T_1 is the kinetic energy of the DWMR without the wheels, T_2 is the kinetic energy of the actuated wheels in the plane and T_3 is the kinetic energy of all the wheels considering the orthogonal plane.

The determination of T_1 can be obtained considering the DWMR velocities in the horizontal (υ_x) and vertical (υ_y), i.e., from Eq. (1.27) one has:

$$\begin{aligned} \upsilon_{\mathrm{x}} &= \dot{x} + \mathrm{d}\dot{\theta}\sin(\theta), \\ \upsilon_{\mathrm{y}} &= \dot{y} - \mathrm{d}\dot{\theta}\cos(\theta), \end{aligned} \tag{1.37}$$

therefore,

$$T_1 = \frac{1}{2}\mathrm{m_c}(\upsilon_{\mathrm{x}}^2 + \upsilon_{\mathrm{y}}^2) + \frac{1}{2}\mathrm{I_c}\dot{\theta}^2,$$

replacing,

$$T_1 = \frac{1}{2}\mathrm{m_c}\left(\dot{x}^2 + \dot{y}^2 - 2\dot{y}\mathrm{d}\dot{\theta}\cos(\theta) + 2\dot{x}\mathrm{d}\dot{\theta}\sin(\theta)\right) + \frac{1}{2}\mathrm{m_c}\mathrm{d}^2\dot{\theta}^2 + \frac{1}{2}\mathrm{I_c}\dot{\theta}^2. \tag{1.38}$$

The kinetic energy of the actuated wheels about the plane of the wheels, denoted as T_2, is calculated based on the right and left wheel velocities, ($\dot{x}_{\mathrm{D}}, \dot{y}_{\mathrm{D}}$) and ($\dot{x}_{\mathrm{E}}, \dot{y}_{\mathrm{E}}$), i.e.,

$$\begin{cases} \dot{x}_{\mathrm{D}} = \upsilon_{\mathrm{x}} + \mathrm{a}\dot{\theta}\cos(\theta) \\ \dot{y}_{\mathrm{D}} = \upsilon_{\mathrm{y}} + \mathrm{a}\dot{\theta}\sin(\theta) \end{cases}, \quad \begin{cases} \dot{x}_{\mathrm{E}} = \upsilon_{\mathrm{x}} - \mathrm{a}\dot{\theta}\cos(\theta) \\ \dot{y}_{\mathrm{E}} = \upsilon_{\mathrm{y}} - \mathrm{a}\dot{\theta}\sin(\theta) \end{cases},$$

soon,

$$T_2 = \frac{1}{2}\mathrm{m_w}(\dot{x}_{\mathrm{D}}^2 + \dot{y}_{\mathrm{D}}^2) + \frac{1}{2}\mathrm{m_w}(\dot{x}_{\mathrm{E}}^2 + \dot{y}_{\mathrm{E}}^2) + \frac{1}{2}\mathrm{I_{mD}}\dot{\theta}^2 + \frac{1}{2}\mathrm{I_{mE}}\dot{\theta}^2,$$

and after replacing terms,

$$T_2 = \mathrm{m_w}(\dot{x}^2 + \dot{y}^2 + 2\dot{x}\mathrm{d}\dot{\theta}\sin(\theta) - 2\dot{y}\mathrm{d}\dot{\theta}\cos(\theta)) + (\mathrm{m_w}(\mathrm{d}^2 + \mathrm{a}^2) + \mathrm{I_m})\dot{\theta}^2. \tag{1.39}$$

where $\mathrm{I_m} = \mathrm{I_m D} = \mathrm{I_m E}$.

Considering the moment of inertia of the wheels in relation to the axis that passes through them, the parcel T_3 results in:

$$T_3 = \frac{1}{2}\mathrm{I_w}\dot{\varphi}_r^2 + \frac{1}{2}\mathrm{I_w}\dot{\varphi}_l^2 = \frac{1}{2}\mathrm{I_w}(\dot{\varphi}_r^2 + \dot{\varphi}_l^2). \tag{1.40}$$

Thus, the total kinetic energy T is:

$$T = \frac{1}{2}\mathrm{m_t}(\dot{x}^2 + \dot{y}^2) + \mathrm{m_t}(\dot{x}\mathrm{d}\dot{\theta}\sin(\theta) - \dot{y}\mathrm{d}\dot{\theta}\cos(\theta)) + \frac{1}{2}\mathrm{I}\dot{\theta}^2 + \frac{1}{2}\mathrm{I_w}(\dot{\varphi}_r^2 + \dot{\varphi}_l^2), \tag{1.41}$$

where $\mathrm{m_t} = \mathrm{m_c} + 2\mathrm{m_w}$, $\mathrm{I} = \mathrm{m_c d^2} + \mathrm{I_c} + 2\mathrm{m_w}(\mathrm{d^2} + \mathrm{a^2}) + 2\mathrm{I_m}$ and noting that the $\dot{\theta} = \omega$, as in Eq. (1.8).

From Eqs. (1.31), (1.33), and (1.41), the configuration dynamic model as well as, neglecting the mass and inertia of the wheels and motors, the posture dynamic model is obtained.

1.2.2.2 Configuration Dynamic Model

In this model, the matrix $\boldsymbol{A(q)}$ is defined by Eq. (1.14), which summarizes the movement constraints of the DWMR assuming there is no slipping and the DWMR moves only in the direction of the symmetry axis.

The matrices $\boldsymbol{H(q)}$, $\boldsymbol{C(q, \dot{q})}$ and $\boldsymbol{E(q)}$ are given by:

$$\boldsymbol{H(q)} = \begin{bmatrix} \mathrm{m_t} & 0 & \mathrm{m_t d}\sin(\theta) & 0 & 0 \\ 0 & \mathrm{m_t} & -\mathrm{m_t d}\cos(\theta) & 0 & 0 \\ \mathrm{m_t d}\sin(\theta) & -\mathrm{m_t d}\cos(\theta) & \mathrm{I} & 0 & 0 \\ 0 & 0 & 0 & \mathrm{I_w} & 0 \\ 0 & 0 & 0 & 0 & \mathrm{I_w} \end{bmatrix}, \tag{1.42}$$

$$\boldsymbol{C(q, \dot{q})q} = \begin{bmatrix} \mathrm{m_t d}\dot{\theta}^2\cos(\theta) \\ \mathrm{m_t d}\dot{\theta}^2\sin(\theta) \\ 0 \\ 0 \\ 0 \end{bmatrix}, \tag{1.43}$$

$$\boldsymbol{E(q)} = \begin{bmatrix} 0 & 0 \\ 0 & 0 \\ 0 & 0 \\ 1 & 0 \\ 0 & 1 \end{bmatrix}. \tag{1.44}$$

Furthermore, generalized coordinate vector $\boldsymbol{q}$, velocity vector $\boldsymbol{v}$, and the Jacobian matrix $\boldsymbol{S(q)}$ are provided by Eqs. (1.12), (1.15) and (1.23), respectively.

Finally, the vector of torques $\boldsymbol{\tau}$ is given by:

$$\boldsymbol{\tau} = \begin{bmatrix} \tau_r \\ \tau_l \end{bmatrix}, \tag{1.45}$$

where τ_r and τ_l represent, respectively, the torques on the right and left wheels of the DWMR.

1.2.2.3 Posture Dynamic Model

In this model, the mass and inertia of all wheels and motors are disregarded. Due to this reason, the matrix $\boldsymbol{A}(\boldsymbol{q})$ is reduced and given by Eq. (1.17), so that the displacements occur only in the direction of the symmetry axis.

Matrices $\boldsymbol{H}(\boldsymbol{q})$, $\boldsymbol{C}(\boldsymbol{q}, \dot{\boldsymbol{q}})$ and $\boldsymbol{E}(\boldsymbol{q})$ are defined as:

$$\boldsymbol{H}(\boldsymbol{q}) = \begin{bmatrix} m_c & 0 & m_c d \sin(\theta) \\ 0 & m_c & -m_c d \cos(\theta) \\ m_c d \sin(\theta) & -m_c d \cos(\theta) & I \end{bmatrix}, \tag{1.46}$$

$$\boldsymbol{C}(\boldsymbol{q}, \dot{\boldsymbol{q}})\boldsymbol{q} = \begin{bmatrix} m_c d \dot{\theta}^2 \cos(\theta) \\ m_c d \dot{\theta}^2 \sin(\theta) \\ 0 \end{bmatrix}, \tag{1.47}$$

$$\boldsymbol{E}(\boldsymbol{q}) = \begin{bmatrix} \frac{\cos(\theta)}{r} & \frac{\cos(\theta)}{r} \\ \frac{\sin(\theta)}{r} & \frac{\sin(\theta)}{r} \\ \frac{a}{r} & -\frac{a}{r} \end{bmatrix}. \tag{1.48}$$

The velocity vector $\boldsymbol{v}$, the generalized coordinate vector $\boldsymbol{q}$, and the Jacobian matrix $\boldsymbol{S}(\boldsymbol{q})$ are each given by Eqs. (1.18), (1.19), and (1.25), respectively.

In the case of using angular velocity of the right and left wheels, Eq. (1.15), the Jacobian matrix $\boldsymbol{S}(\boldsymbol{q})$ is given by:

$$\boldsymbol{S}(\boldsymbol{q}) = \begin{bmatrix} b(a\cos(\theta) - d\sin(\theta)) & b(a\cos(\theta) + d\sin(\theta)) \\ b(a\sin(\theta) + d\cos(\theta)) & b(a\sin(\theta) - d\cos(\theta)) \\ b & -b \end{bmatrix}, \quad \text{for } C \neq P, \tag{1.49}$$

which is obtained from the multiplication of Eq. (1.25) by the matrix $\boldsymbol{\Omega}^{-1}$ of Eq. (1.10).

The vector of torques $\boldsymbol{\tau}$ is given by:

$$\boldsymbol{\tau} = \begin{bmatrix} \tau_r \\ \tau_l \end{bmatrix}, \tag{1.50}$$

representing, as in [21], the torques in the right and left wheels of the DWMR (τ_r and τ_l). However, these control signals are converted to torques in the directions of x and y, and a component perpendicular to the direction of displacement of the DWMR, since this model disregards the effect of the wheels. That is, as variables φ_r and φ_l are not mentioned, the torques applied τ_r and τ_l are converted to the unity of force and projected in the directions of x e y.

1.2.3 Inclusion of Actuator Dynamics

Robots are driven by actuators that can be electrical, hydraulic, pneumatic, etc. [15]. In the case of DWMRs, the wheels are driven by DC motors that have both mechanical and electrical dynamics. Considering a dynamic system with nonholonomic constraints including the actuator dynamics without neglecting the armature inductance, the dynamics can be described by the following formulation:

$$\dot{\boldsymbol{q}} = \boldsymbol{S}(\boldsymbol{q})\boldsymbol{v}, \tag{1.51}$$

$$\boldsymbol{H}(\boldsymbol{q})\ddot{\boldsymbol{q}} + \boldsymbol{C}(\boldsymbol{q},\dot{\boldsymbol{q}})\dot{\boldsymbol{q}} + \boldsymbol{\tau}_{\mathrm{d}} = -\boldsymbol{A}^{\mathrm{T}}(\boldsymbol{q})\boldsymbol{\rho} + \boldsymbol{E}(\boldsymbol{q})\boldsymbol{\tau}, \tag{1.52}$$

$$\mathbf{L}_{\mathrm{a}}\frac{d\boldsymbol{i}_{\mathrm{a}}}{dt} + \mathbf{R}_{\mathrm{a}}\boldsymbol{i}_{\mathrm{a}} + \mathbf{K}_{\mathrm{e}}\dot{\boldsymbol{\varphi}}_{\mathrm{o}} + \boldsymbol{\tau}_{\mathrm{e}} = \boldsymbol{u}, \tag{1.53}$$

where $\boldsymbol{q}$, $\boldsymbol{H}(\boldsymbol{q})$, $\boldsymbol{C}(\boldsymbol{q},\dot{\boldsymbol{q}})$, $\boldsymbol{A}(\boldsymbol{q})$, $\boldsymbol{\rho}$, $\boldsymbol{E}(\boldsymbol{q})$ and $\boldsymbol{\tau}$ are defined as in Eq. (1.35); $\boldsymbol{i}_{\mathrm{a}} = [i_{\mathrm{a}_1} \ \dots \ i_{\mathrm{a}_\mathrm{m}}]^{\mathrm{T}}$ denotes the vector of armature currents; $\mathrm{L_a} = \mathrm{diag}([\mathrm{L}_{\mathrm{a}_1} \ \dots \ \mathrm{L}_{\mathrm{a}_\mathrm{m}}])$ represent the armature inductances; $\mathrm{R_a} = \mathrm{diag}([\mathrm{R}_{\mathrm{a}_1} \ \dots \ \mathrm{R}_{\mathrm{a}_\mathrm{m}}])$ are armature resistances; $\mathrm{K_e} = \mathrm{diag}([\mathrm{K}_{\mathrm{e}_1} \ \dots \ \mathrm{K}_{\mathrm{e}_\mathrm{m}}])$ back electromotive force constants; $\dot{\boldsymbol{\varphi}}_{\mathrm{o}} = [\dot{\varphi}_{\mathrm{o}_1} \ \dots \ \dot{\varphi}_{\mathrm{o}_\mathrm{m}}]^{\mathrm{T}}$ motor angular velocities; $\boldsymbol{\tau}_{\mathrm{e}} = [\tau_{\mathrm{e}_1} \ \dots \ \tau_{\mathrm{e}_\mathrm{m}}]^{\mathrm{T}}$ electrical disturbances; and $\boldsymbol{u} = [u_1 \ \dots \ u_\mathrm{m}]^{\mathrm{T}}$ control input voltages. In Fig. 1.3 the configuration of this system is shown. To fully operate the nonholonomic system, $\boldsymbol{E}(\boldsymbol{q})$ is assumed to be a full rank matrix and $\mathrm{m} \geq \mathrm{n} - \mathrm{p}$.

The angular velocities of the motors $\dot{\boldsymbol{\varphi}}_{\mathrm{o}}$ and the corresponding angular velocities of the wheels $\dot{\boldsymbol{\varphi}}$ are related by the transmission rate of the gears N as:

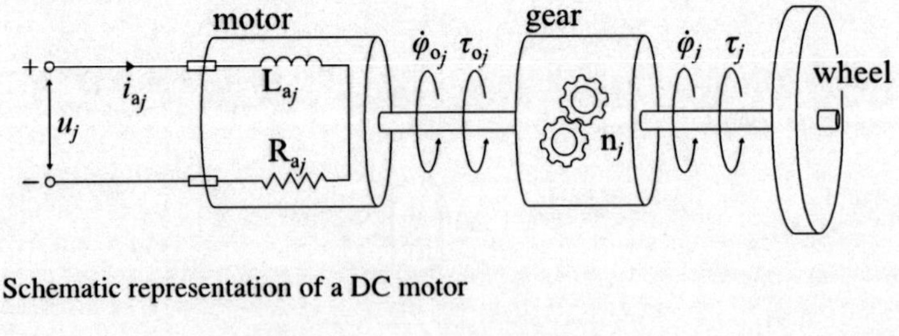

Fig. 1.3 Schematic representation of a DC motor

$$\dot{\boldsymbol{\varphi}} = \frac{\dot{\boldsymbol{\varphi}}_{\mathbf{o}}}{\mathbf{N}}, \tag{1.54}$$

where $\mathbf{N} = \mathrm{diag}([\mathrm{n}_1\ \ldots\ \mathrm{n}_\mathrm{m}])$ and the torques generated by the motors $\boldsymbol{\tau}_\mathrm{o}$ are related to the torques of the wheels $\boldsymbol{\tau}$ as:

$$\boldsymbol{\tau} = \mathbf{N}\boldsymbol{\tau}_\mathrm{o} = \mathbf{N}\mathbf{K}_\tau \boldsymbol{i}_\mathrm{a}, \tag{1.55}$$

where $\mathbf{K}_\tau$ is a positive definite diagonal matrix that characterizes the electromechanical conversion between current and torque (motor torque constant).

The relationship between the angular velocities of the wheels $\dot{\boldsymbol{\varphi}}$ and the velocities of the DWMR $\boldsymbol{v}$ is dependent on the type of mechanical system and can generally be expressed as:

$$\dot{\boldsymbol{\varphi}} = \boldsymbol{\Omega}\boldsymbol{v}. \tag{1.56}$$

The structure of $\boldsymbol{\Omega}$ depends on the mechanical system to be controlled. In the case of a DWMR model, $\boldsymbol{\Omega}$ is the same as in Eq. (1.9). Eliminating $\dot{\boldsymbol{\varphi}}_\mathrm{m}$ from the actuator dynamics Eq. (1.53) by substituting Eqs. (1.54) and (1.56), as well as by substituting Eqs. (1.55) and (1.52) to obtain the following dynamics:

$$\dot{\boldsymbol{q}} = \boldsymbol{S}(\boldsymbol{q})\boldsymbol{v}, \tag{1.57}$$

$$\boldsymbol{H}(\boldsymbol{q})\ddot{\boldsymbol{q}} + \boldsymbol{C}(\boldsymbol{q}, \dot{\boldsymbol{q}})\dot{\boldsymbol{q}} + \boldsymbol{\tau}_\mathrm{d} = -\boldsymbol{A}^\mathrm{T}(\boldsymbol{q})\boldsymbol{\rho} + \boldsymbol{E}(\boldsymbol{q})\mathbf{N}\mathbf{K}_\tau \boldsymbol{i}_\mathrm{a}, \tag{1.58}$$

$$\mathbf{L}_\mathrm{a}\frac{d\boldsymbol{i}_\mathrm{a}}{dt} + \mathbf{R}_\mathrm{a}\boldsymbol{i}_\mathrm{a} + \mathbf{K}_\mathrm{e}\mathbf{N}\boldsymbol{\Omega}\boldsymbol{v} + \boldsymbol{\tau}_\mathrm{e} = \boldsymbol{u}. \tag{1.59}$$

Frequently, the electrical time constants of the armature circuits of the motors are faster (or much smaller) than the mechanical time constants, i.e., $\frac{\mathbf{L}_\mathrm{a}}{\mathbf{R}_\mathrm{a}} \approx 0$. In this case, it is appropriate to neglect the armature inductances of the motors to rewrite the dynamics as:

$$\dot{\boldsymbol{q}} = \boldsymbol{S}(\boldsymbol{q})\boldsymbol{v}, \tag{1.60}$$

$$\boldsymbol{H}(\boldsymbol{q})\ddot{\boldsymbol{q}} + \boldsymbol{C}(\boldsymbol{q}, \dot{\boldsymbol{q}})\dot{\boldsymbol{q}} + \boldsymbol{\tau}_\mathrm{d} = -\boldsymbol{A}^\mathrm{T}(\boldsymbol{q})\boldsymbol{\rho} + \boldsymbol{E}(\boldsymbol{q})\mathbf{N}\mathbf{K}_\tau \boldsymbol{i}_\mathrm{a}, \tag{1.61}$$

$$\mathbf{R}_\mathrm{a}\boldsymbol{i}_\mathrm{a} + \mathbf{K}_\mathrm{e}\mathbf{N}\boldsymbol{\Omega}\boldsymbol{v} + \boldsymbol{\tau}_\mathrm{e} = \boldsymbol{u}, \tag{1.62}$$

or

$$\dot{\boldsymbol{q}} = \boldsymbol{S}(\boldsymbol{q})\boldsymbol{v}, \tag{1.63}$$

$$\begin{aligned} \boldsymbol{H}(\boldsymbol{q})\ddot{\boldsymbol{q}} + \boldsymbol{C}(\boldsymbol{q}, \dot{\boldsymbol{q}})\dot{\boldsymbol{q}} + \boldsymbol{\tau}_\mathrm{d} = \boldsymbol{E}(\boldsymbol{q})\tfrac{\mathbf{N}\mathbf{K}_\tau}{\mathbf{R}_\mathrm{a}}\boldsymbol{u} - \tfrac{\mathbf{N}^2\mathbf{K}_\tau\mathbf{K}_\mathrm{e}}{\mathbf{R}_\mathrm{a}}\boldsymbol{E}(\boldsymbol{q})\boldsymbol{\Omega}\boldsymbol{v} \\ -\tfrac{\mathbf{N}\mathbf{K}_\tau}{\mathbf{R}_\mathrm{a}}\boldsymbol{E}(\boldsymbol{q})\boldsymbol{\tau}_\mathrm{e} - \boldsymbol{A}^\mathrm{T}(\boldsymbol{q})\boldsymbol{\rho}. \end{aligned} \tag{1.64}$$

1.2.4 Formulation of the Dynamic State-Space Model

The dynamic system of Eq. (1.35) can suffer a transformation of coordinates to obtain a more appropriate representation for control purposes. Differentiating Eq. (1.21), substituting this result into Eq. (1.35), and then pre-multiplying by $\boldsymbol{S}^{\mathrm{T}}(\boldsymbol{q})$, one can eliminate the term containing the constraint matrix, i.e., $\boldsymbol{A}(\boldsymbol{q})\boldsymbol{\rho}$. Therefore, the motion equation of the DWMR is given by:

$$\begin{aligned}\boldsymbol{S}^{\mathrm{T}}(\boldsymbol{q})\boldsymbol{H}(\boldsymbol{q})\boldsymbol{S}(\boldsymbol{q})\dot{\boldsymbol{v}} + \boldsymbol{S}^{\mathrm{T}}(\boldsymbol{q})\left[\boldsymbol{H}(\boldsymbol{q})\dot{\boldsymbol{S}}(\boldsymbol{q}) + \boldsymbol{C}(\boldsymbol{q},\dot{\boldsymbol{q}})\boldsymbol{S}(\boldsymbol{q})\right]\boldsymbol{v} \\ +\boldsymbol{S}^{\mathrm{T}}(\boldsymbol{q})\boldsymbol{\tau}_{\mathrm{d}} = \boldsymbol{S}^{\mathrm{T}}(\boldsymbol{q})\boldsymbol{E}(\boldsymbol{q})\boldsymbol{\tau}.\end{aligned} \tag{1.65}$$

By appropriate definitions, one can rewrite the Eq. (1.65) as follows:

$$\bar{\boldsymbol{H}}(\boldsymbol{q})\dot{\boldsymbol{v}} + \bar{\boldsymbol{C}}(\boldsymbol{q},\dot{\boldsymbol{q}})\boldsymbol{v} + \bar{\boldsymbol{\tau}}_{\mathrm{d}} = \bar{\boldsymbol{E}}(\boldsymbol{q})\boldsymbol{\tau} = \bar{\boldsymbol{\tau}}. \tag{1.66}$$

Equation (1.66) can be represented as:

$$\dot{\boldsymbol{\chi}} = \boldsymbol{f}(\boldsymbol{\chi}) + \boldsymbol{g}(\boldsymbol{\chi})\boldsymbol{\tau},$$

explicitly,

$$\dot{\boldsymbol{\chi}} = \begin{bmatrix} \boldsymbol{S}(\boldsymbol{q})\boldsymbol{v} \\ -\bar{\boldsymbol{H}}^{-1}(\boldsymbol{q})\left(\bar{\boldsymbol{C}}(\boldsymbol{q},\dot{\boldsymbol{q}})\boldsymbol{v} + \bar{\boldsymbol{\tau}}_{\mathrm{d}}\right)\end{bmatrix} + \begin{bmatrix}\boldsymbol{0} \\ \bar{\boldsymbol{H}}^{-1}(\boldsymbol{q})\bar{\boldsymbol{E}}(\boldsymbol{q})\end{bmatrix}\boldsymbol{\tau}, \tag{1.67}$$

where $\boldsymbol{\chi} = [\boldsymbol{q}\ \boldsymbol{v}]^{\mathrm{T}}$.

Equation (1.66) describes the behavior of the nonholonomic system in a new set of local coordinates, i.e., $\boldsymbol{S}(\boldsymbol{q})$ is a Jacobian matrix that transforms the velocities $\boldsymbol{v}$ in coordinates of the mobile base to the velocities $\dot{\boldsymbol{q}}$ in Cartesian coordinates. Therefore, the properties of the original dynamics are maintained for the new set of coordinates [15], i.e.,

- *Property 1.3: Limitation*—The matrix $\bar{\boldsymbol{H}}(\boldsymbol{q})$, the norm of matrix $\bar{\boldsymbol{C}}(\boldsymbol{q},\dot{\boldsymbol{q}})$ and the vector $\bar{\boldsymbol{\tau}}_{\mathrm{d}}$ are all bounded;
- *Property 1.4: Skew-symmetry*—The matrix $\dot{\bar{\boldsymbol{H}}}(\boldsymbol{q}) - 2\bar{\boldsymbol{C}}(\boldsymbol{q},\dot{\boldsymbol{q}})$ is skew-symmetric.

Proof: The derivative of the inertia matrix and the matrix of Coriolis and centripetal torques are given by:

$$\begin{aligned}\dot{\bar{\boldsymbol{H}}}(\boldsymbol{q}) &= \dot{\boldsymbol{S}}^{\mathrm{T}}(\boldsymbol{q})\boldsymbol{H}(\boldsymbol{q})\boldsymbol{S}(\boldsymbol{q}) + \boldsymbol{S}^{\mathrm{T}}(\boldsymbol{q})\dot{\boldsymbol{H}}(\boldsymbol{q})\boldsymbol{S}(\boldsymbol{q}) + \boldsymbol{S}^{\mathrm{T}}(\boldsymbol{q})\boldsymbol{H}(\boldsymbol{q})\dot{\boldsymbol{S}}(\boldsymbol{q}), \\ \bar{\boldsymbol{C}}(\boldsymbol{q},\dot{\boldsymbol{q}}) &= \boldsymbol{S}^{\mathrm{T}}(\boldsymbol{q})\boldsymbol{H}(\boldsymbol{q})\dot{\boldsymbol{S}}(\boldsymbol{q}) + \boldsymbol{S}^{\mathrm{T}}(\boldsymbol{q})\boldsymbol{C}(\boldsymbol{q},\dot{\boldsymbol{q}})\boldsymbol{S}(\boldsymbol{q}).\end{aligned}$$

Since $\dot{\boldsymbol{H}}(\boldsymbol{q}) - 2\boldsymbol{C}(\boldsymbol{q},\dot{\boldsymbol{q}})$ is skew-symmetric, it is easy to show directly that Eq. (1.68) is also skew-symmetric:

$$\dot{\bar{H}}(q) - 2\bar{C}(q,\dot{q}) = \dot{S}^{T}(q)H(q)S(q) - \left[\dot{S}^{T}(q)H(q)S(q)\right]^{T} + S^{T}(q)\left[\dot{H}(q) - 2C(q,\dot{q})\right]S(q). \tag{1.68}$$

Using the same procedure for the inclusion of actuator dynamics, one can reformulate the dynamics [(Eqs. (1.58) and (1.64)] as:

$$\bar{H}(q)\dot{v} + \bar{C}(q,\dot{q})v + \bar{\tau}_{d} = \bar{E}\tau = \bar{E}(q)\mathrm{N}\tau_{o} = \bar{E}(q)\mathrm{NK}_{\tau}i_{a}, \tag{1.69}$$

$$\bar{H}(q)\dot{v} + \bar{C}(q,\dot{q})v + \bar{\tau}_{d} = \frac{\mathrm{NK}_{\tau}}{\mathrm{R}_{a}}\bar{E}(q)u - \frac{\mathrm{N}^{2}\mathrm{K}_{\tau}\mathrm{K}_{e}}{\mathrm{R}_{a}}\bar{E}(q)\Omega v - \frac{\mathrm{NK}_{\tau}}{\mathrm{R}_{a}}\bar{E}(q)\tau_{e}. \tag{1.70}$$

It is noted that for the dynamics Eqs. (1.69) and (1.70), properties 1.1 and 1.2 of the original dynamics are also maintained.

Equations (1.69) and (1.70) can be represented respectively as:

$$\begin{aligned}\dot{\chi} &= f(\chi) + g(\chi)u,\\ &= \begin{bmatrix} S(q)v \\ \bar{H}^{-1}(q)\left(\bar{E}(q)\mathrm{NK}_{\tau}i_{a} - \bar{C}(q,\dot{q})v - \bar{\tau}_{d}\right) \\ -\left(\frac{\mathrm{K}_{e}\mathrm{N}\Omega v + \mathrm{R}_{a}i_{a} + \tau_{e}}{\mathrm{L}_{a}}\right)\end{bmatrix} + \begin{bmatrix} 0 \\ 0 \\ \frac{1}{\mathrm{L}_{a}}\end{bmatrix}u,\end{aligned} \tag{1.71}$$

with $\chi = \left[q\ v\ i_{a}\right]^{T}$, and

$$\begin{aligned}\dot{\chi} &= f(\chi) + g(\chi)u,\\ &= \begin{bmatrix} S(q)v \\ -\bar{H}^{-1}(q)\left(\bar{C}(q,\dot{q})v + \frac{\mathrm{N}^{2}\mathrm{K}_{\tau}\mathrm{K}_{e}}{\mathrm{R}_{a}}\bar{E}(q)\Omega v + \bar{\tau}_{d} + \frac{\mathrm{NK}_{\tau}}{\mathrm{R}_{a}}\bar{E}(q)\tau_{e}\right)\end{bmatrix} \\ &\quad + \begin{bmatrix} 0 \\ \bar{H}^{-1}(q)\frac{\mathrm{NK}_{\tau}}{\mathrm{R}_{a}}\bar{E}(q)\end{bmatrix}u,\end{aligned} \tag{1.72}$$

with $\chi = \left[q\ v\right]^{T}$.

1.2.5 Controllability of the Dynamic Model

A system described by Eq. (1.67) or (1.71) or (1.72) is said to be controllable when any state can be reached under the action of piecewise continuous inputs [18]. A system satisfies the rank condition of Lie algebra (or controllability rank condition or rank accessibility condition) in the state x_0 if the rank of the linear combination

$\boldsymbol{g}$ and all their recursively computed Lie brackets[2] is equal to the dimension of the state-space [2].

In [11], it is known that the posture vector is provided by Eq. (1.18) as well as their velocity vector of the DWMR is given by Eq. (1.19). As the DWMR has nonholonomic constraint described by Eq. (1.11), where the matrix $\boldsymbol{A}(\boldsymbol{q})$ is defined as in Eq. (1.16), therefore, this system has three-dimensional configuration space ($\mathrm{n} = 3$) and a constraint ($\mathrm{p} = 1$) which leads to a velocity vector of dimension $\mathrm{m} = \mathrm{n} - \mathrm{p} = 2$. Choosing $\boldsymbol{v} = [\upsilon\ \omega]^{\mathrm{T}}$, Eq. (1.19), as the internal state variables, ones obtain Eq. (1.21) with the matrix $\boldsymbol{S}(\boldsymbol{q})$ defined in Eq. (1.24).

Then, one has the smooth and linearly independent vector fields $\boldsymbol{g}$ [12] as:

$$\boldsymbol{g}_1 = \begin{bmatrix} \cos(\theta) \\ \sin(\theta) \\ 0 \end{bmatrix}, \quad \boldsymbol{g}_2 = \begin{bmatrix} 0 \\ 0 \\ 1 \end{bmatrix}. \tag{1.73}$$

The rank of the matrix formed by ($\boldsymbol{g}_1,\ \boldsymbol{g}_2$) is two and inferior to the configuration space, therefore, it is necessary to generate a Lie bracket $[\boldsymbol{g}_1,\ \boldsymbol{g}_2]$ to verify if this system is holonomic and the possibility of completing the distribution to verify if this system is controllable. Soon,

$$[\boldsymbol{g}_1,\ \boldsymbol{g}_2] = \boldsymbol{g}_3 = \frac{\partial \boldsymbol{g}_2}{\partial \boldsymbol{q}}\boldsymbol{g}_1 - \frac{\partial \boldsymbol{g}_1}{\partial \boldsymbol{q}}\boldsymbol{g}_2 = \begin{bmatrix} \sin(\theta) \\ -\cos(\theta) \\ 0 \end{bmatrix}, \tag{1.74}$$

and the Frobenius theorem [12], if this system of Eq. (1.21) is completely integrable (holonomic), then it is involutive, i.e.,

$$\operatorname{rank}\left(\boldsymbol{g}_1\ \boldsymbol{g}_2\right) = \operatorname{rank}\left(\boldsymbol{g}_1\ \boldsymbol{g}_2\ [\boldsymbol{g}_1,\ \boldsymbol{g}_2]\right). \tag{1.75}$$

Verification leads to:

$$\operatorname{rank}\left\{\begin{bmatrix} \cos(\theta) & 0 & \sin(\theta) \\ \sin(\theta) & 0 & -\cos(\theta) \\ 0 & 1 & 0 \end{bmatrix}\right\} = 3. \tag{1.76}$$

With this, one can conclude that this system is nonholonomic. By Chow's theorem (rank condition of Lie algebra or controllability rank condition or rank accessibility condition) [12], one has that the rank of this system plus $\boldsymbol{g}_3$ is equal to the rank of the configuration space, so this system is controllable.

It is important to emphasize that a controllability verification based on the configuration kinematic model, Eq. (1.23), is performed by [6].

[2] The Lie bracket of $\boldsymbol{f}$ and $\boldsymbol{g}$ is defined as $\boldsymbol{ad}_{\mathrm{fg}} = [\boldsymbol{f},\ \boldsymbol{g}] = \frac{\partial \boldsymbol{g}}{\partial x}\boldsymbol{f} - \frac{\partial \boldsymbol{f}}{\partial x}\boldsymbol{g}$, where $\boldsymbol{f}$ and $\boldsymbol{g}$ are two vector fields of $\Re^{\mathrm{n}}$ [13].

References

1. Alexander, J.C., Maddocks, J.H.: On the kinematics of wheeled mobile robots. Int. J. Robot. Res. **8**(5), 15–27 (1989)
2. Barraquand, J., Latombe, J.: Nonholonomic multibody mobile robots: controllability and motion planning in the presence of obstacles. Algorithmica **10**(2), 121 (1993)
3. Campion, G., Bastin, G., d'Aandrea Novel, B.: Structural properties and classification of kinematic and dynamic models of wheeled mobile robots. IEEE Trans. Robot. Autom. **12**(1), 47–62 (1996)
4. Campion, G., Chung, W.: Wheeled Robots, chapter Part B 17, pp. 391–410. Springer (2008)
5. Campion, G., d'Andrea-Novel, B., Bastin, G.: Modelling and state feedback control of nonholonomic mechanical systems. In: IEEE Conference on Decision and Control, pp. 1184–1189 (1991)
6. Coelho, P., Nunes, U.: Lie algebra application to mobile robot control: a tutorial. Robotica **21**(05), 483–493 (2003)
7. Coelho, P., Nunes, U.: Path-following control of mobile robots in presence of uncertainties. IEEE Trans. Robot. **21**(2), 252–261 (2005)
8. d'Andrea-Novel, B., Bastin, G., Campion, G.: Modelling and control of non-holonomic wheeled mobile robots. In: IEEE International Conference on Robotics and Automation, pages 1130–1135 (1991)
9. Canudas de Wit, C., Bastin, G., Siciliano, B.: Theory of Robot Control. Springer, Berlin, Heidelberg (1996)
10. Dhaouadi, R., Abu Hatab, A.: Dynamic modelling of differential-drive mobile robots using Lagrange and Newton-Euler methodologies: a unified framework. Adv. Robot. Autom. **2**(2), 576–586 (2013)
11. Fierro, R., Lewis, F.L.: Control of a nonholonomic mobile robot using neural networks. IEEE Trans. Neural Netw. **9**(4), 589–600 (1998)
12. Figueiredo, L.C., Jota, F.G.: Introdução ao controle de sistemas não-holonômicos. Revista de Controle & Automação **15**(3), 9 (2004)
13. Khalil, H.K.: Nonlinear Systems. Pearson, 31 edition (2001)
14. Killough, S.M., Pin, F.G.: Design of an omnidirectional and holonomic wheeled platform prototype. In: IEEE International Conference on Robotics and Automation, pages 84–90 (1992)
15. Lewis, F.L., Dawson, D.M., Abdallah, C.T.: Robot Manipulator Control: Theory and Practice, 2nd edn. Marcel Dekker, Inc. (2003)
16. Muir, P.F., Neuman, C.P.: Kinematic modeling of wheeled mobile robots. J. Robot. Syst. **4**(2), 281–340 (1987)
17. Muir, P.F., Neuman, C.P.: Kinematic Modeling for Feedback Control of an Omnidirectional Wheeled Mobile Robot, pages 25–31. Springer New York (1990)
18. Samson, C., Ait-Abderrahim, K.: Feedback control of a nonholonomic wheeled cart in Cartesian space. In: Proceedings of 1991 IEEE International Conference on Robotics and Automation, pp. 1136–1141 (1991)
19. Thanjavur, K., Rajagopalan, R.: Ease of dynamic modelling of wheeled mobile robots (WMRs) using Kane's approach. In: Proceedings of International Conference on Robotics and Automation, pages 2926–2931 (1997)
20. Vos, D.W., Von Flotow, A.H.: Dynamics and nonlinear adaptive control of an autonomous unicycle: theory and experiment. In: 29th IEEE Conference on Decision and Control, pages 182–187 (1990)
21. Yamamoto, Y., Yun, X.: Coordinating locomotion and manipulation of a mobile manipulator. IEEE Trans. Autom. Control **39**(6), 1326–1332 (1994)

Chapter 2
Background on Control Problems and Systems

2.1 Trajectory Tracking Problem

For the DWMR to arrive at a point in the plane, control is necessary. Sciavicco et al. [5] define robotics as the science that studies the intelligent connection between perception and action (coordination may be included for multi-agent robotics). The basic components of a robotic system are actuators, sensors, and control. The ability to exert an action is provided by the actuators, which move the mechanical components of the DWMR. The perception capacity is given by the sensing system, which acquires information on the internal state of the mechanical system and also on the external environment. The ability to intelligently connect action with perception is provided by the control system, which commands the execution of the action taking into account the objectives defined by the task planning technique.

In general, the purpose of the control is to eliminate the error between where you want the DWMR to be and their actual position. The main control problems of DWMR presented in the literature are [6]:

- **Stabilization or regulation**: it is to achieve a desired final posture for the DWMR, which is defined by a location in the Cartesian space and a given orientation, from any initial configuration of the DWMR, i.e., from the initial location of the DWMR, at the end of the task should be presented exactly as the reference regardless of the path taken to reach it. In Fig. 2.1 there is an illustration of this type of task where possible paths are tracked by the DWMR. In other words, stabilizing a system corresponds to bringing the state of the system to a given equilibrium point. This type of task is not an attractive solution for real environments, precisely because there is no predictability about the trajectory that the DWMR will perform during their movement, and unexpected collisions can occur.
- **Trajectory tracking**: it consists of making the DWMR reach and follow a certain trajectory in the Cartesian space or a geometric path with an associated temporal

N. A. Martins and D. W. Bertol, *Wheeled Mobile Robot Control*, Studies in Systems, Decision and Control 380, https://doi.org/10.1007/978-3-030-77912-2_2

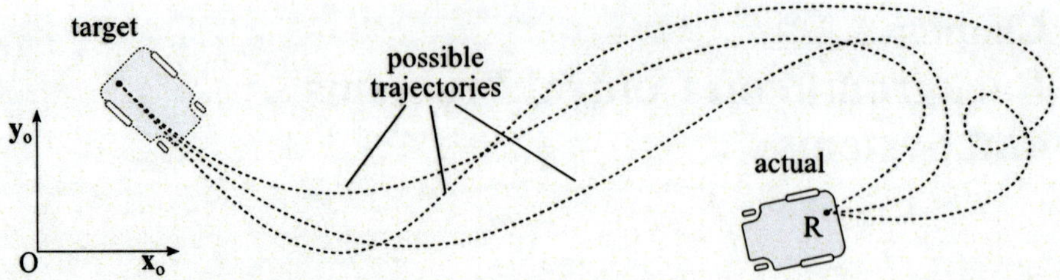

Fig. 2.1 Example of stabilization task

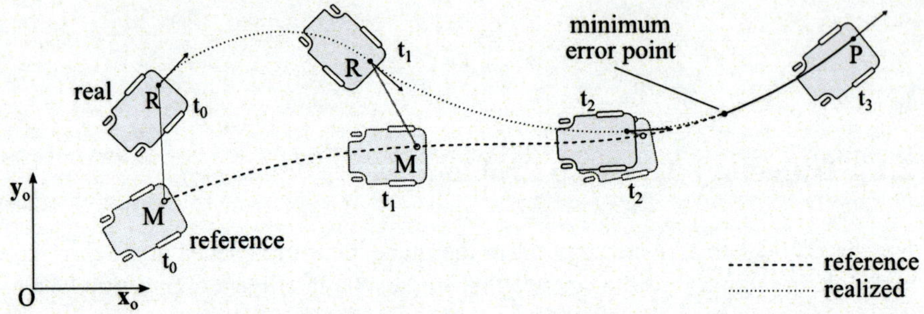

Fig. 2.2 Example of trajectory tracking task

law, starting from a certain initial posture, inside or outside the trajectory. In short, it corresponds to following a reference trajectory as a function of time. Figure 2.2 illustrates an example of a trajectory tracking task where a DWMR, R, pursues the desired location, M, with temporal requirements, about a reference curve. Trajectory tracking is used, for example, to avoid collisions in a controlled environment, where the DWMR must respect certain posture at a given time.

- **Path tracking**: it consists of a trajectory tracking problem, but with the absence of the temporal law. Thus, the DWMR must only follow a certain geometric path without time constraints, i.e., the DWMR must follow a reference path as a function of time-independent parameters. Figure 2.3 illustrates a path tracking task where a DWMR, R, follows a reference curve and obtains as a target the point M of the curve closest to the DWMR, without temporal requirements.
- **Fault-Tolerant**: Fault tolerance can be defined as the ability of a system to complete a given task in the presence of hardware or software defects. This is done by reconfiguring the control system according to the isolated fault.

This book has as main focus the synthesis of controllers in the resolution of the trajectory tracking problem in a flat and regular environment for a DWMR of medium size, subject to limitations of performance, uncertainties, and disturbances, that make the hypothesis of perfect velocity tracking is not maintained. Thus, to meet the control purposes, the trajectory tracking control problem for DWMRs is posed as follows.

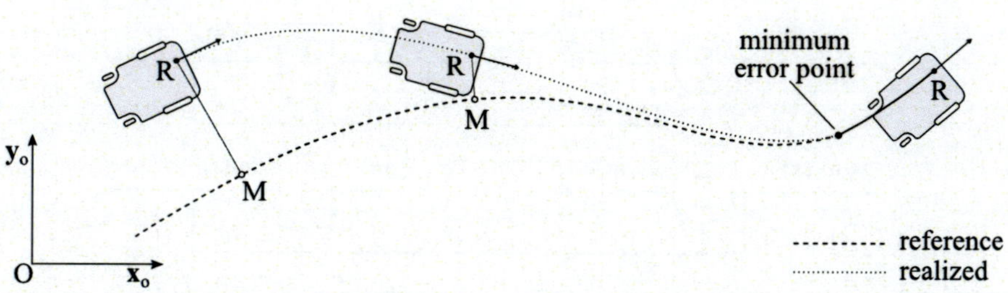

Fig. 2.3 Example of a path tracking task

Let there be a reference DWMR that is prescribed, in which the structure of DWMR is given and the reference posture and velocities are known, with $\boldsymbol{v}_r > 0$ for all t, find a smooth velocity control $\boldsymbol{v}_c$ such that $\lim_{t\to\infty}(\boldsymbol{q}_e) = 0$, where $\boldsymbol{q}_e = \boldsymbol{q}_r - \boldsymbol{q}$ and $\boldsymbol{v}_r$ are the posture tracking errors (or posture error) and the reference velocity vector, respectively. Then compute the control torques $\boldsymbol{\tau}$ for Eq. (1.67) or control input voltages $\boldsymbol{u}$ for Eqs. (1.71) and (1.72), such that $\boldsymbol{v} \to \boldsymbol{v}_c$ as $t \to \infty$ [2].

2.2 PD Dynamic Control

The objective of the dynamic controller is to compensate the known torques and forces described in Eq. (1.66) and ensure fast auxiliary velocity tracking errors $\boldsymbol{v}_e = \boldsymbol{v}_c - \boldsymbol{v}$ (see Fig. 2.4). As the uncertainties and disturbances are unknown, they are set to zero for this design and will be considered just for adjusting control gain and the design of the kinematic controller.

The solution presented by Spong et al. [7] about the calculus of the PD control is considered to act as the dynamic controller. Thus it has to change the applied wheels torque control to the body torques:

$$\boldsymbol{\tau} = \bar{\boldsymbol{E}}(\boldsymbol{q})^{-1}\,\bar{\boldsymbol{u}}, \tag{2.1}$$

and apply to the system, Eq. (1.66), the vector $\bar{\boldsymbol{u}} = [\bar{u}_\upsilon \ \bar{u}_\omega]^T$ as a new control input that will be designed as the PD control to achieve fast convergence of $\boldsymbol{v}_e$. The desired control is detailed in Fig. 2.5, in the frequency domain. Thus, the control signals $\bar{u}_\upsilon(s)$ and $\bar{u}_\omega(s)$ are generated by the PD controllers as follows:

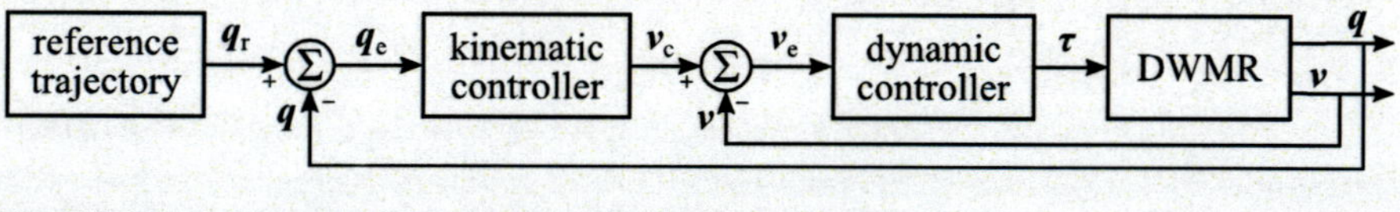

Fig. 2.4 Block diagram representation of the control structure

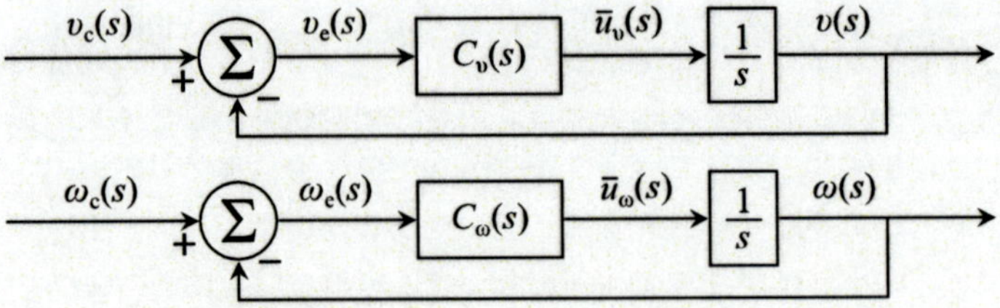

Fig. 2.5 Block diagram representation of the closed-loop PD control

$$
\begin{aligned}
C_\upsilon(s) &= \frac{\upsilon_e(s)}{\bar{u}_\upsilon(s)} = \mathrm{k}_{\mathrm{p}_\upsilon} + \frac{\mathrm{k}_{\mathrm{d}_\upsilon}\mathrm{n}_\upsilon}{1+\frac{\mathrm{n}_\upsilon}{s}} \\
C_\omega(s) &= \frac{\omega_e(s)}{\bar{u}_\omega(s)} = \mathrm{k}_{\mathrm{p}_\omega} + \frac{\mathrm{k}_{\mathrm{d}_\omega}\mathrm{n}_\omega}{1+\frac{\mathrm{n}_\omega}{s}}
\end{aligned}
\tag{2.2}
$$

with the proportional gains, $\mathrm{k}_{\mathrm{p}_\upsilon}$ and $\mathrm{k}_{\mathrm{p}_\omega}$, the derivative gains, $\mathrm{k}_{\mathrm{d}_\upsilon}$ and $\mathrm{k}_{\mathrm{d}_\omega}$, and the derivative filter parameter gains n_υ and n_ω being positive and adjusted to achieve stability with good time response performance. It is necessary to consider that adjusting of n_υ and n_ω plays an important role to accelerate the system response despite neglected dynamics and to avoid excitation and the chattering phenomenon [1].

This dynamic control architecture ensures fast auxiliary velocity tracking errors, being the following stability analysis of the closed-loop control system proved by means Lyapunov theory [3] and similar to [7].

Given the control velocity $\boldsymbol{v}_c(t) \in \Re^{nxm}$, the auxiliary velocity tracking error is defined as:

$$\boldsymbol{v}_e = \boldsymbol{v}_c - \boldsymbol{v} \Rightarrow \boldsymbol{v} = \boldsymbol{v}_c - \boldsymbol{v}_e, \tag{2.3}$$

and, their derivative as

$$\dot{\boldsymbol{v}}_e = \dot{\boldsymbol{v}}_c - \dot{\boldsymbol{v}} \Rightarrow \dot{\boldsymbol{v}} = \dot{\boldsymbol{v}}_c - \dot{\boldsymbol{v}}_e. \tag{2.4}$$

Using Eqs. (2.3) and (2.4) as well as the DWMR dynamics, Eq. (1.66), disregarding the effect of the uncertainties and disturbances,

$$\bar{\boldsymbol{H}}(\boldsymbol{q})\dot{\boldsymbol{v}} + \bar{\boldsymbol{C}}(\boldsymbol{q},\dot{\boldsymbol{q}})\boldsymbol{v} = \bar{\boldsymbol{\tau}} = \bar{\boldsymbol{E}}(\boldsymbol{q})\boldsymbol{\tau}, \tag{2.5}$$

one obtains the closed-loop system error dynamics in terms of the auxiliary velocity tracking error as

$$
\begin{aligned}
\bar{\boldsymbol{H}}(\boldsymbol{q})[\dot{\boldsymbol{v}}_c - \dot{\boldsymbol{v}}_e] + \bar{\boldsymbol{C}}(\boldsymbol{q},\dot{\boldsymbol{q}})[\boldsymbol{v}_c - \boldsymbol{v}_e] &= \bar{\boldsymbol{H}}(\boldsymbol{q})\dot{\boldsymbol{v}}_c - \bar{\boldsymbol{H}}(\boldsymbol{q})\dot{\boldsymbol{v}}_e + \bar{\boldsymbol{C}}(\boldsymbol{q},\dot{\boldsymbol{q}})\boldsymbol{v}_c - \bar{\boldsymbol{C}}(\boldsymbol{q},\dot{\boldsymbol{q}})\boldsymbol{v}_e, \\
&= \bar{\boldsymbol{E}}(\boldsymbol{q})\boldsymbol{\tau}, \\
\bar{\boldsymbol{H}}(q)\dot{\boldsymbol{v}}_e &= \bar{\boldsymbol{H}}(q)\dot{\boldsymbol{v}}_c + \bar{\boldsymbol{C}}(\boldsymbol{q},\dot{\boldsymbol{q}})\boldsymbol{v}_c - \bar{\boldsymbol{C}}(\boldsymbol{q},\dot{\boldsymbol{q}})\boldsymbol{v}_e - \bar{\boldsymbol{E}}(q)\boldsymbol{\tau}.
\end{aligned}
\tag{2.6}
$$

Applying the property of linearity in the parameters [3], the error dynamics results in:

$$\begin{aligned}\bar{H}(q)\dot{v}_c + \bar{C}(q,\dot{q})v_c = \bar{\gamma}(q,\dot{q},v_c,\dot{v}_c)\bar{\varphi},\\ \bar{H}(q)\dot{v}_e = \bar{\gamma}(q,\dot{q},v_c,\dot{v})\bar{\varphi} - \bar{C}(q,\dot{q})v_e - \bar{E}(q)\tau.\end{aligned} \tag{2.7}$$

Stability analysis is realized by considering the following Lyapunov function candidate:

$$V = \frac{1}{2}v_e^T\bar{H}(q)v_e + \frac{1}{2}v_e^T\mathbf{K}_d v_e. \tag{2.8}$$

It can be verified that V is positive definite. Differentiating Eq. (2.8) results in:

$$\dot{V} = \frac{1}{2}\left[\dot{v}_e^T\bar{H}(q)v_e + v_e^T\dot{\bar{H}}(q)v_e + v_e^T\bar{H}(q)\dot{v}_e + \dot{v}_e^T\mathbf{K}_d v_e + v_e^T\dot{\mathbf{K}}_d v_e + v_e^T\mathbf{K}_d\dot{v}_e\right],$$

$$\dot{V} = v_e^T\bar{H}(q)\dot{v}_e + \frac{1}{2}v_e^T\dot{\bar{H}}(q)v_e + v_e^T\mathbf{K}_d\dot{v}_e. \tag{2.9}$$

Substituting Eq. (2.7) into Eq. (2.9) yields to:

$$\dot{V} = -v_e^T\bar{E(q)}\tau + v_e^T\mathbf{K}_d\dot{v}_e + v_e^T\bar{\varphi}^T\bar{\gamma}^T + \frac{1}{2}v_e^T\left(\dot{\bar{H}}(q) - 2\bar{C}(q,\dot{q})\right)v_e. \tag{2.10}$$

Using the skew-symmetry property, $\dot{\bar{H}}(q) - 2\bar{C}(q,\dot{q}) = 0$, one has:

$$\dot{V} = -v_e^T\bar{E}(q)\tau + v_e^T\mathbf{K}_d\dot{v}_e^T + v_e^T\bar{\varphi}^T\bar{\gamma}^T. \tag{2.11}$$

The control law, Eqs. (2.1) and (2.2), can be rewritten in the time domain as:

$$\tau = \bar{E}(q)^{-1}\left(\mathbf{K}_d\dot{v}_e + \mathbf{K}_p v_e\right). \tag{2.12}$$

Using the PD control law, Eq. (2.12), in Eq. (2.11), $\dot{V}$ results in:

$$\dot{V} = -v_e^T\bar{E}(q)\bar{E}(q)^{-1}\left(\mathbf{K}_d\dot{v}_e + \mathbf{K}_p v_e\right) + v_e^T\mathbf{K}_d\dot{v}_e + v_e^T\bar{\varphi}^T\bar{\gamma}^T,$$
$$\dot{V} = -v_e v^T\mathbf{K}_d\dot{v}_e - v_e^T\mathbf{K}_p v_e + v_e^T\mathbf{K}_d\dot{v}_e + v_e^T\bar{\varphi}^T\bar{\gamma}^T,$$

$$\dot{V} = -v_e^T\mathbf{K}_p v_e + v_e^T\bar{\varphi}^T\bar{\gamma}^T. \tag{2.13}$$

By the Rayleigh-Ritz theorem [3], Eq. (2.13) is rewritten as:

$$\dot{V} \leq -\lambda_{\min}\left\{\mathbf{K}_p\right\}||v_e||^2 + ||\bar{\varphi}^T\bar{\gamma}^T v_e||,$$

$$\dot{V} \leq -||v_e||\left(\lambda_{\min}\left\{\mathbf{K}_p\right\}||v_e|| - ||\bar{\varphi}^T\bar{\gamma}^T||\right). \tag{2.14}$$

where $\lambda_{\min}\left\{\mathbf{K}_{\mathrm{p}}\right\}$ is the minimum singular value of $\mathbf{K}_{\mathrm{p}}$.

Thus, $\dot{V}$ is guaranteed negative definite while the term in parentheses in Eq. (2.14) is positive. Also, Eq. (2.14) can be obtained from a condition sufficient for $\dot{V}$ to be negative definite, i.e.:

$$||\boldsymbol{\nu}_{\mathrm{e}}|| > \frac{||\bar{\boldsymbol{\varphi}}^{\mathrm{T}}\bar{\boldsymbol{\gamma}}^{\mathrm{T}}||}{\lambda_{\min}\left\{\mathbf{K}_{\mathrm{p}}\right\}}. \tag{2.15}$$

Therefore, $\dot{V}$ is a negative definite in a region bounded by a constant. According to the standard theory of Lyapunov and the La Salle extension [3], it is shown that $||\boldsymbol{\nu}_{\mathrm{e}}||$ is uniformly ultimately bounded (UUB). It is observed that $||\boldsymbol{\nu}_{\mathrm{e}}||$ can be kept arbitrarily small by increasing or adding the gain $\lambda_{\min}\left\{\mathbf{K}_{\mathrm{p}}\right\}$ in Eq. (2.15). Finally, the right side of Eq. (2.15) can be considered as the practical bound of $\boldsymbol{\nu}_{\mathrm{e}}(t)$. Also, Eq. (2.15) represents the worst case one can have.

2.3 Posture Error Dynamics

In this book, the syntheses of kinematic controllers are developed based on the posture error dynamics of the closed-loop system required to solve the trajectory tracking control problem.

Thus, the objective of kinematic control is to determine the linear and angular velocities necessary for the actual DWMR, C, (Fig. 1.1), to track a given trajectory, determined by a reference DWMR, R, described in posture coordinates, as can be in Fig. 2.6. The kinematic behavior of actual DWMR posture is given by the Eqs. (1.18), (1.19), (1.21), and (1.25).

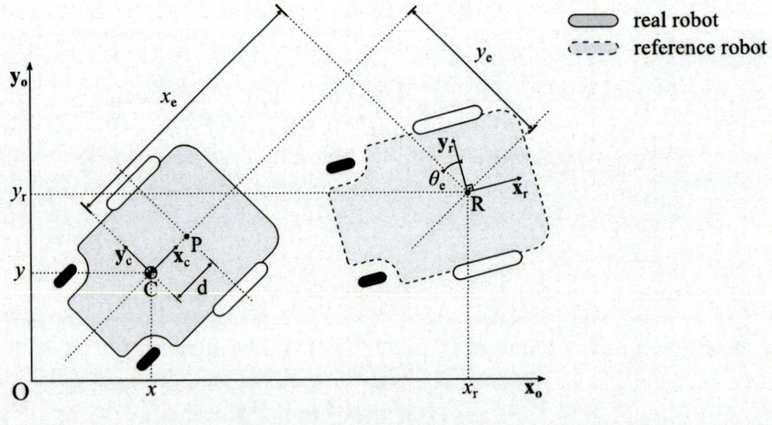

Fig. 2.6 Control strategy for the trajectory tracking: definition of the posture tracking errors between the actual DWMR, C, and reference DWMR, R

The reference DWMR must have the same kinematic characteristics as the actual DWMR [4]. However, because the reference DWMR is commonly described with C = P (Fig. 2.6) in the literature for the elaboration of reference trajectories, the reference kinematic model considered is formulated as:

$$\dot{\boldsymbol{q}}_{\mathrm{r}} = \boldsymbol{S}_{\mathrm{r}}(\boldsymbol{q}_{\mathrm{r}})\boldsymbol{v}_{\mathrm{r}}, \tag{2.16}$$

in which:

$$\boldsymbol{S}_{\mathrm{r}}(\boldsymbol{q}_{\mathrm{r}}) = \begin{bmatrix} \cos(\theta_{\mathrm{r}}) & 0 \\ \sin(\theta_{\mathrm{r}}) & 0 \\ 0 & 1 \end{bmatrix}, \quad \boldsymbol{q}_{\mathrm{r}} = \begin{bmatrix} x_{\mathrm{r}} \\ y_{\mathrm{r}} \\ \theta_{\mathrm{r}} \end{bmatrix}, \quad \boldsymbol{v}_{\mathrm{r}} = \begin{bmatrix} \upsilon_{\mathrm{r}} \\ \omega_{\mathrm{r}} \end{bmatrix}, \tag{2.17}$$

being $\boldsymbol{S}_{\mathrm{r}}(\boldsymbol{q}_{\mathrm{r}})$ the reference Jacobian matrix, $\boldsymbol{q}_{\mathrm{r}}$ the reference posture vector, and $\boldsymbol{v}_{\mathrm{r}}$ the reference velocity vector.

The posture error $\boldsymbol{q}_{\mathrm{e}}$ is defined as the transformation to the local coordinate system of the DWMR, with origin in $\mathrm{CX_CY_C}$, from the difference between the reference posture and the actual posture as shown in Fig. 2.6. The posture error $\boldsymbol{q}_{\mathrm{e}}$ is given by:

$$\boldsymbol{q}_{\mathrm{e}} = \begin{bmatrix} x_{\mathrm{e}} \\ y_{\mathrm{e}} \\ \theta_{\mathrm{e}} \end{bmatrix} = \begin{bmatrix} \cos(\theta) & \sin(\theta) & 0 \\ -\sin(\theta) & \cos(\theta) & 0 \\ 0 & 0 & 1 \end{bmatrix} \begin{bmatrix} x_{\mathrm{r}} - x \\ y_{\mathrm{r}} - y \\ \theta_{\mathrm{r}} - \theta \end{bmatrix} = \boldsymbol{R}(\theta)\left(\boldsymbol{q}_{\mathrm{r}} - \boldsymbol{q}\right), \tag{2.18}$$

with the posture error being described in the DWMR reference system and $\boldsymbol{R}(\theta)$, Eq. (1.30), is the rotation matrix between inertial coordinate system $\mathrm{OX_OY_O}$ and mass center coordinate system of the actual DWMR $\mathrm{CX_CY_C}$ (Figs. 1.1 and 2.6).

The posture error dynamics of the closed-loop system is obtained from the definition of posture error of the DWMR, Eq. (2.18), and their time derivative, using the kinematic model together with the simplification due to the perfect tracking velocity ($[\upsilon\ \omega]^{\mathrm{T}} = [\upsilon_{\mathrm{c}}\ \omega_{\mathrm{c}}]^{\mathrm{T}}$).

The derivative of x_{e} is given by

$$\dot{x}_{\mathrm{e}} = (\dot{x}_{\mathrm{r}} - \dot{x})\cos(\theta) + (\dot{y}_{\mathrm{r}} - \dot{y})\sin(\theta) - (x_{\mathrm{r}} - x)\dot{\theta}\sin(\theta) + (y_{\mathrm{r}} - y)\dot{\theta}\cos(\theta).$$

But from Eqs. (1.7), (1.8) and (2.18), leads to:

$$\dot{x}_{\mathrm{e}} = y_{\mathrm{e}}\omega - \upsilon + \dot{x}_{\mathrm{r}}\cos(\theta) + \dot{y}_{\mathrm{r}}\sin(\theta).$$

Again, from Eq. (2.18), results in

$$\dot{x}_{\mathrm{e}} = y_{\mathrm{e}}\omega - \upsilon + \dot{x}_{\mathrm{r}}\cos(\theta_{\mathrm{r}} - \theta_{\mathrm{e}}) + \dot{y}_{\mathrm{r}}\sin(\theta_{\mathrm{r}} - \theta_{\mathrm{e}}).$$

Expanding the terms in sine and cosine, it follows that

$$\begin{aligned} \dot{x}_{\mathrm{e}} = {} & y_{\mathrm{e}}\omega - \upsilon + \dot{x}_{\mathrm{r}}(\cos(\theta_{\mathrm{r}})\cos(\theta_{\mathrm{e}}) + \sin(\theta_{\mathrm{r}})\sin(\theta_{\mathrm{e}})) \\ & + \dot{y}_{\mathrm{r}}(\sin(\theta_{\mathrm{r}})\cos(\theta_{\mathrm{e}}) - \cos(\theta_{\mathrm{r}})\sin(\theta_{\mathrm{e}})). \end{aligned}$$

The terms $\cos(\theta_e)$ and $\sin(\theta_e)$ are highlighted, and one obtains

$$\begin{aligned}\dot{x}_e = y_e\omega - \upsilon + \cos(\theta_e)(\dot{x}_r\cos(\theta_r) + \dot{y}_r\sin(\theta_r))\\ + \sin(\theta_e)(\dot{x}_r\sin(\theta_r) - \dot{y}_r\cos(\theta_r)).\end{aligned}$$

Using the relation $\dot{x}_r\sin(\theta_r) = \dot{y}_r\cos(\theta_r)$, deduced from the Eqs. (2.16) and (2.17), finally it arrives to

$$\dot{x}_e = y_e\omega - \upsilon + \upsilon_r\cos(\theta_e). \tag{2.19}$$

The derivative of y_e is provided by

$$\dot{y}_e = -(\dot{x}_r - \dot{x})\sin(\theta) + (\dot{y}_r - \dot{y})\cos(\theta) - (x_r - x)\dot{\theta}\cos(\theta) - (y_r - y)\dot{\theta}\sin(\theta).$$

From Eq. (2.18) and the use of the Eqs. (1.21) and (1.25), one can write

$$\dot{y}_e = -(\mathrm{d} + x_e)\omega - \dot{x}_r\sin(\theta) + \dot{y}_r\cos(\theta).$$

Again, from Eq. (2.18), it follows

$$\dot{y}_e = -(\mathrm{d} + x_e)\omega - \dot{x}_r\sin(\theta_r - \theta_e) + \dot{y}_r\cos(\theta_r - \theta_e).$$

With the expansion of the terms in sine and cosine, one obtains

$$\begin{aligned}\dot{y}_e = -(\mathrm{d} + x_e)\omega - \dot{x}_r(\sin(\theta_r)\cos(\theta_e) - \cos(\theta_r)\sin(\theta_e))\\ + \dot{y}_r(\cos(\theta_r)\cos(\theta_e) + \sin(\theta_r)\sin(\theta_e)).\end{aligned}$$

Evidencing the terms of $\cos(\theta_e)$ and $\sin(\theta_e)$, one can write

$$\begin{aligned}\dot{y}_e = -(\mathrm{d} + x_e)\omega + \sin(\theta_e)(\dot{x}_r\cos(\theta_r) + \dot{y}_r\sin(\theta_r))\\ + \cos(\theta_e)(\dot{y}_r\cos(\theta_r) - \dot{x}_r\sin(\theta_r)).\end{aligned}$$

By using the relation $\dot{x}_r\sin(\theta_r) = \dot{y}_r\cos(\theta_r)$, deduced from Eqs. (2.16) and (2.17), it results in

$$\dot{y}_e = -(\mathrm{d} + x_e)\omega + \upsilon_r\sin(\theta_e). \tag{2.20}$$

The derivative of θ_e is given by

$$\dot{\theta}_e = \dot{\theta}_r - \dot{\theta} = \omega_r - \omega. \tag{2.21}$$

Therefore, posture error dynamics of the closed-loop system $\dot{q}_e$ is provided by

$$\dot{\boldsymbol{q}}_e = \begin{bmatrix}\dot{x}_e\\ \dot{y}_e\\ \dot{\theta}_e\end{bmatrix} = \begin{bmatrix}\omega y_e - \upsilon + \upsilon_r\cos(\theta_e)\\ -\omega(\mathrm{d} + x_e) + \upsilon_r\sin(\theta_e)\\ \omega_r - \omega\end{bmatrix}. \tag{2.22}$$

In the case of using the posture kinematic model of the actual DWMR, Eqs. (1.21) and (1.24), where $C = P$ or $d = 0$, the posture error dynamics of the closed-loop system $\dot{q}_e$, Eq. (2.22), yields to

$$\dot{\boldsymbol{q}}_e = \begin{bmatrix} \dot{x}_e \\ \dot{y}_e \\ \dot{\theta}_e \end{bmatrix} = \begin{bmatrix} \omega y_e - \upsilon + \upsilon_r \cos(\theta_e) \\ -\omega x_e + \upsilon_r \sin(\theta_e) \\ \omega_r - \omega \end{bmatrix}. \tag{2.23}$$

2.4 Robustness Considerations

Under robustness considerations, in practical situations, velocities and tracking errors are not equal to zero. The best that can be done is to guarantee that the error converges to a neighborhood of the origin. If uncertainties and disturbances (e.g., modeling imprecisions; surface friction; external disturbances; bounded unknown disturbances; unmodeled and unstructured dynamics of actuators, sensors, and other electronic devices; integration errors; noises; operational, sampling, and physical limitations) drive the system away from the convergence compact set, the derivative of the Lyapunov function become negative and the energy of the system decreases uniformly; therefore, the error becomes small again [2].

As "perfect velocity tracking" does not hold into practice, the dynamic controller generates auxiliary velocity tracking errors $\boldsymbol{\nu}_e$, Eq. (2.3), which is bounded by some known constant. This tracking error can be seen as uncertainties and disturbances for the kinematic model, see Fig. 2.7. Based on Eqs. (1.24) and (1.25), the closed-loop kinematic model becomes:

$$\begin{cases} \dot{x} = (\upsilon_c + \upsilon_e)\cos(\theta) \\ \dot{y} = (\upsilon_c + \upsilon_e)\sin(\theta) \\ \dot{\theta} = (\omega_c + \omega_e) \end{cases}, \text{ for } C = P, \tag{2.24}$$

$$\begin{cases} \dot{x} = (\upsilon_c + \upsilon_e)\cos(\theta) - (\omega_c + \omega_e)d\sin(\theta) \\ \dot{y} = (\upsilon_c + \upsilon_e)\sin(\theta) + (\omega_c + \omega_e)d\cos(\theta) \\ \dot{\theta} = (\omega_c + \omega_e) \end{cases}, \text{ for } C \neq P, \tag{2.25}$$

where $\boldsymbol{\nu}_e = [\upsilon_e \ \omega_e]^T$ and $\boldsymbol{\nu}_c = [\upsilon_c \ \omega_c]^T$ denote the auxiliary velocity tracking errors and the desired velocity control inputs, respectively. The uncertainties and disturbances, given by $\boldsymbol{\nu}_e$, satisfy the matching conditions i.e., the nonholonomic constraints, Eqs. (1.1) and (1.2), are not violated. Afterward, by using standard Lyapunov methods, it can be shown that along with a system's solution, $||\boldsymbol{q}_e||$ is bounded, and thus $||\dot{\boldsymbol{q}}_e||$ is also bounded. The norm of the auxiliary velocity tracking errors affects directly the norm of the posture tracking errors. Note that the norm of the auxiliary velocity tracking errors $||\boldsymbol{\nu}_e||$ depends on the proportional gains, k_{p_υ} and k_{p_ω}, the derivative gains, k_{d_υ} and k_{d_ω}, and the derivative filter parameter gains n_υ and n_ω,

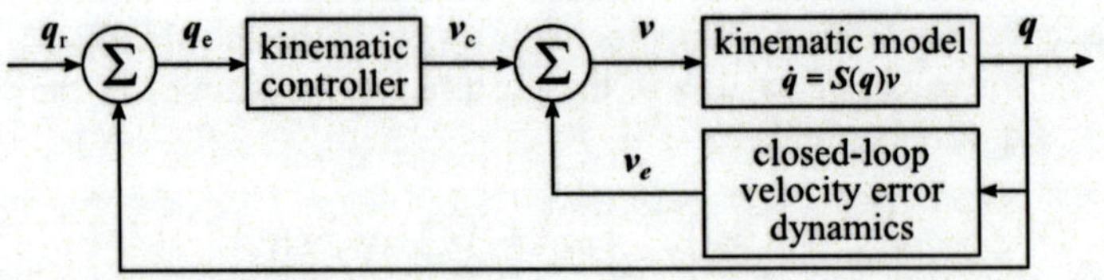

Fig. 2.7 Block diagram of the closed-loop control system: υ_e as uncertainties and disturbances for the kinematic model (similar to [2])

Eq. (2.2). Since $||\boldsymbol{v}_e||$ can be made arbitrarily small, then $||\boldsymbol{q}_e||$ can also be made arbitrarily small [2].

The effect of the uncertainties and disturbances (e.g., modeling imprecisions; surface friction; external disturbances; bounded unknown disturbances; unmodeled and unstructured dynamics of actuators, sensors, and other electronic devices; integration errors; noises; operational, sampling, and physical limitations) that affect the system can be considered, after mathematical manipulations, as a term added in the right side of error dynamics, Eqs. (2.22) and (2.23), i.e.,

$$\dot{\boldsymbol{q}}_e = \begin{bmatrix} \dot{x}_e \\ \dot{y}_e \\ \dot{\theta}_e \end{bmatrix} = \begin{bmatrix} (\omega_c + \omega_e)y_e - (\upsilon_c + \upsilon_e) + \upsilon_r \cos(\theta_e) \\ -(\omega_c + \omega_e)x_e + \upsilon_r \sin(\theta_e) \\ \omega_r - (\omega_c + \omega_e) \end{bmatrix}, \text{ for C} = \text{P}, \tag{2.26}$$

$$\dot{\boldsymbol{q}}_e = \begin{bmatrix} \dot{x}_e \\ \dot{y}_e \\ \dot{\theta}_e \end{bmatrix} = \begin{bmatrix} (\omega_c + \omega_e)y_e - (\upsilon_c + \upsilon_e) + \upsilon_r \cos(\theta_e) \\ -(\omega_c + \omega_e)(d + x_e) + \upsilon_r \sin(\theta_e) \\ \omega_r - (\omega_c + \omega_e) \end{bmatrix}, \text{ for C} \neq \text{P}. \tag{2.27}$$

2.5 Generic Model for Nonlinear Systems

The derivation of the kinematic controllers treated in this book and their properties are made directly for an important class of nonlinear systems, whose model, in the form of state equations, is given by:

$$\dot{\tilde{z}} = \boldsymbol{A}(\tilde{z}, \boldsymbol{p}, t) + \boldsymbol{B}(\tilde{z}, \boldsymbol{p}, t)\boldsymbol{v}_c(\tilde{z}, t) + \boldsymbol{d}_b(t), \tag{2.28}$$

where $\tilde{z}(t)$ is the vector of states; $\boldsymbol{A}(\tilde{z}, \boldsymbol{p}, t) = \boldsymbol{A}_0(\tilde{z}, t) + \Delta\boldsymbol{A}(\tilde{z}, \boldsymbol{p}, t)$ is the vector of nonlinear functions; $\boldsymbol{B}(\tilde{z}, \boldsymbol{p}, t) = \boldsymbol{B}_0(\tilde{z}, t) + \Delta\boldsymbol{B}(\tilde{z}, \boldsymbol{p}, t)$ is the matrix of nonlinear functions; $\boldsymbol{v}_c(\tilde{z}, t)$ is the vector of control inputs; $\boldsymbol{d}_b(t)$ is the vector of uncertainties and disturbances; $\Delta\boldsymbol{A}(\tilde{z}, \boldsymbol{p}, t)$ and $\Delta\boldsymbol{B}(\tilde{z}, \boldsymbol{p}, t)$ are respectively, the vector and the matrix representing the uncertainties and disturbances in the system arising from the parametric uncertainties; and $\boldsymbol{A}_0(\tilde{z}, \boldsymbol{p}, t)$ and $\boldsymbol{B}_0(\tilde{z}, \boldsymbol{p}, t)$ refers to the vector and the matrix of nominal parameters, respectively.

The aim of this study is the derivation of a robust kinematic controller to the current uncertainties and disturbances in the kinematic model, Eqs. (2.24) and (2.25). To ensure the robustness of the controller, the uncertainties and disturbances must be bounded, the matrix $\boldsymbol{B}(\tilde{z}, \boldsymbol{p}, t)$ must be nonsingular and the following conditions must be satisfied [8]:

$$\begin{cases} \Delta \boldsymbol{A}(\tilde{z}, \boldsymbol{p}, t) = \boldsymbol{B}_0(\tilde{z}, t)\tilde{\boldsymbol{a}} \\ \Delta \boldsymbol{B}(\tilde{z}, \boldsymbol{p}, t) = \boldsymbol{B}_0(\tilde{z}, t)\tilde{\boldsymbol{b}} \\ \boldsymbol{d}_{\mathrm{b}}(t) = \boldsymbol{B}_0(\tilde{z}, t)\tilde{\boldsymbol{d}}_0 \end{cases}, \tag{2.29}$$

which means that $\Delta \boldsymbol{A}(\tilde{z}, \boldsymbol{p}, t)$, $\Delta \boldsymbol{B}(\tilde{z}, \boldsymbol{p}, t)$, and $\boldsymbol{d}_{\mathrm{b}}(t)$ must belong to the image of $\boldsymbol{B}_0(\tilde{z}, t)$; $\tilde{\boldsymbol{a}}$ and $\tilde{\boldsymbol{b}}$ are respectively, the vector and the matrix that incorporate the parametric uncertainties; $\tilde{\boldsymbol{d}}_0$ represents uncertainties and disturbances. So, the error dynamics in Eqs. (2.26) and (2.27) can be rewritten, based on Eqs. (2.28) and (2.29), as:

$$\dot{\tilde{z}} = \boldsymbol{A}_0(\tilde{z}, t) + \boldsymbol{B}_0(\tilde{z}, t)\boldsymbol{v}_{\mathrm{c}}(\tilde{z}, t) + \boldsymbol{d}_{\mathrm{b}}(t). \tag{2.30}$$

since there are not parametric uncertainties ($\Delta \boldsymbol{A} = 0$, $\Delta \boldsymbol{B} = 0$).

References

1. Elyoussef, E.S., Martins, N.A., De Pieri, E.R., Moreno, U.F.: PD-super-twisting second order sliding mode tracking control for a nonholonomic wheeled mobile robot. In: Proceedings of the 19th World Congress of the International Federation of Automatic Control—IFAC World Congress, vol. 11, pp. 3827–3832 (2014)
2. Fierro, R., Lewis, F.L.: Control of a nonholonomic mobile robot using neural networks. IEEE Trans. Neural Netw. **9**(4), 589–600 (1998)
3. Lewis, F.L., Dawson, D.M., Abdallah, C.T.: Robot Manipulator Control: Theory and Practice, 2nd edn. Marcel Dekker, Inc. (2003)
4. Oriolo, G., Cefalo, M., Vendittelli, M.: Repeatable motion planning for redundant robots over cyclic tasks. IEEE Trans. Robot. **33**(5), 1170–1183 (2017)
5. Sciavicco, L..., Villani, L., Oriolo, G., Siciliano, B.: Robotics: Modelling, Planning and Control. Springer, London (2009)
6. Silva-Ortigoza, R., Marcelino-Aranda, M., Ortigoza, G.S., Guzman, V.M.H., Molina-Vilchis, M.A., Saldana-Gonzalez, G., Herrera-Lozada, J.C., Olguin-Carbajal, M.: Wheeled mobile robots: a review. Latin America Trans. IEEE (Revista IEEE America Latina) **10**(6), 2209–2217 (2012)
7. Spong, M.W., Hutchinson, S., Vidyasagar, M.: Robot Modeling and Control. Wiley, New York (2006)
8. Utkin, V.I., Guldner, J., Shi, J.: Sliding Mode Control in Electro-Mechanical Systems. CRC Press, 2nd edn. (2009)

Chapter 3
Background on Simulation and Experimentation Environments

3.1 Implementation Environment

To verify the performance of the controllers treated in this book they are implemented in Matlab/Simulink software, version R2019b, and evaluated for the trajectory tracking control problem using: simulation using the DWMR configuration model determined by the Eq. (1.71) (Ideal scenario); DWMR simulator—the MobileSim (Realistic scenario); and application in the PowerBot DWMR (Experimental scenario).

The executions using Matlab/Simulink software are performed with the total simulation time of 50 s and the Dormand-Prince integration method with a sampling time of 1 ms. The choice of this value of sampling time is due to the existing internal PD controller, which generates the control torques, in the closed robotic platform, in this case, the PowerBot DWMR. As consequence, this value is also considered in the implementation of the kinematic controllers treated in an application running on the embedded computer. Thus, only the PD controller, as described in Sect. 2.2, is considered as a dynamic controller of the PowerBot DWMR in the three scenarios. Moreover, the sampling time of the kinematic and dynamic configuration state-space models with the actuator dynamics or PowerBot DWMR is 0.1 ms to allow the calculation of the temporal response of the entire system. It is important to emphasize that these sampling times must be considered in the simulation and experimental scenarios, as can be seen in Fig. 3.1.

With the use of the PowerBot DWMR in all scenarios, it is believed necessary to provide information about the features of this DWMR to test the controllers described in this book.

N. A. Martins and D. W. Bertol, *Wheeled Mobile Robot Control*, Studies in Systems, Decision and Control 380, https://doi.org/10.1007/978-3-030-77912-2_3

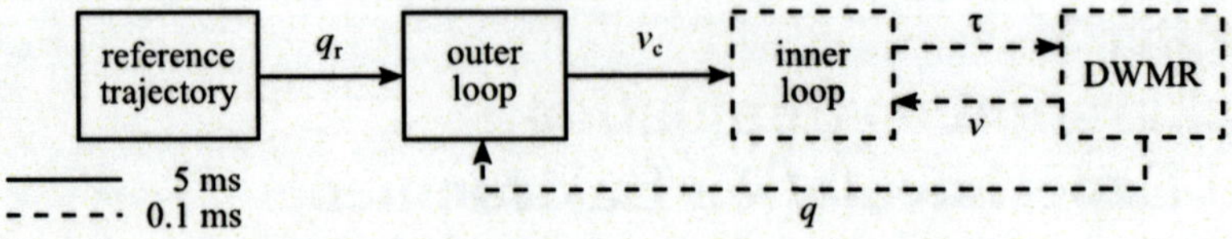

Fig. 3.1 Diagram of sampling time of the PowerBot DWMR control loops

3.1.1 PowerBot DWMR

The PowerBot DWMR (Fig. 3.2), manufactured by the MobileRobots Inc., is composed of two conventional fixed wheels that have independent actuation and two passive caster wheels used as support for platform balancing. In its composition are frontal and rear sonars; accelerometer; gyroscope; compass; odometry; camera with position control; front and rear bumpers with touch sensors; an internal computer with video capture card; and wireless communication (802.11b/g) (Fig. 3.3).

The location of the PowerBot DWMR in the environment is provided by their firmware using the dead reckoning technique, with error correction through the data fusion of the accelerometer, gyroscope, and compass.

Figure 3.3 shows the block diagram of the hardware architecture of the PowerBot DWMR. Note that the actuators and part of the sensing are only accessible by the main controller of the DWMR.

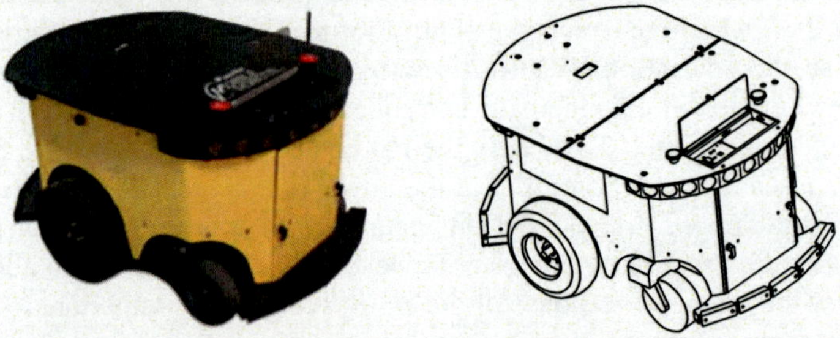

Fig. 3.2 PowerBot DWMR

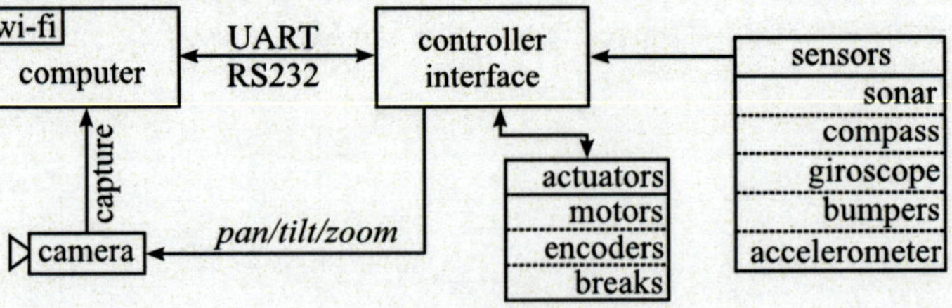

Fig. 3.3 Hardware architecture of the PowerBot DWMR

Table 3.1 PowerBot DWMR specifications: physical parameters

Parameter	Value
r	0.135 m
c	0.9 m (DWMR length)
2R	0.66 m (DWMR width)
h	0.48 m (DWMR height)
d	0.1 m

All data obtained from the sensors are prepossessed and encapsulated in an information packet that is sent via RS-232 serial communication to a selectable frequency 1 Hz and 1 kHz.

The control inputs are in terms of their velocities (body or wheels), which are treated as references of the internal control loop, composed of PID controllers at a frequency 200 Hz, with configurable gains in firmware. Direct access to the drive of the motors is not possible, which prevents the application of torque references.

The manufacturer provides the Aria library, which contains methods for connecting to the DWMR, handling the data packet, and managing DWMR control tasks.

The solution of the control loop of the library uses the frequency of sending the data packets by the DWMR to start each controller code execution cycle. Although it inherits the frequency of arrival of the data packets, the original solution of the PowerBot DWMR does not provide temporal guarantees of execution, i.e., although the DWMR was sending packets at a guaranteed frequency, the control software might not be able to receive them, losing control steps.

Thus, to have greater predictability about the periodicity of the control task, a control loop was implemented using the real-time toolbox of the Matlab/Simulink software in conjunction with the connection and communication methods of the Aria library. Thus it was possible to implement a control system with a frequency 200 Hz and with the capacity to analyze deadline losses.

By completing the information, the specifications and the detailed schematic of the dimensions of the PowerBot DWMR, respectively, are found in Tables 3.1, 3.2, 3.3, 3.4 and Fig. 3.4.

3.1.2 Trajectory Adopted

The trajectory used as a reference trajectory in the simulations and experiments is an eight-shape trajectory [2]. The mathematical formulation of this trajectory is given by Eq. (3.1) as,

Table 3.2 PowerBot DWMR specifications: kinematic and dynanic parameters

Parameter	Value
m_t	120.0 kg
m_c	110.0 kg
m_w	5.0 kg
I	15.0656 kg m^2
I_c	11.4180 kg ^{2}
I_w	0.0456 kg ^{2}
I_m	0.7293 kg ^{2}
m_{load}	100 kg (load capacity or payload)
υ_{max}	2.1 m/s (maximum linear velocity)
ω_{max}	5.24 rad/s (maximum angular velocity)
$\tau_{m_{max}}$	20.45 N m (maximum motor torque)

Table 3.3 PowerBot DWMR specifications: mechanical and electrical parameters of the actuators

Parameter	Value
N	22.3
K_e	0.02 rad/s/V
K_τ	0.2247 Nm/A
L_a	0.01 H
R_a	6.0 Ω
u_{max}	24.0 V_{DC} (maximum motor voltage)
$I_{a_{max}}$	4.0 A (maximum motor current)

Table 3.4 PowerBot DWMR specifications: internal PID control parameters

Parameter	Value
k_{p_υ}	40.0
k_{p_ω}	40.0
k_{i_υ}	0.0
k_{i_ω}	0.0
k_{d_υ}	20.0
k_{d_ω}	20.0
N_υ	1.0
N_ω	1.0

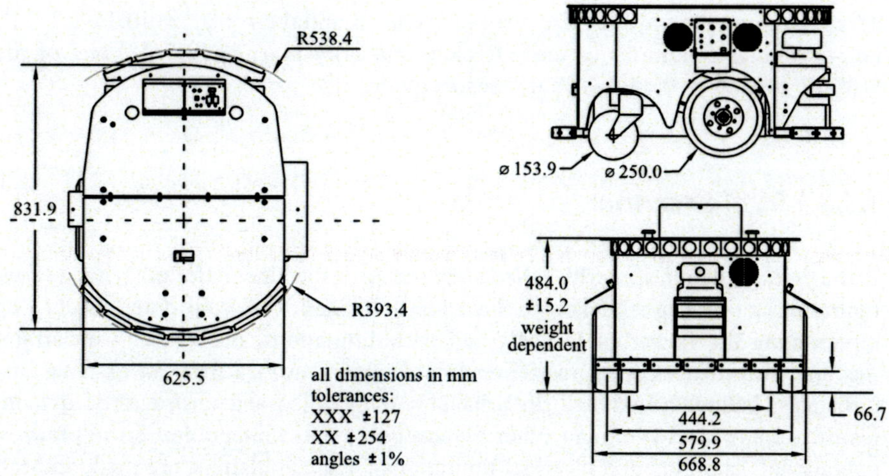

Fig. 3.4 PowerBot DWMR specifications

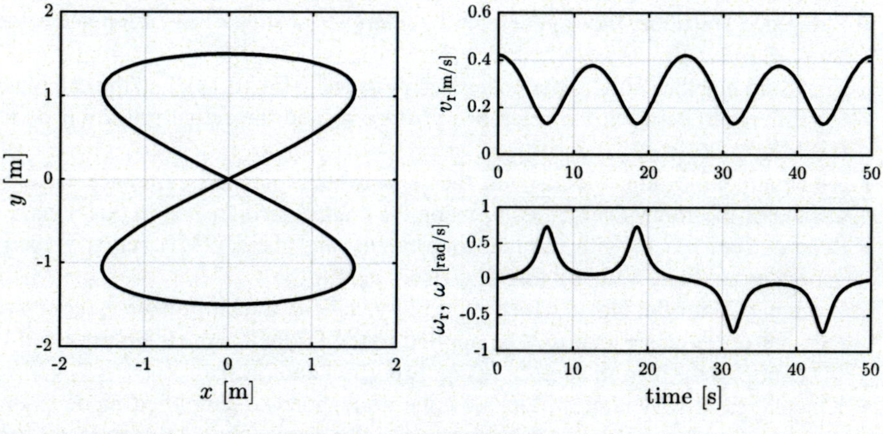

Fig. 3.5 Eight-shape reference trajectory and their velocities

$$\dot{\boldsymbol{q}}_\mathrm{r} = \begin{bmatrix} \dot{x}_\mathrm{r} \\ \dot{y}_\mathrm{r} \\ \dot{\theta}_\mathrm{r} \end{bmatrix} = \begin{bmatrix} \frac{3\pi}{50} \sin\left(\frac{2\pi}{50}\left(t + \frac{50}{4}\right)\right) \\ -\frac{6\pi}{50} \cos\left(\frac{4\pi}{50}\left(t + \frac{50}{4}\right)\right) \\ \dfrac{\ddot{y}_\mathrm{r}\dot{x}_\mathrm{r} - \ddot{x}_\mathrm{r}\dot{y}_\mathrm{r}}{\dot{x}_\mathrm{r}^2 + \dot{y}_\mathrm{r}^2} \end{bmatrix}, \quad \boldsymbol{v}_\mathrm{r} = \begin{bmatrix} v_\mathrm{r} \\ \omega_\mathrm{r} \end{bmatrix} = \begin{bmatrix} \sqrt{\dot{x}_\mathrm{r}^2 + \dot{y}_\mathrm{r}^2} \\ \dot{\theta}_\mathrm{r} \end{bmatrix}. \tag{3.1}$$

as well as this trajectory and their velocities can be visualized as in Fig. 3.5.

Analyzing Fig. 3.5 and Eq. (3.1), it is verified that this trajectory is more complex, considering the deceleration and acceleration, with the linear velocity varying between 0.15 m/s and 0.45 m/s and the angular velocity varying between −0.7 and

0.7 rad/s along the trajectory, as well as it has an initial error $\boldsymbol{q}_{e_0} = [0.15\ 0.2\ \frac{\pi}{2}]^T$, i.e., the initial conditions of the reference trajectory is $\boldsymbol{q}_{r_0} = [0\ 0\ \frac{\pi}{2}]^T$ and of the DWMR is $\boldsymbol{q}_0 = [-0.15\ \ -0.2\ 0]^T$, respectively.

3.1.3 Ideal Scenario

In the first simulation scenario, called ideal scenario, the kinematic and dynamic configuration state-space models with the actuator dynamics are considered, Eq. (1.71), repressnting the Powerbot DWMR and their limitations, disregarding uncertainties and disturbances (e.g., modeling imprecisions; surface friction; external disturbances; bounded unknown disturbances; unmodeled and unstructured dynamics of actuators, sensors, and other electronic devices; integration errors; noises; operational, sampling and physical limitations). The simulations are made in Matlab/Simulink, following the block diagram presented in Fig. 3.6.

An explanation of the blocks corresponding to Fig. 3.6 is as follows:

- Reference Trajectory block generates the reference posture $\boldsymbol{q}_r$ and reference velocity $\boldsymbol{v}_r$, Eq. (3.1);
- First Sum block calculates the posture error $\boldsymbol{q}_e$, as in Eq. (2.18), i.e., the calculates the difference between the reference posture $\boldsymbol{q}_r$ and the actual posture $\boldsymbol{q}$ of the DWMR;
- Kinematic Controller block, with the posture error and the reference velocity, calculates the control velocities $\boldsymbol{v}_c$ using the control laws treated in this book;
- Second Sum block, with the control velocity and the DWMR actual velocity, generates auxiliary velocity tracking errors $\boldsymbol{v}_e$, Eq. 2.3);
- Dynamic Controller block, with the auxiliary velocity tracking errors $\boldsymbol{v}_e$, generates the control torques $\boldsymbol{\tau}$ that will be applied to the DWMR, which are provided by PD control described in Sect. 2.2 using Eqs. (2.1) and (2.2);
- Kinematics and Dynamics blocks, which correspond to simplified models, simulate the PowerBot DWMR behavior, using the kinematic and dynamic configuration state-space models with the actuator dynamics, Eq. (1.71), with parameters provided by the manufacturer of the Powerbot DWMR. This results in the velocities $\boldsymbol{v}$ and $\dot{\boldsymbol{q}}$, and finally;
- Integration block obtains the actual posture $\boldsymbol{q}$ of the DWMR, Eq. (1.12), which is used to calculate the posture error $\boldsymbol{q}_e$, Eq. (2.18).

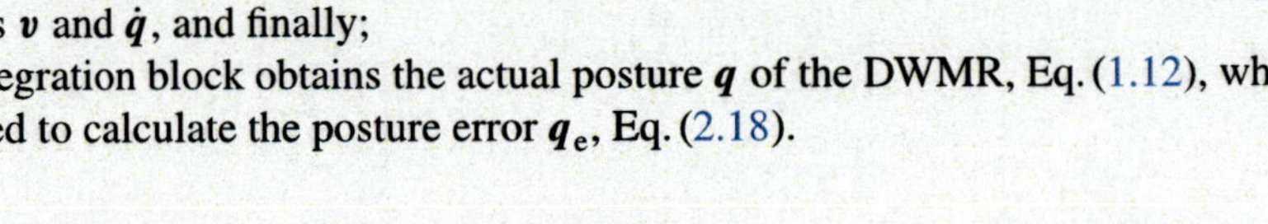

Fig. 3.6 Block diagram of simulation or ideal scenario

To perform the simulations in this scenario, matrices $\bar{\boldsymbol{H}}(\boldsymbol{q})$, $\bar{\boldsymbol{C}}(\boldsymbol{q}, \dot{\boldsymbol{q}})$ and $\bar{\boldsymbol{E}}(\boldsymbol{q})$ of the configuration state-space model with the actuator dynamics, Eq. (1.71), are defined as:

$$\bar{\boldsymbol{H}}(\boldsymbol{q}) = \begin{bmatrix} \bar{\mathrm{H}}_{11} & \bar{\mathrm{H}}_{12} \\ \bar{\mathrm{H}}_{21} & \bar{\mathrm{H}}_{22} \end{bmatrix}, \bar{\boldsymbol{C}}(\boldsymbol{q}, \dot{\boldsymbol{q}}) = \begin{bmatrix} 0 & 0 \\ 0 & 0 \end{bmatrix}, \bar{\boldsymbol{E}}(\boldsymbol{q}) = \begin{bmatrix} 1 & 0 \\ 0 & 1 \end{bmatrix}, \tag{3.2}$$

being

$$\begin{aligned} \bar{\mathrm{H}}_{11} = \bar{\mathrm{H}}_{22} &= \frac{\mathrm{m_t r^2 R^2 + 4R^2 I_w - m_t r^2 d^2 + r^2 I}}{4\mathrm{R}^2}, \\ \bar{\mathrm{H}}_{12} = \bar{\mathrm{H}}_{21} &= \frac{\mathrm{r^2 \left(m_t R^2 + m_t d^2 - I\right)}}{4\mathrm{R}^2}. \end{aligned} \tag{3.3}$$

with $\boldsymbol{q}$, $\boldsymbol{v}$, $\boldsymbol{S}(\boldsymbol{q})$, $\boldsymbol{H}(\boldsymbol{q})$, $\boldsymbol{C}(\boldsymbol{q}, \dot{\boldsymbol{q}})$, $\boldsymbol{E}(\boldsymbol{q})$ and $\boldsymbol{\tau}$ given by the Eqs.(1.12), (1.15), (1.23) and (1.42) to (1.45), respectively.

3.1.4 Realistic Scenario

The second simulation scenario is called the realistic scenario, which instead of using the kinematics and dynamics, Eq. (1.71), uses the MobileSim simulator to represent the behavior of the PowerBot DWMR.

MobileSim is a software designed to simulate the behavior of Mobile Robot platforms produced by Mobile Robots Inc. for debugging and experimentation. To establish communication between the controller in Matlab/Simulink and the MobileSim simulator, ARIA is used. ARIA (Advanced Robot Interface for Applications) is a library for all Mobile Robot platforms capable of dynamically controlling the DWMRs velocity, heading, relative heading, and other motion parameters. ARIA also receives position estimation, sonar readings, and all other current operating data sent by the robotic platform. Figure 3.7 shows a simulation in the MobileSim after the Powerbot DWMR tracks the eight-shape trajectory.

The block diagram presented in Fig. 3.8 shows how ARIA and MobileSim simulator are used to make the simulations in the realistic scenario.

In Fig. 3.8 the Reference Trajectory, Sum, and Kinematic Controller blocks work as in Fig. 3.6. Now, in the realistic scenario, since the DWMR is no longer represented by an equation in Matlab/Simulink, the velocity calculated by the controller passes to the ARIA Function block. This block establishes the communication between Matlab/Simulink and the MobileSim simulator. The MobileSim block simulates the DWMR behavior, representing the Dynamic Controller, Dynamics, and Kinematics blocks of Fig. 3.6, and gives the actual posture $\boldsymbol{q}$ of the DWMR. Again, an ARIA Function block is needed to receive the posture information from the MobileSim simulator, which is used to calculate the posture error.

The main difference is that the MobileSim simulator considers the uncertainties and disturbances, such as modeling imprecisions; surface friction; external

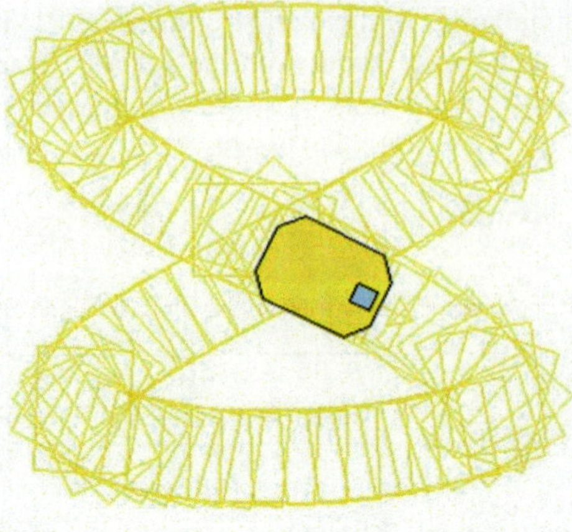

Fig. 3.7 MobileSim interface—simulation of the eight-shape trajectory

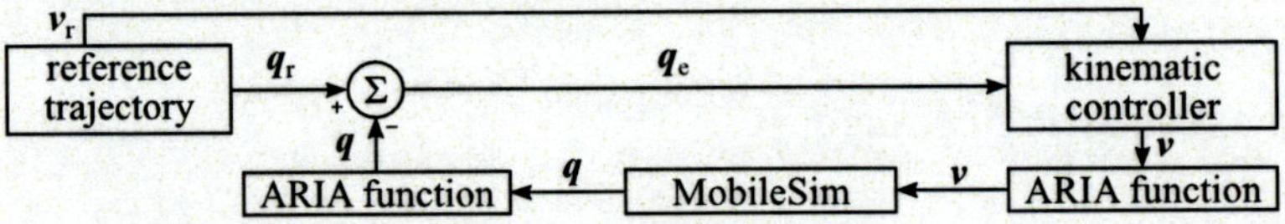

Fig. 3.8 Block diagram of the simulations in the realistic scenario

disturbances; bounded unknown disturbances; unmodeled and unstructured dynamics of actuators, sensors, and other electronic devices; integration errors; noises; operational, sampling and physical limitations. Thus, the auxiliary velocity tracking errors $\boldsymbol{v}_e$ happen, seen as uncertainties and disturbances for the kinematic model and given by the difference between the control velocities $\boldsymbol{v}_c$ and the actual velocities of the DWMR $\boldsymbol{v}$. Therefore, the differences in the simulation results from the ideal to realistic scenarios are due to these uncertainties and disturbances of the DWMR.

3.1.5 Experimental Scenario

After the simulations in the ideal and realistic scenarios, the controllers treated in this book are tested in the Powerbot DWMR, which is a high-payload differential drive robotic platform for research and rapid prototyping. It is an ideal platform for laboratory and research tasks involving delivery, navigation, and manipulation. It is an automated guided vehicle, specially designed and equipped for autonomous, intelligent delivery and handling of large payloads. Powerbot DWMR is a member of the MobileRobots Pioneer family of mobile robots, which are research and development platforms that share a common architecture, foundation software, and employ intelligence based on client-server robotics control [1].

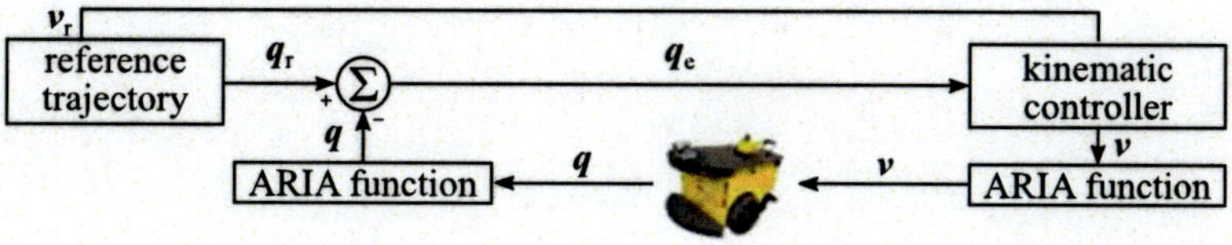

Fig. 3.9 Block diagram of the implementation in a real-time or experimental scenario

Figure 3.9 shows the block diagram of the execution as the MobileSim simulator is replaced by the Powerbot DWMR. ARIA is still used for the communication between the Matlab/Simulink software and the Powerbot DWMR. This communication is carried out by a serial port.

With the internal control system of the PowerBot DWMR, it is possible to obtain experimental results of the integration of the dynamic controller, i.e., internal PD control with kinematic controllers approached here. Simulations in ideal and realistic scenarios are done before of realization of the experimental tests to verify the suitability of the kinematic controllers with the internal control loop and to study the tuning of the gains. Thus, these simulation results provided a preview of what can be expected in the experimental tests, which are developed with the same trajectory and parameter adjustments of the controllers.

References

1. Filipescu, A., Minzu, V., Dumitrascu, B., Minca, E.: Trajectory tracking and discrete-time sliding-mode control of wheeled mobile robots. In *Proceedings of the 2011 IEEE International Conference on Information and Automation—ICIA'2011*, pp. 27–32. IEEE (2011)
2. Oriolo, G., De Luca, A., Vendittelli, M.: WMR control via dynamic feedback linearization: design, implementation, and experimental validation. IEEE Trans. Control Syst. Technol. **10**(6), 835–852 (2002)

Chapter 4
Backstepping Control

4.1 Introduction

In this chapter, the backstepping control technique in Cartesian coordinates (CCC) proposed by [6, 7] used to solve the reference trajectory tracking problem for the kinematics of DWMRs is presented. The CCC is based on the kinematic model expressed in Cartesian coordinates and the backstepping control law is deduced from the stability proof using the Lyapunov-like analysis, which guarantees the asymptotic convergence of the posture error. Moreover, simulations and real-time results are presented and discussed to verify the performance of the CCC.

4.2 Control Design

DWMRs are characterized by nonholonomic systems, a class of systems that cannot be stabilized by smooth static state feedback laws. Backstepping control is one of the most important methods for stabilizing a nonholonomic system. The control problem for nonlinear systems has been researched [2].

Backstepping is a recursive procedure that breaks a design problem for the whole system into a sequence of design problems for lower order systems. Backstepping for nonlinear systems was investigated by several researchers [8, 18].

The method proved to be effective for the trajectory tracking control problem of DWMRs [1–3, 6, 7, 14], i.e., the backstepping controller is the most commonly used approach for a DWMR to track the desired trajectory.

Electronic supplementary material The online version of this chapter (https://doi.org/10.1007/978-3-030-77912-2_4) contains supplementary material, which is available to authorized users.

N. A. Martins and D. W. Bertol, *Wheeled Mobile Robot Control*, Studies in Systems, Decision and Control 380, https://doi.org/10.1007/978-3-030-77912-2_4

The backstepping controller denominated as CCC based on the kinematic model of the DWMR was first suggested by Kanayama et al. [6, 7]. This method is preferable due to its simple structure and suitable for applications that have small tracking errors. The inputs to the CCC are the reference velocities and the desired position trajectories, whereas the CCC outputs are the DWMR linear and angular velocities. It is noteworthy that the DWMR velocity via the backstepping method is directly related to the state tracking errors and this CCC law is widely used and recommended in various studies [2, 14].

As described in Sects. 2.1 and 2.3, the problem of tracking a trajectory consists of defining the linear and angular velocities ($\upsilon_{\rm r}$ and $\omega_{\rm r}$) of the reference trajectory described by a fictional (ideal) DWMR, having the same kinematic characteristics of the real DWMR, moments of inertia and other dynamic properties. The kinematic model is then used to obtain the reference trajectory, which should be followed by the real DWMR.

For the CCC are considered the kinematic model of the actual DWMR with $\mathrm{C} = \mathrm{P}$, Eqs. (1.18), (1.19), (1.21) and (1.24), and the reference kinematic model, Eqs. (2.16) and (2.17), in treatment of the trajectory tracking problem, as can be seen in Fig. 2.2.

The posture error vector $\boldsymbol{q}_{\rm e}$ expressed on frame $\mathrm{CX_cY_c}$ (Fig. 1.1) is defined as in Eq. (2.18). Differentiating the Eq. (2.18) yields the posture error dynamics of the closed-loop system $\dot{q}_{\rm e}$, Eq. (2.22).

By using the Barbalat Lemma [10], one can deduce the CCC law proposed by Kanayama et al. [6, 7] which guarantees system stability and asymptotic convergence of posture error. Assuming that $\theta_{\rm e} \in [-\pi, \pi]$, consider the following scalar function as Lyapunov function candidate:

$$V(\boldsymbol{q}_{\rm e}) = \frac{1}{2}(x_{\rm e}^2 + y_{\rm e}^2) + \frac{1}{\mathrm{k_y}}\Big(1 - \cos(\theta_{\rm e})\Big) \geq 0, \tag{4.1}$$

with $\mathrm{k_y} > 0$. If $\boldsymbol{q}_{\rm e} = \boldsymbol{0}$, $V(\boldsymbol{q}_{\rm e}) = \boldsymbol{0}$. If $\boldsymbol{q}_{\rm e} \neq 0$, $V(\boldsymbol{q}_{\rm e}) > 0$. Differentiating the Eq. (4.1) in relation to time, one obtains:

$$\dot{V}(\boldsymbol{q}_{\rm e}) = \dot{x}_{\rm e} x_{\rm e} + \dot{y}_{\rm e} y_{\rm e} + \frac{1}{\mathrm{k_y}} \dot{\theta}_{\rm e} \sin(\theta_{\rm e}). \tag{4.2}$$

Substituting the Eq. (2.23) into Eq. (4.2), it results:

$$\begin{aligned} \dot{V}(\boldsymbol{q}_{\rm e}) = {} & \Big(\omega y_{\rm e} - \upsilon + \upsilon_{\rm r}\cos(\theta_{\rm e})\Big)x_{\rm e} + \Big(\upsilon_{\rm r}\sin(\theta_{\rm e}) - \omega x_{\rm e}\Big)y_{\rm e} \\ & + \tfrac{1}{\mathrm{k_y}}\left(\omega_{\rm r} - \omega\right)\sin(\theta_{\rm e}). \end{aligned} \tag{4.3}$$

Consider the following CCC law:

$$\boldsymbol{\upsilon} = \boldsymbol{\upsilon}_{\rm c} = \begin{bmatrix} \upsilon_{\rm c} \\ \omega_{\rm c} \end{bmatrix} = \begin{bmatrix} \upsilon \\ \omega \end{bmatrix} = \begin{bmatrix} \upsilon_{\rm r}\cos(\theta_{\rm e}) + \upsilon_{\rm c_1}^* \\ \omega_{\rm r} + \upsilon_{\rm c_2}^* \end{bmatrix}, \tag{4.4}$$

where

$$
\begin{aligned}
v_{c_1}^* &= k_x x_e, \\
v_{c_2}^* &= v_r\Big(k_y y_e + k_\theta \sin(\theta_e)\Big),
\end{aligned} \tag{4.5}
$$

with k_x, k_y, and k_θ being positive constants defined in the CCC design.

Substituting Eq. (4.4) into Eq. (4.3), $\dot{V}$ leads to:

$$
\dot{V}(\boldsymbol{q}_e) = -k_x x_e^2 - \frac{v_r k_\theta \sin^2(\theta_e)}{k_y} \leq 0, \tag{4.6}
$$

being, therefore, negative semidefinite for $v_r > 0$. Therefore, it can be concluded that the point $\boldsymbol{q}_e = 0$ is a stable equilibrium point if the reference velocity $v_r > 0$.

Differentiating from Eq. (4.6), one obtains:

$$
\ddot{V}(\boldsymbol{q}_e) = -2k_x x_e \dot{x}_e - \frac{2v_r k_\theta \dot{\theta}_e \sin(\theta_e)\cos(\theta_e)}{k_y}. \tag{4.7}
$$

From Eq. (4.6) it is verified that x_e, y_e and θ_e are bounded, and since $\dot{x}_e$ and $\dot{y}_e$ are functions of x_e, y_e, θ_e, and of v_r, ω_r, k_x, k_y, k_θ, bounded by definition, it can be deduced that $\ddot{V}(\boldsymbol{q}_e)$ is bounded. Thus, from the Barbalat Lemma [10], it can be inferred that $\dot{V}(\boldsymbol{q}_e)$ is uniformly continuous, and therefore $\dot{V}(\boldsymbol{q}_e) \to 0$. Thus, it is immediate that both x_e and θ_e converge asymptotically to zero ($x_e \to 0$ and $\theta_e \to 0$). It remains to prove the convergence of y_e. From Eq. (2.23), it is found that $\dot{y}_e \to 0$, and as y_e is bounded, it converges to a constant. Equation (4.4) ensures that $v \to v_r$. Since $x_e \to 0$ and $\theta_e \to 0$, and the uniform continuity of $\dot{x}_e$ and $\dot{\theta}_e$ in Eq. (2.23), one can conclude that $\omega \to \omega_r$ and $y_e \to 0$ for $\omega \neq 0$. Equation (4.4) also ensures that $y_e \to 0$ for $v_r \neq 0$.

Moreover, [6, 7] makes the following proposition, "*It is assumed that (a) v_r and ω_r are continuously differentiable and bounded, (b) there is a positive constant δ_r such that $v_r \geq \delta_r$ for all $t \geq 0$, (c)* k_x, k_y *and* k_θ *are positive constants, and (d) $\dot{v}_r$ and $\dot{\omega}_r$ are sufficiently small. Under these conditions, $\boldsymbol{q}_e = 0$ from* Eqs. (2.23) *and* (4.4) *is a uniformly, asymptotically stable point on* $[0, \infty)$.", whose the proof of this proposition as well as more details about the stability proof of this CCC law can be found in [6, 7].

4.3 Simulations Using Matlab and/or MobileSim Simulator

Following some results are presented, obtained by simulations, to verify the effectiveness of the CCC law presented in Eq. (4.4). The reference trajectory, the linear and angular reference velocities as well as initial conditions of posture (reference trajectory and DWMR) are founded in Sect. 3.1.2. The gains selected for this CCC are provided in Table 4.1.

Table 4.1 CCC gains

Controller	k_x	k_y	k_θ
CCC	0.9	8.0	3.0

Table 4.2 RMS of the errors—simulation results in the ideal scenario

Controller	x_e (m)	y_e (m)	θ_e (rad)
CCC	0.0161	0.0357	0.0463

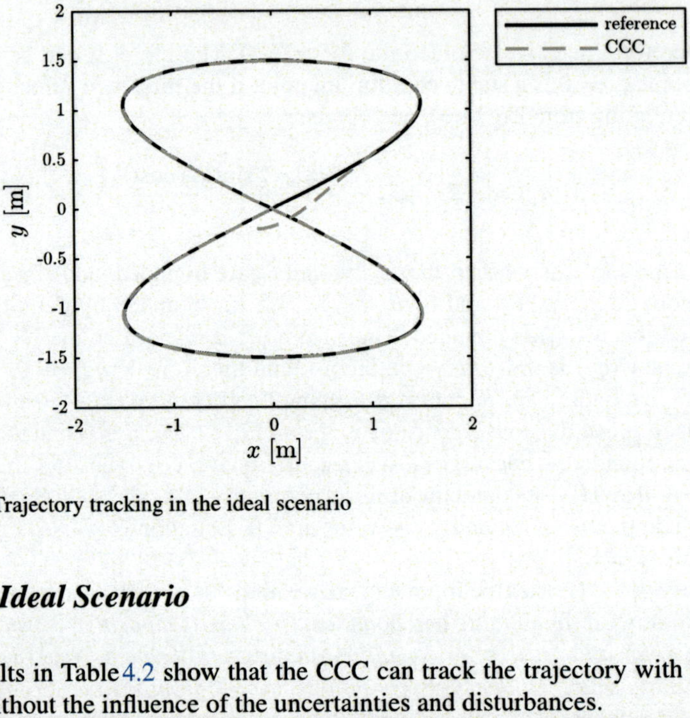

Fig. 4.1 Trajectory tracking in the ideal scenario

4.3.1 Ideal Scenario

The results in Table 4.2 show that the CCC can track the trajectory with minimum errors without the influence of the uncertainties and disturbances.

From the observation of Figs. 4.1 and 4.2, it can be seen that the CCC performs the trajectory tracking satisfactorily, as well as the posture trajectories (x, y and θ) tend to converge to their desired values (x_r, y_r, and θ_r).

According to Fig. 4.3, posture tracking errors oscillate very close to zero during the entire DWMR movement, since, for this trajectory, the curvature is variable, consequently the velocities, especially the angular, are also variable.

Figure 4.4 shows the control velocities generated by the CCC while Fig. 4.5 shows that the behavior of the DWMR linear and angular velocity signals produced by the CCC are similar and converge to the linear and angular reference velocities, whose profiles are not constant during the entire tracking.

Observing Fig. 4.6, it is verified that the control velocities v_c^*, Eq. (4.5), were necessary to compensate the auxiliary velocity tracking errors, noting that there is

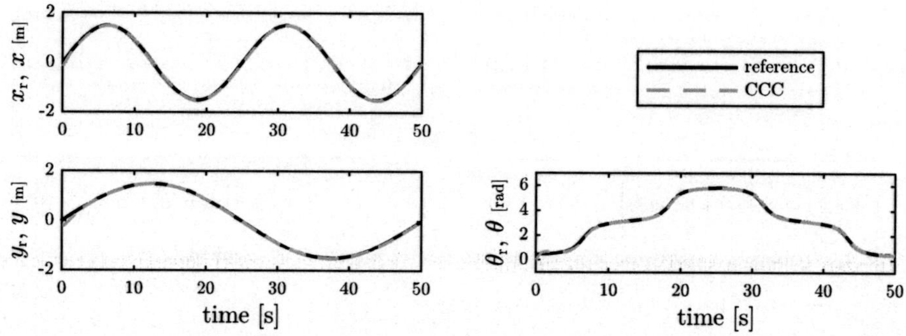

Fig. 4.2 Trackings (x, y, and θ) in the ideal scenario

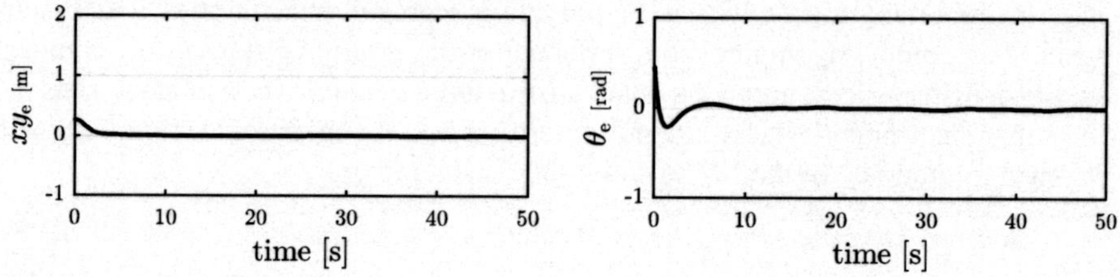

Fig. 4.3 Posture tracking errors, with $xy_e = \sqrt{x_e^2 + y_e^2}$, in the ideal scenario

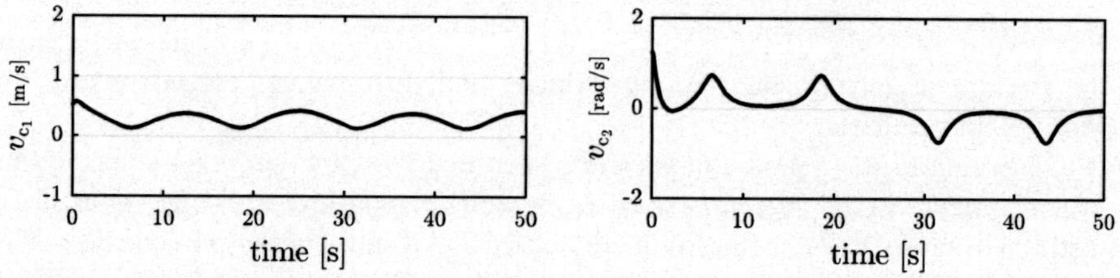

Fig. 4.4 Control velocities in the ideal scenario

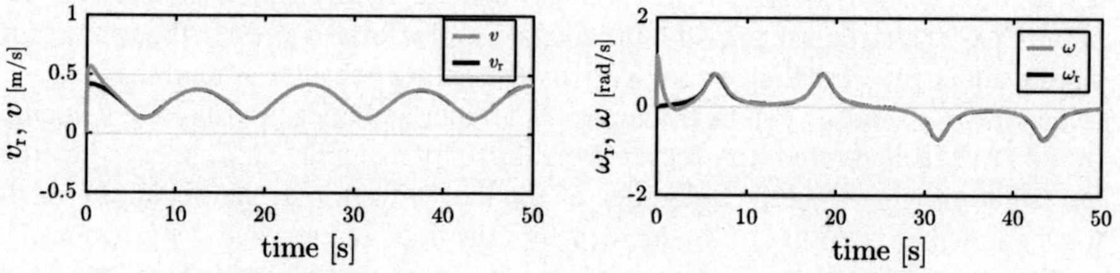

Fig. 4.5 Velocities of the DWMR and reference velocities in the ideal scenario

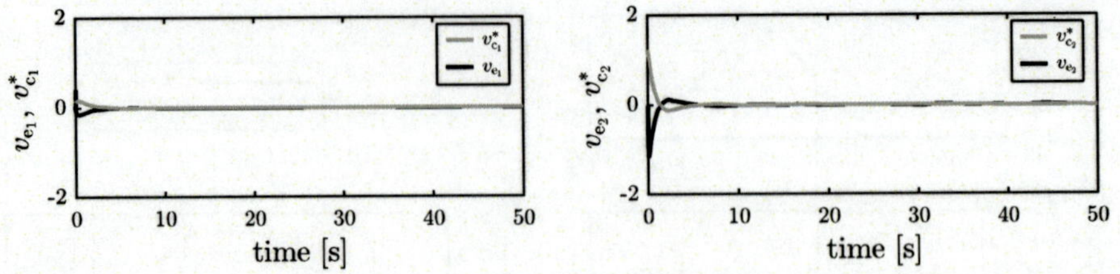

Fig. 4.6 Auxiliary velocity tracking errors and control velocities υ_c^* in the ideal scenario

no influence of the uncertainties and disturbances, even with the integration of the kinematic (CCC) and dynamic (PD) controls.

It is important to recall that in the ideal scenario uncertainties and disturbances (e.g., modeling imprecision; surface friction; external disturbances; bounded unknown disturbances; unmodeled and unstructured dynamics of actuators, sensors, and other electronic devices; integration errors; noises; operational, sampling, and physical limitations) of the DWMR are not considered.

4.3.2 *Realistic Scenario*

The results of RMS of the errors in Table 4.3 suffered a significant increase in the tracking trajectory, due to uncertainties and disturbances considered by the MobileSim simulator.

Observing Figs. 4.7 and 4.8, it can be seen that the CCC suffers a deterioration of performance in the trajectory tracking as well as in posture trackings. This fact is justified because the incidence of uncertainties and disturbances is not considered in the design of this CCC, Eq. (4.4).

Regarding posture tracking errors, Fig. 4.9, note that the influence of uncertainties and disturbances on the kinematic model of posture visibly affects the performance of the CCC, which fails to provide sufficient control efforts to reject it, thus presenting significant posture tracking errors with an alternating behavior around the zero.

Figures 4.10 and 4.11 show the control velocities and linear and angular velocities of the DWMR and reference, respectively. Still, by using the CCC, these velocities are slightly higher than the velocities of the ideal scenario at the beginning of the transient regime, because of the incidence of the uncertainties and disturbances.

Verifying Fig. 4.12, the control velocities υ_c^*, Eq. (4.5), apparently compensate the auxiliary velocity tracking errors. Although it is barely perceptible in the visualization

Table 4.3 RMS of the errors—simulation results in the realistic scenario

Controller	x_e (m)	y_e (m)	θ_e (rad)
CCC	0.0190	0.0356	0.0451

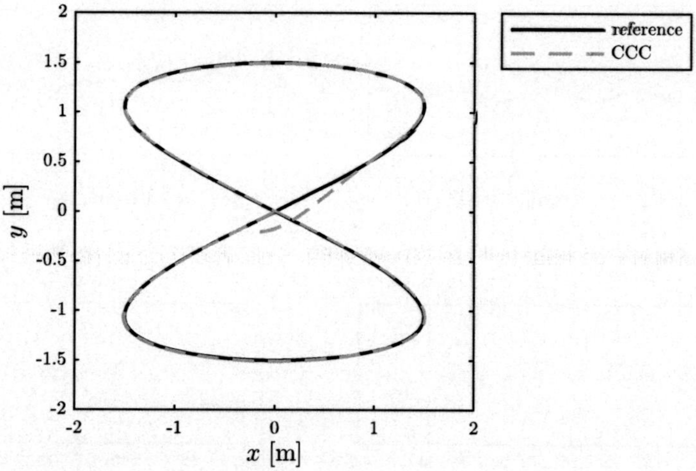

Fig. 4.7 Trajectory tracking in the realistic scenario

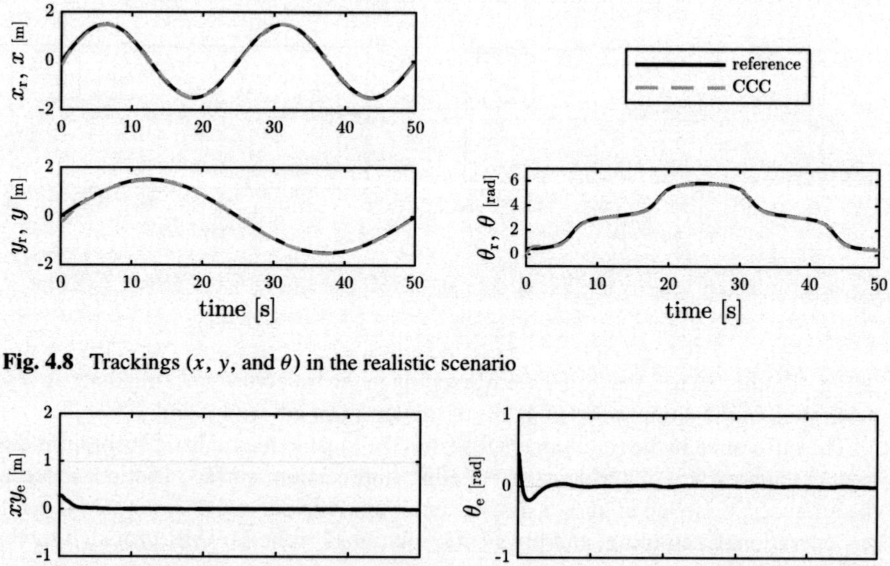

Fig. 4.8 Trackings (x, y, and θ) in the realistic scenario

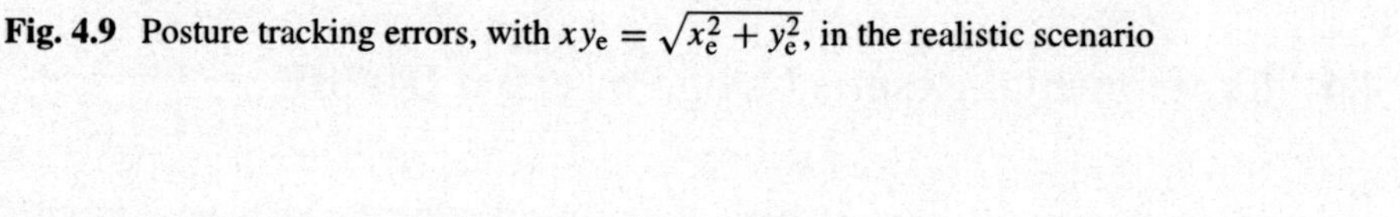

Fig. 4.9 Posture tracking errors, with $xy_e = \sqrt{x_e^2 + y_e^2}$, in the realistic scenario

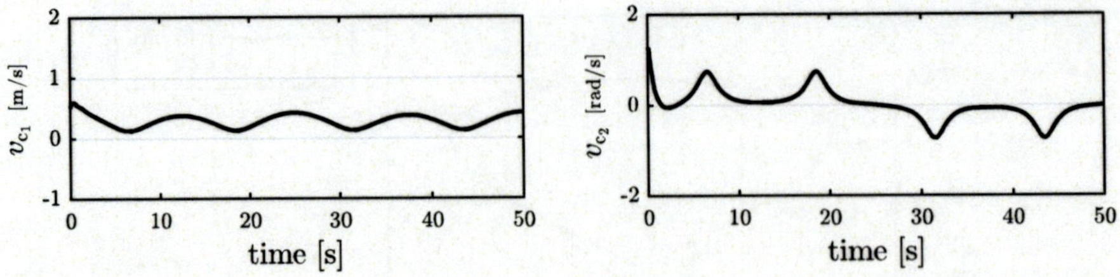

Fig. 4.10 Control velocities in the realistic scenario

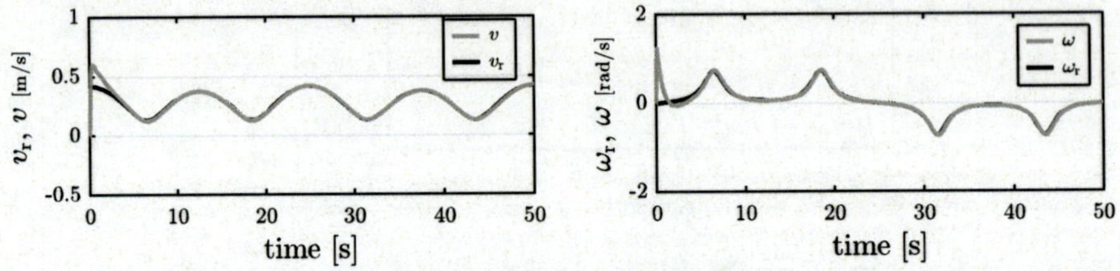

Fig. 4.11 Velocities of the DWMR and reference velocities in the realistic scenario

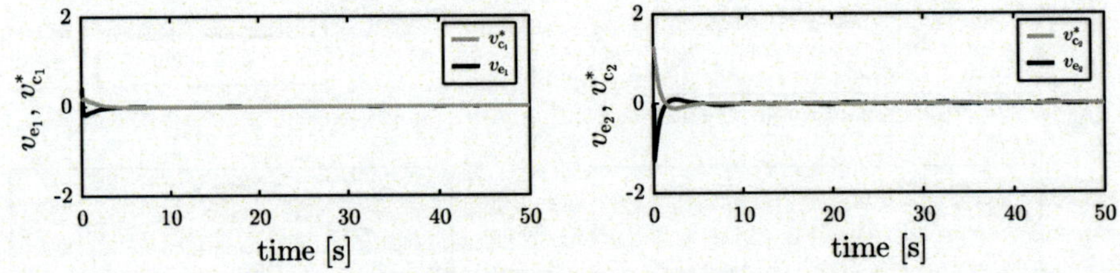

Fig. 4.12 Auxiliary velocity tracking errors and control velocities v_c^* in the realistic scenario

of Fig. 4.7, there was a performance worsening in the trajectory tracking that can also be verified in the comparison of RMS of the errors in Tables 4.2 and 4.3.

The difference in the tracking results from the ideal to the realistic scenario is due to uncertainties and disturbances (modeling imprecision; surface friction; external disturbances; bounded unknown disturbances; unmodeled, and unstructured dynamics; operational, sampling, and physical limitations) of the DWMR provided by the MobileSim simulator.

4.4 Experimental Results Using PowerBot DWMR

The tests performed on the Powerbot DWMR took into account the CCC law Eq. (4.4), the reference trajectory in Sect. 3.1.2, and the controller gains contained in Table 4.1 for obtaining the experimental results.

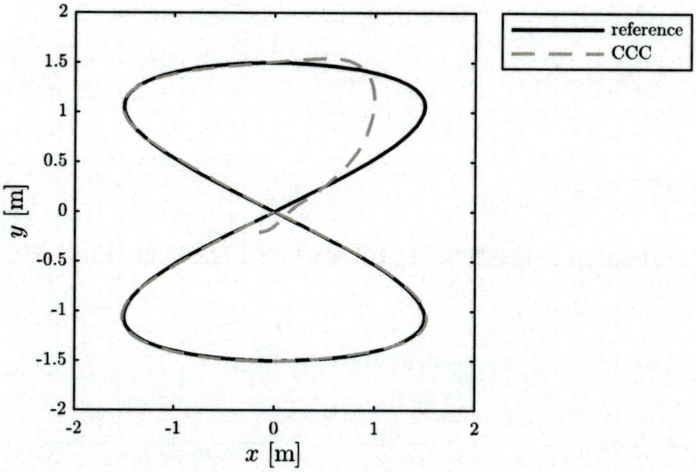

Fig. 4.13 Trajectory tracking in the experimental scenario

Comparing the RMS error values using Tables 4.2, 4.3, and 4.4, it can be seen that PowerBot DWMR performed with a significant deterioration in the experimental tests compared to the simulations performed in the ideal and realistic scenarios.

Such performance deterioration can also be seen in Figs. 4.13, 4.14, and 4.15, where PowerBot DWMR does not tracks the desired trajectory, as well as posture tracking errors, are significant within the first 15 s, which justifies the higher RMS error values compared to the ideal and realistic scenarios. After this time interval, PowerBot DWMR behaves similarly to the realistic scenario.

Observing Fig. 4.16, it is verified that the control velocity profile has similar behavior and close to the reference velocity profile.

Based on Fig. 4.17, velocities of the PowerBot DWMR in the first 15 s do not converge to reference velocities. Also, both PowerBot DWMR and reference velocity profiles are very similar as can be viewed after this time interval.

Regarding the behavior of the control velocities υ_c^*, Eq. (4.5), and the auxiliary velocity tracking errors, Fig. 4.18, the same statements considered in the realistic scenario must be valid in the experimental scenario.

Finally, intrinsic uncertainties and disturbances (modeling imprecision; surface friction; external disturbances; bounded unknown disturbances; unmodeled and unstructured dynamics; operational, sampling, and physical limitations) of the

Table 4.4 RMS of the errors—results in the experimental scenario

Controller	x_e (m)	y_e (m)	θ_e (rad)
CCC	0.2904	0.1809	0.2299

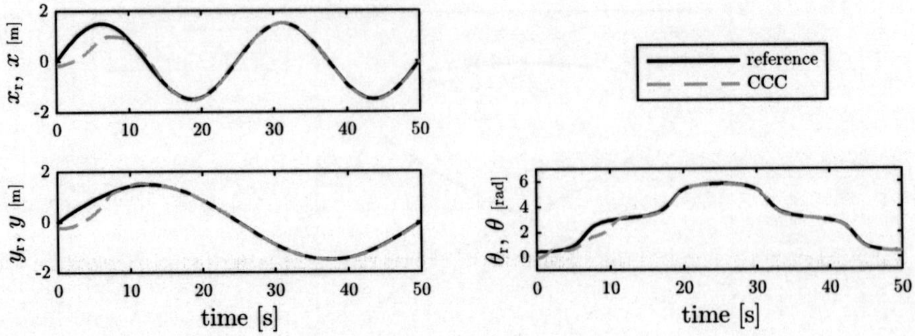

Fig. 4.14 Trackings (x, y, and θ) in the experimental scenario

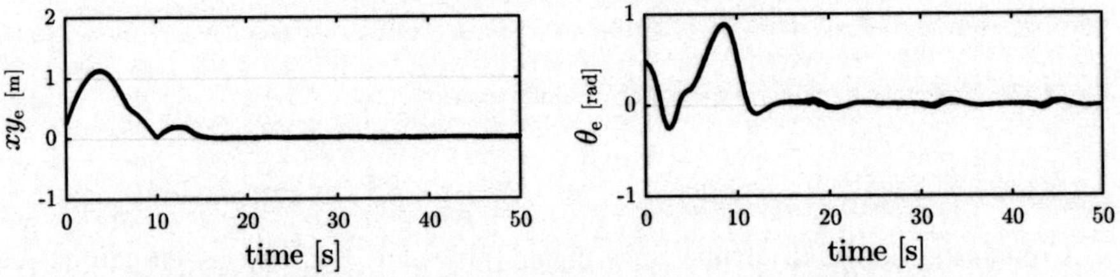

Fig. 4.15 Posture tracking errors, with $xy_e = \sqrt{x_e^2 + y_e^2}$, in the experimental scenario

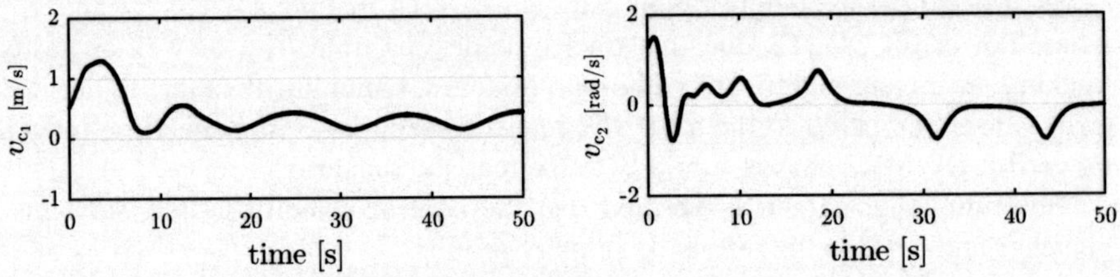

Fig. 4.16 Control velocities in the experimental scenario

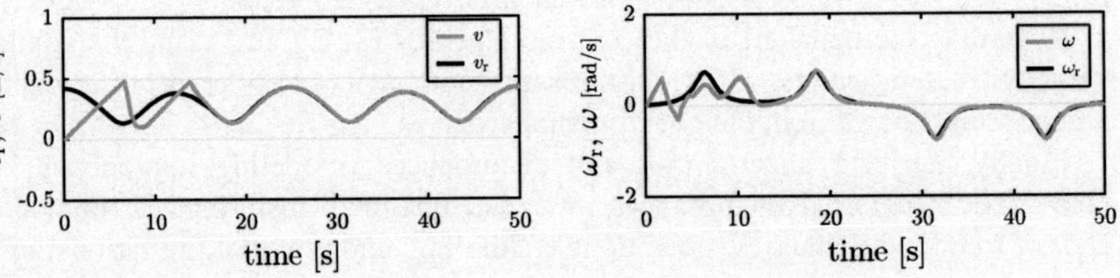

Fig. 4.17 Velocities of the DWMR and reference velocities in the experimental scenario

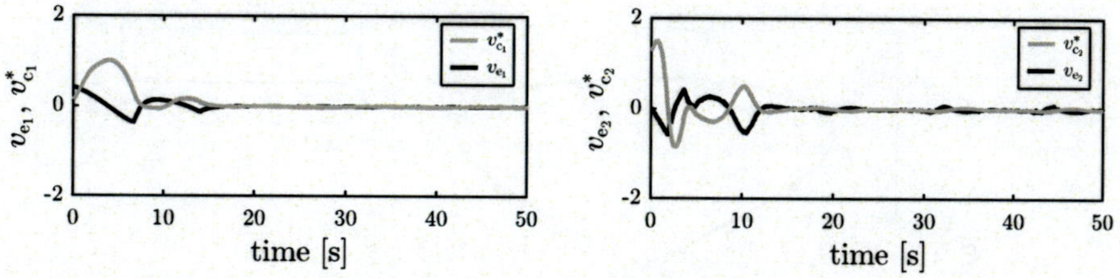

Fig. 4.18 Auxiliary velocity tracking errors and control velocities v_c^* in the experimental scenario

PowerBot DWMR during experimental tests should be taken into account in obtaining the results, mainly as can be seen in the first 15 s of the experiments.

4.5 Analysis and Discussion of Results

The analysis and discussion are carried out by comparing the results of simulations obtained in the ideal scenario with or without the incidence of uncertainties and disturbances. Therefore, this comparative study considers the CCC simulation for the nominal case without uncertainties and disturbances described in Sect. 4.3.1 and the CCC simulation for the case with uncertainties and disturbances addressed in this section.

To verify the damaging effects, the simulation results in the ideal scenario are shown with the performance of this controller for the case with uncertainties and disturbances in the dynamic model, which, consequently, causes auxiliary velocity tracking errors in the kinematic model of the DWMR. The simulations in the ideal scenario are performed considering load variations (mass and moment of inertia), unmodeled dynamics introduced in the form of friction, time-varying external disturbances with unknown upper bounds in the dynamic model. These uncertainties and disturbances are similar to [11, 12, 16, 17].

Analyzing Table 4.5, which shows the RMS error values for the CCC under the incidence of uncertainties and disturbances in the ideal scenario, it is verified that the DWMR does not track the trajectory adequately due to these values are significantly higher than those in Table 4.2.

Observing Fig. 4.19, it appears that the CCC does not present a satisfactory performance in the trajectory tracking. A possible justification for this fact is due to the

Table 4.5 RMS of the errors—results in the ideal scenario under the incidence of uncertainties and disturbances

Controller	x_e (m)	y_e (m)	θ_e (rad)
CCC	0.0801	0.0754	0.1112

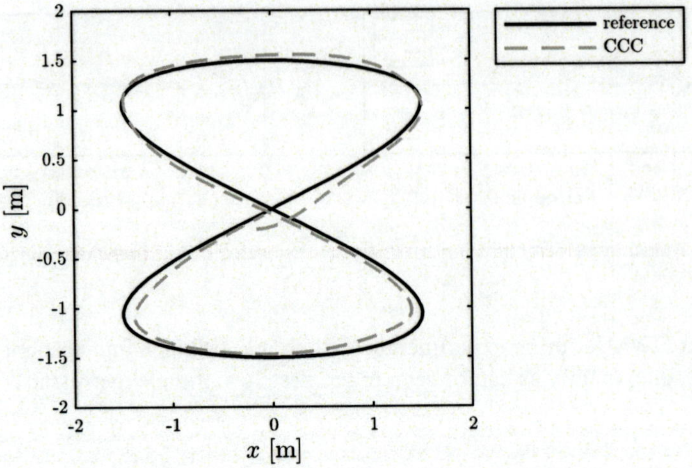

Fig. 4.19 Trajectory tracking in the ideal scenario under the incidence of uncertainties and disturbances

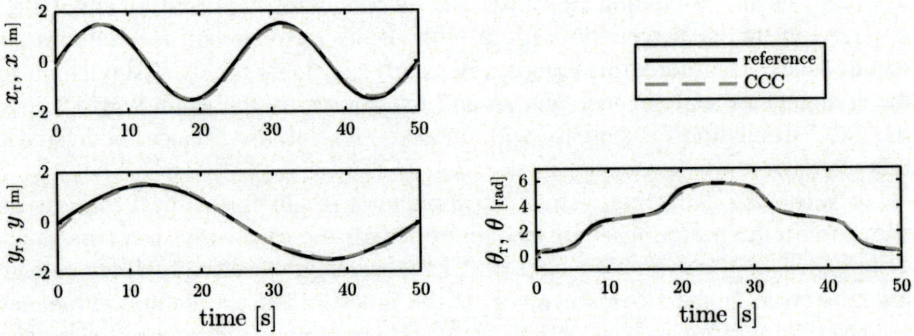

Fig. 4.20 Trackings (x, y, and θ) in the ideal scenario under the incidence of uncertainties and disturbances

incidence of uncertainties and disturbances, whose treatment is not considered in the design of this controller, Eq. (4.4). In the observation of Fig. 4.20, the influence of uncertainties and disturbances is almost imperceptible, but there is a small difference in behavior between the posture tracking performed by the DWMR and the reference ones.

Concerning the posture tracking errors, Fig. 4.21, it is noted that the influence of uncertainties and disturbances on the kinematic posture model visibly affects the performance of the CCC, thus presenting posture tracking errors significant with alternating behavior around zero.

Furthermore, due to the incidence of uncertainties and disturbances, from Fig. 4.22, it is possible to verify that, by using the CCC, the control velocities are significantly higher when compared to the control velocities Fig. 4.4, as well as the linear and angu-

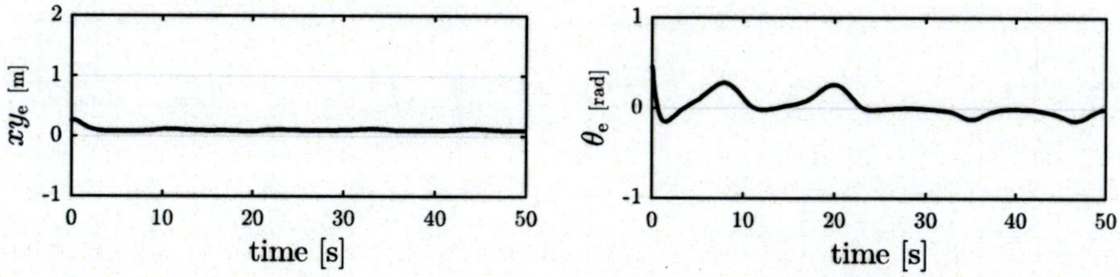

Fig. 4.21 Posture tracking errors, with $xy_e = \sqrt{x_e^2 + y_e^2}$, in the ideal scenario under the incidence of uncertainties and disturbances

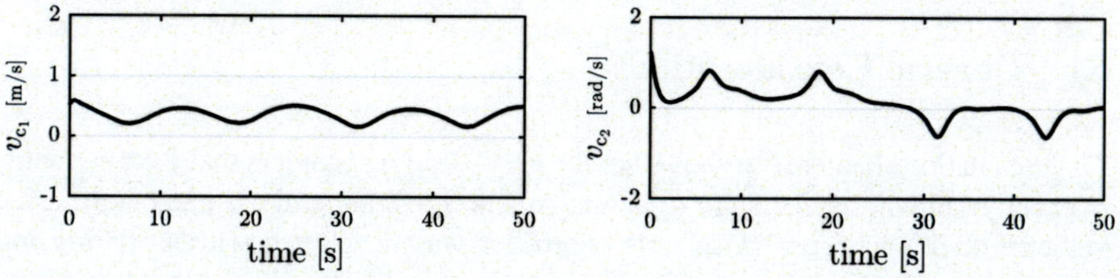

Fig. 4.22 Control velocities in the ideal scenario under the incidence of uncertainties and disturbances

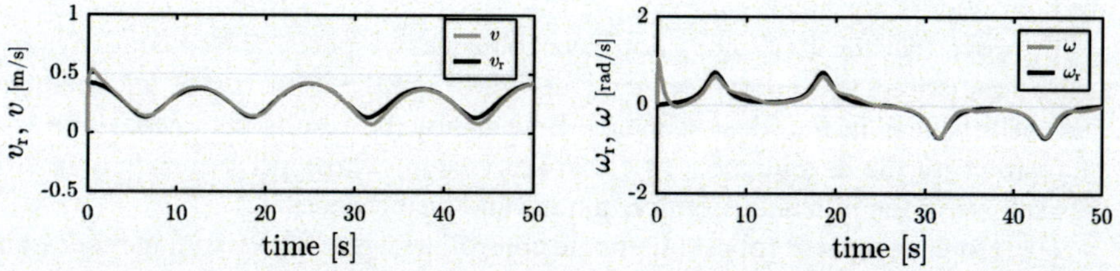

Fig. 4.23 Velocities of the DWMR and reference velocities in the ideal scenario under the incidence of uncertainties and disturbances

lar velocities of the DWMR, shown in Fig. 4.23, present slightly different behaviors compared to the linear and angular velocities of the DWMR, as well as with the linear and angular reference speeds of Fig. 4.5.

Concerning the behavior of the control velocities v_c^*, Eq. (4.5), and the auxiliary velocity tracking errors, Fig. 4.24, it is verified that control efforts provided by v_c^* are more than sufficient to compensate for such errors, so that the control velocities v_c, Eq. (4.4), exceed in their magnitudes necessary for adequate compensation, which consequently results in the application of unnecessary control efforts to the DWMR actuators, in addition to providing an increase in posture tracking errors, as already seen in the Fig. 4.21.

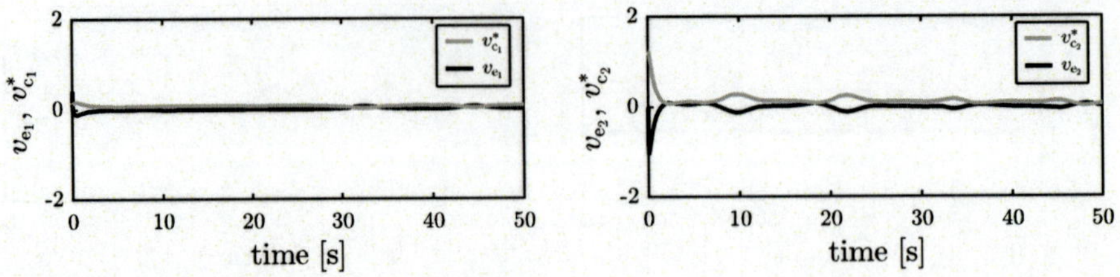

Fig. 4.24 Auxiliary velocity tracking errors and control velocities v_c^* in the ideal scenario under the incidence of uncertainties and disturbances

4.6 General Considerations

The idea in this chapter is to show that the CCC design considers that the kinematics is exactly known, neglects the dynamics of the DWMR, and assumes that perfect velocity tracking occurs. Also, this controller was developed without taking into account aspects of robustness. However, the statement of perfect velocity tracking does not remain in practice, since, in real situations, the uncertainties and disturbances in the kinematic and dynamic models can compromise the accuracy of the results, if the controller is not robust [4, 5, 9, 13, 15].

To prove that the CCC does not have robustness aspects, a comparative study using simulations was carried out considering or not the influence of uncertainties and disturbances in the ideal scenario. Because of the results obtained, there is a motivation for the development of controller design, whose performance is free or less subject to the influence of uncertainties and disturbances.

Therefore, designs of robust kinematic controllers, specifically sliding mode controllers, to deal with problems of uncertainties and disturbances in the kinematic model in rectangular coordinates are developed in later chapters.

References

1. Amin, S.H.M., Husain A.R., Sanhoury, I.M.H.: Trajectory tracking of steering system mobile robot. In: International Conference on Mechatronics: Integrated Engineering for Industrial and Societal Development, pp. 1–5. ICOM'11. IEEE, Kuala Lumpur (2011)
2. Dumitrascu, B., Filipescu, A., Minzu, V., Filipescu, A.: Backstepping control of wheeled mobile robots. In: 15th International Conference on System Theory, Control and Computing, pages 1–6 (2011)
3. Fierro, R., Lewis, F.L.: Control of a nonholonomic mobile robot: backstepping kinematics into dynamics. J. Robot. Syst. **14**(3), 149–163 (1997)
4. Fierro, R., Lewis, F.L.: Control of a nonholonomic mobile robot using neural networks. IEEE Trans. Neural Netw. **9**(4), 589–600 (1998)
5. Figueiredo, L.C., Jota, F.G.: Introdução ao controle de sistemas não-holonômicos. Revista de Controle & Automação **15**(3), 9 (2004)

6. Kanayama, Y., Kimura, Y., Miyazaki, F., Noguchi, T.: A stable tracking control method for an autonomous mobile robot. In: IEEE International Conference on Robotics and Automation, pages 384–389, 1990
7. Kanayama, Y., Kimura, Y., Miyazaki, F., Noguchi, T.: A stable tracking control method for a non-holonomic mobile robot. In: Proceedings of the IEEE/RSJ International Workshop on Intelligent Robots and Systems (IROS'1991), pages 1236–1241. IEEE (1991)
8. Khalil, H.K.: Nonlinear Systems, 31 edn. Pearson (2001)
9. Kolmanovsky, I., Harris McClamroch, N.: Developments in nonholonomic control problems. IEEE Control Syst. **15**(6), 20–36 (1995)
10. Lewis, F.L., Dawson, D.M., Abdallah, C.T.: Robot Manipulator Control: Theory and Practice, 2nd edn. Marcel Dekker, Inc. (2003)
11. Li, Y.D., Zhu, L., Sun, M.: Adaptive neural-network control of mobile robot formations including actuator dynamics. In:Sensors, Measurement and Intelligent Materials, volume 303 of Applied Mechanics and Materials, pages 1768–1773. Trans Tech Publications (2013)
12. Li, Y.D., Zhu, L., Sun, M.: Adaptive RBFNN formation control of multi-mobile robots with actuator dynamics. Indo. J. Electr. Eng. **11**(4), 1797–1806 (2013)
13. Morin, P., Samson, C.: Motion control of wheeled mobile robots. In: Siciliano, B., Khatib, O., (eds) Handbook of Robotics, chapter Part E 34, pages 799–826. Springer (2008)
14. Obaid, M.A.M., Husain, A.R.: Time varying backstepping control for trajectory tracking of mobile robot. Int. J. Comput. Vis. Robot. **7**(1/2), 172–181 (2017)
15. Oriolo, G.: Wheeled robots. In: Encyclopedia of Systems and Control, pages 1–9. Springer (2014)
16. Park, B.S., Yoo, S.J., Park, J.B., Choi, Y.H.: Adaptive neural sliding mode control of nonholonomic wheeled mobile robots with model uncertainty. IEEE Trans. Control Syst. Technol. **17**(1), 207–214 (2009)
17. Seok Park, B., Park, J.B., Choi, Y.H.: Robust formation control of electrically driven nonholonomic mobile robots via sliding mode technique. Int. J. Control Autom. Syst. **9**(5), 888–894 (2011)
18. Shi, H.: A novel scheme for the design of backstepping control for a class of nonlinear systems. Appl. Math. Model. **35**(4), 1893–1903 (2011)

Chapter 5
Robust Control: First-Order Sliding Mode Control Techniques

5.1 Introduction

In this chapter, the problem of control of trajectory tracking of DWMRs based only on their kinematic model described in rectangular coordinates is investigated. This model, which is a simplified representation of the actual DWMR, takes into account uncertainties and disturbances during its movement, so that the statement of perfect velocity tracking (auxiliary velocity tracking error is null) is not maintained in the practice. Thus, to follow the trajectories, as well as to deal with these uncertainties and disturbances in the kinematic model, controller variants are treated based on sliding mode control theory (SMC) to provide robustness, being the stability analysis deduced by the method of Lyapunov. The performance of these variants termed first-order sliding mode controls (FOSMCs) is demonstrated employing simulations and real-time experiments.

5.2 Control Design

The term sliding mode control (SMC) is due to the fact that the control structure is altered intentionally, satisfying some criterion or condition. A simple example is the switching between different gain values in the feedback loop, obeying a rule that satisfies the control objectives. The SMC is a feedback control with high-velocity switching, whose task is to drive the trajectory of states of the system (linear or nonlinear) to a region of the state-space chosen by the designer, keeping it there for all subsequent time. In general, this region is defined as a hypersurface of states of the

Electronic supplementary material The online version of this chapter (https://doi.org/10.1007/978-3-030-77912-2_5) contains supplementary material, which is available to authorized users.

N. A. Martins and D. W. Bertol, *Wheeled Mobile Robot Control*, Studies in Systems, Decision and Control 380, https://doi.org/10.1007/978-3-030-77912-2_5

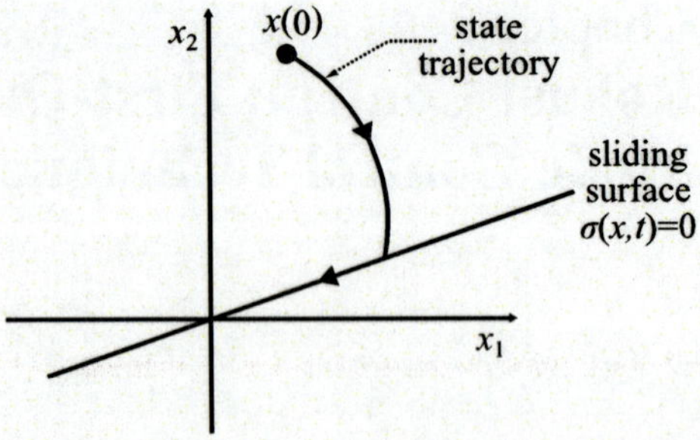

Fig. 5.1 Trajectory of states of a system under the action of an SMC

system, known as a sliding surface. When constrained to this surface, the trajectory of states of the controlled system will depend on the dynamic properties of the surface and will become insensitive to uncertainties and disturbances [7, 14].

The control structure is changed according to the relative position of the trajectory of states related to the sliding surface. If the trajectory of states is above this surface, an appropriate gain value is assigned; if it is below it, another gain is chosen, such that the trajectory of states always directed towards the sliding surface.

The property of the trajectory of states remaining on the sliding surface once the intercept is called sliding mode. A sliding mode will exist for a system if, in the vicinity of the sliding surface, the velocity vector of states (derived from the vector of states) is directed to that surface. For a clear understanding of these concepts, consider Fig. 5.1, in which a trajectory of states intercepts the sliding surface and slides on it under an ideal SMC.

Figure 5.1 shows the existence of two phases in the SMC: the reaching phase and the sliding phase. The first phase is to make the trajectory of states of the system reach the sliding surface in finite time from any initial conditions. During this period, trajectories are sensitive to uncertainties and disturbances. In the second phase, the control effort is directed to maintain the trajectory of states of the system by sliding on the surface towards the control objective, which is obtained using the switched control.

5.2.1 Control Technique

In this subsection, an SMC is derived based on Eq. (2.30) to act on the trajectory of the states in the reaching and sliding phases, even in the presence of uncertainties and disturbances. In the sliding phase, the trajectories of the states are restricted to the sliding surface, so the behavior of the SMC depends on the shape and parameters of the chosen surface.

The sliding surface, $\boldsymbol{\sigma}$, should aim at the convergence of the system errors $\tilde{z}$ to zero, i.e.:

$$\boldsymbol{\sigma}(\tilde{z}, t) = \boldsymbol{\Lambda}^{\mathrm{T}}\tilde{z} = \mathbf{0}, \tag{5.1}$$

thus, in the sliding phase, the errors $\tilde{z}$ will tend exponentially to zero according to a standard determined by the matrix $\boldsymbol{\Lambda}$ containing positive constants.

Also, to influence in the process of reaching the sliding surfaces, the control $\boldsymbol{v}_c(\tilde{z}, t)$ will be chosen in such a way that it imposes $\boldsymbol{\sigma}(\tilde{z}, t)$ to have the dynamics given by the following first-order differential equation [7, 14]:

$$\dot{\boldsymbol{\sigma}}(\tilde{z}, t) = -\mathbf{G}\ \text{sign}(\boldsymbol{\sigma}) - \mathbf{K}\, h(\boldsymbol{\sigma}), \tag{5.2}$$

where $\mathbf{G} = \text{diag}\left(\left[\mathrm{g}_1\ \mathrm{g}_2\ \ldots\ \mathrm{g}_n\right]\right)$ and $\mathbf{K} = \text{diag}\left([\mathrm{k}_1\ \mathrm{k}_2\ \ldots\ \mathrm{k}_i]\right)$ are positive definite diagonal matrices, $h(\boldsymbol{\sigma})$ is a vector that can be a function since $\boldsymbol{\sigma}^{\mathrm{T}} h(\boldsymbol{\sigma}) > 0$ and $\text{sign}(\boldsymbol{\sigma})$ is a componentwise discontinuous function defined as:

$$\text{sign}(\boldsymbol{\sigma}) = \boldsymbol{\sigma} \oslash |\boldsymbol{\sigma}|_\circ = \left[\text{sign}(\sigma_1)\ \ldots\ \text{sign}(\sigma_{\mathrm{n}})\right]^{\mathrm{T}}, \tag{5.3}$$

where $\oslash$ represents the Hadamard (element-wise) division and $|\cdot|_\circ$ is the element wise module of the elements of a vector. So the ith element of $\text{sign}(\boldsymbol{\sigma})$ is defined as:

$$\text{sign}(\sigma_i) = \frac{\sigma_i}{|\sigma_i|} = \frac{|\sigma_i|}{\sigma_i} = \begin{cases} 1 \text{ if } \sigma_i > 0 \\ 0 \text{ if } \sigma_i = 0 \\ -1 \text{ if } \sigma_i < 0 \end{cases}.$$

In Eq. (5.2) the term $\mathbf{K}\, h(\boldsymbol{\sigma})$ contributes to accelerating the convergence process to the sliding surface, whereas the discontinuous term $\mathbf{G}\ \text{sign}(\boldsymbol{\sigma})$ takes the vector of errors $\tilde{z}$ to zero in an asymptotic way [7, 14].

Time differentiating Eq. (5.1), substituting Eq. (2.30), and taking Eq. (5.2), one obtains:

$$\begin{aligned} \dot{\boldsymbol{\sigma}}(\tilde{z}, t) &= \frac{\partial \boldsymbol{\sigma}(\tilde{z}, t)}{\partial \tilde{z}} \dot{\tilde{z}} + \frac{\partial \boldsymbol{\sigma}(\tilde{z}, t)}{\partial t} \\ &= \frac{\partial \boldsymbol{\sigma}}{\partial \tilde{z}} (\boldsymbol{A}_0 + \boldsymbol{B}_0 \boldsymbol{v}_{\mathrm{c}} + \boldsymbol{d}_{\mathrm{b}}) + \frac{\partial \boldsymbol{\sigma}}{\partial t} \\ &= \frac{\partial \boldsymbol{\sigma}}{\partial \tilde{z}} (\boldsymbol{A}_0 + \boldsymbol{B}_0 \boldsymbol{v}_{\mathrm{c}}) + \frac{\partial \boldsymbol{\sigma}}{\partial \tilde{z}} \boldsymbol{d}_{\mathrm{b}} + \frac{\partial \boldsymbol{\sigma}}{\partial t}, \end{aligned} \tag{5.4}$$

from which the following control law is derived:

$$\boldsymbol{v}_{\mathrm{c}} = -(\boldsymbol{B}_{0_\sigma})^{-1} \left(\boldsymbol{A}_{0_\sigma} + \frac{\partial \boldsymbol{\sigma}}{\partial t} + \mathbf{G}\ \text{sign}(\boldsymbol{\sigma}) + \mathbf{K}\, h(\boldsymbol{\sigma}) \right), \tag{5.5}$$

where

$$\boldsymbol{A}_{0_\sigma} = \frac{\partial \boldsymbol{\sigma}}{\partial \tilde{z}} \boldsymbol{A}_0, \qquad \boldsymbol{B}_{0_\sigma} = \frac{\partial \boldsymbol{\sigma}}{\partial \tilde{z}} \boldsymbol{B}_0, \qquad \frac{\partial \boldsymbol{\sigma}}{\partial \tilde{z}} = \boldsymbol{\Lambda}^{\mathrm{T}}.$$

Defining

$$\boldsymbol{v}_{\mathrm{c}}^* = -\mathbf{G}\ \text{sign}(\boldsymbol{\sigma}) - \mathbf{K}\, h(\boldsymbol{\sigma}), \tag{5.6}$$

and substituting Eqs. (5.5) and (5.6) in Eq. (5.4), results in:

$$\begin{aligned}\dot{\boldsymbol{\sigma}} &= \boldsymbol{A}_{0_\sigma} - \boldsymbol{B}_{0_\sigma}\boldsymbol{B}_{0_\sigma}^{-1}\left(\boldsymbol{A}_{0_\sigma} + \frac{\partial\sigma}{\partial t} - \boldsymbol{v}_c^*\right) + \boldsymbol{d}_\sigma + \frac{\partial\sigma}{\partial t} \\ &= -\mathbf{G}\,\text{sign}(\boldsymbol{\sigma}) - \mathbf{K}\,h(\boldsymbol{\sigma}) + \boldsymbol{\psi},\end{aligned} \tag{5.7}$$

being $\boldsymbol{B}_{0_\sigma}\boldsymbol{B}_{0_\sigma}^{-1} = \mathbf{I}_n$ and $\boldsymbol{\psi}$ the disturbances in the system defined as:

$$\boldsymbol{d}_\sigma = \frac{\partial\sigma}{\partial\tilde{z}}\boldsymbol{d}_b = \frac{\partial\sigma}{\partial\tilde{z}}\boldsymbol{B}_0\tilde{\boldsymbol{d}}_0,$$

by using Eq. (2.29).

5.2.2 *Stability Analysis*

Choosing the Lyapunov function candidate in the form:

$$V = \frac{1}{2}\boldsymbol{\sigma}^{\mathrm{T}}\boldsymbol{\sigma}, \tag{5.8}$$

which is positive definite, the sliding surface will be attractive since the control law, Eq. (5.5), ensures that $\dot{V} = \boldsymbol{\sigma}^{\mathrm{T}}\dot{\boldsymbol{\sigma}}$ is negative definite. Using the result described by Eq. (5.7), an expression for $\dot{V}$ is immediately obtained, i.e.,

$$\dot{V} = \boldsymbol{\sigma}^{\mathrm{T}}\dot{\boldsymbol{\sigma}} = -\boldsymbol{\sigma}^{\mathrm{T}}\mathbf{G}\,\text{sign}(\boldsymbol{\sigma}) - \boldsymbol{\sigma}^{\mathrm{T}}\mathbf{K}\,h(\boldsymbol{\sigma}) + \boldsymbol{\sigma}^{\mathrm{T}}\boldsymbol{\psi}. \tag{5.9}$$

As $\boldsymbol{\sigma}^{\mathrm{T}}\mathbf{K}\,h(\boldsymbol{\sigma}) \geq 0$, $\dot{V} \leq 0$ is guaranteed by:

$$\boldsymbol{\sigma}^{\mathrm{T}}\mathbf{G}\,\text{sign}(\boldsymbol{\sigma}) \geq \boldsymbol{\sigma}^{\mathrm{T}}\boldsymbol{\psi}, \tag{5.10}$$

which is satisfied if the diagonal elements of **G** meet the following condition:

$$\lambda_{\min}\{\mathbf{G}\} > |\boldsymbol{\psi}| > \overline{\psi}_{\max}, \tag{5.11}$$

with $\lambda_{\min}\{\mathbf{G}\}$ being the minimum singular value of **G** and $\overline{\psi}_{\max}$ representing the maximum effect of the uncertainties and disturbances. Meeting the condition of Eq. (5.11), then $\dot{V} \leq 0\,(\dot{V} = 0$ only when $V = 0)$, which implies that V may decrease to V = 0 exponentially [2–5].

For the existence and reachability of a sliding mode, the Lyapunov function candidate must be positive definite so that the sliding surface will be attractive if the control law, Eq. (5.5), ensures $\dot{V} \leq 0$. Then, a nonsingular matrix $\boldsymbol{B}_{0_\sigma}$ is necessary. As **G**, in the Eq. (5.5), is a positive definite diagonal matrix, the sliding mode can be forced to the condition where the matrix $\boldsymbol{B}_{0_\sigma}$ is also a positive definite, and the values of **G** are large enough. However, in Eq. (5.5) the matrix $\boldsymbol{B}_{0_\sigma}$ is only nonsingular. To solve this problem, a method of diagonalization was used, requiring new sliding surfaces $\boldsymbol{\sigma}^*(\tilde{z}, t)$ [6, 10]:

$$\boldsymbol{\sigma}^*(\tilde{z}, t) = \boldsymbol{\Gamma}(\tilde{z}, t)\boldsymbol{\sigma}(\tilde{z}, t) = \boldsymbol{\Gamma}(\tilde{z}, t)\boldsymbol{\Lambda}^{\mathrm{T}}\tilde{z}, \tag{5.12}$$

where $\boldsymbol{\Gamma}(\tilde{z}, t) \in \mathbb{R}^{\mathrm{m}\times\mathrm{m}}$ is a suitable nonsingular transformation, which is defined as follows:

$$\boldsymbol{\Gamma}(\tilde{z}, t) = \left(\frac{\partial\boldsymbol{\sigma}}{\partial\tilde{z}}\boldsymbol{B}_0\right)^{\mathrm{T}} = \boldsymbol{B}_{0_\sigma}^{\mathrm{T}}. \tag{5.13}$$

This method is based on the fact that the equivalent system is invariant to a nonsingular transformation of the sliding surfaces. Assuming that the original system is given by the Eq. (2.30), with sliding surfaces $\boldsymbol{\sigma}(\tilde{z}, t) = \mathbf{0}$, Eq. (5.1), then the sliding motion (the trajectory of the equivalent system) is invariant to the transformation of the sliding surfaces $\boldsymbol{\sigma}^* = \boldsymbol{\Gamma}(\tilde{z}, t)\boldsymbol{\sigma} = \mathbf{0} \in \mathbb{R}^{\mathrm{m}}$ if $||\dot{\boldsymbol{\Gamma}}||$ and $||\boldsymbol{\Gamma}^{-1}||$ are bounded for all $t, \tilde{z} \in \boldsymbol{\Lambda} \subseteq \mathbb{R} \times \mathbb{R}^{\mathrm{n}}$ [6, 10]. Therefore, using Eq. (5.4) and applying the equivalent control method [14], one obtains:

$$\begin{aligned}\dot{\boldsymbol{\sigma}}(\tilde{z}, t) &= \frac{\partial\boldsymbol{\sigma}(\tilde{z}, t)}{\partial\tilde{z}}\dot{\tilde{z}} + \frac{\partial\boldsymbol{\sigma}(\tilde{z}, t)}{\partial t} \\ &= \frac{\partial\boldsymbol{\sigma}}{\partial\tilde{z}}\left(\boldsymbol{A}_0 + \boldsymbol{B}_0\boldsymbol{v}_{\mathrm{c_{eq}}} + \boldsymbol{d}_{\mathrm{b}}\right) + \frac{\partial\boldsymbol{\sigma}}{\partial t} = \mathbf{0},\end{aligned} \tag{5.14}$$

and

$$\begin{aligned}\dot{\boldsymbol{\sigma}}^*(\tilde{z}, t) &= \boldsymbol{\Gamma}(\tilde{z}, t)\frac{\partial\boldsymbol{\sigma}(\tilde{z}, t)}{\partial\tilde{z}}\dot{\tilde{z}} + \boldsymbol{\Gamma}(\tilde{z}, t)\frac{\partial\boldsymbol{\sigma}(\tilde{z}, t)}{\partial t} + \dot{\boldsymbol{\Gamma}}(\tilde{z}, t)\boldsymbol{\sigma}(\tilde{z}, t) \\ &= \boldsymbol{\Gamma}(\tilde{z}, t)\frac{\partial\boldsymbol{\sigma}}{\partial\tilde{z}}\left(\boldsymbol{A}_0 + \boldsymbol{B}_0\boldsymbol{v}^*_{\mathrm{c_{eq}}} + \boldsymbol{d}_{\mathrm{b}}\right) + \boldsymbol{\Gamma}(\tilde{z}, t)\frac{\partial\boldsymbol{\sigma}}{\partial t} + \dot{\boldsymbol{\Gamma}}(\tilde{z}, t)\boldsymbol{\sigma} = \mathbf{0},\end{aligned} \tag{5.15}$$

Since $\boldsymbol{\Gamma}(\tilde{z}, t)$ is an m × m nonsingular matrix,

$$\boldsymbol{v}_{\mathrm{c_{eq}}} = -\left(\frac{\partial\boldsymbol{\sigma}}{\partial\tilde{z}}\boldsymbol{B}_0\right)^{-1}\left(\frac{\partial\boldsymbol{\sigma}}{\partial\tilde{z}}(\boldsymbol{A}_0 + \boldsymbol{d}_{\mathrm{b}}) + \frac{\partial\boldsymbol{\sigma}}{\partial t}\right), \tag{5.16}$$

and

$$\begin{aligned}\boldsymbol{v}^*_{\mathrm{c_{eq}}} &= -\left(\frac{\partial\boldsymbol{\sigma}}{\partial\tilde{z}}\boldsymbol{B}_0\right)^{-1}\boldsymbol{\Gamma}^{-1}\left(\boldsymbol{\Gamma}\frac{\partial\boldsymbol{\sigma}}{\partial\tilde{z}}(\boldsymbol{A}_0 + \boldsymbol{d}_{\mathrm{b}}) + \boldsymbol{\Gamma}\frac{\partial\boldsymbol{\sigma}}{\partial t} + \dot{\boldsymbol{\Gamma}}\boldsymbol{\sigma}\right) \\ &= -\left(\frac{\partial\boldsymbol{\sigma}}{\partial\tilde{z}}\boldsymbol{B}_0\right)^{-1}\left(\frac{\partial\boldsymbol{\sigma}}{\partial\tilde{z}}(\boldsymbol{A}_0 + \boldsymbol{d}_{\mathrm{b}}) + \frac{\partial\boldsymbol{\sigma}}{\partial t}\right) - \left(\frac{\partial\boldsymbol{\sigma}}{\partial\tilde{z}}\boldsymbol{B}_0\right)^{-1}\boldsymbol{\Gamma}^{-1}\dot{\boldsymbol{\Gamma}}\boldsymbol{\sigma},\end{aligned} \tag{5.17}$$

in which Eq. (5.17) is different from Eq. (5.16) by the term $\left(\frac{\partial\boldsymbol{\sigma}}{\partial\tilde{z}}\boldsymbol{B}_0\right)^{-1}\boldsymbol{\Gamma}^{-1}\dot{\boldsymbol{\Gamma}}\boldsymbol{\sigma}$, however, $\boldsymbol{\sigma} = \mathbf{0}$ in the sliding mode, resulting in:

$$\boldsymbol{v}^*_{c_{eq}} = -\left(\frac{\partial \boldsymbol{\sigma}}{\partial \tilde{z}} \boldsymbol{B}_0\right)^{-1} \left(\frac{\partial \boldsymbol{\sigma}}{\partial \tilde{z}} (\boldsymbol{A}_0 + \boldsymbol{d}_b) + \frac{\partial \boldsymbol{\sigma}}{\partial t}\right) = \boldsymbol{v}_{c_{eq}}, \tag{5.18}$$

i.e., the equivalent systems are identical and the sliding mode motions coincide. Therefore, sliding motion is independent of a nonsingular, possibly time-variant, transformation of the sliding surfaces, and that any nonsingular transformation with bounded derivatives will produce the same equivalent system [6].

Differentiating V (Eq. (5.8)), substituting $\dot{\boldsymbol{\sigma}}$ Eq. (5.4), using the new sliding surfaces Eqs. (5.12) and (5.13) and doing the proper mathematical manipulations, one obtains $\dot{V}$ as:

$$\begin{aligned}
\dot{V} &= \boldsymbol{\sigma}^{\mathrm{T}} \dot{\boldsymbol{\sigma}} \\
&= \boldsymbol{\sigma}^{\mathrm{T}} \boldsymbol{A}_{0_\sigma} + \boldsymbol{\sigma}^{\mathrm{T}} \boldsymbol{B}_{0_\sigma} \boldsymbol{v}_c + \boldsymbol{\sigma}^{\mathrm{T}} \frac{\partial \boldsymbol{\sigma}}{\partial t} + \boldsymbol{\sigma}^{\mathrm{T}} \boldsymbol{B}_{0_\sigma} \tilde{\boldsymbol{d}}_0 \\
&= \boldsymbol{\sigma}^{\mathrm{T}} \boldsymbol{B}_{0_\sigma} \boldsymbol{B}_{0_\sigma}^{-1} \left(\boldsymbol{A}_{0_\sigma} + \frac{\partial \boldsymbol{\sigma}}{\partial t}\right) + \boldsymbol{\sigma}^{\mathrm{T}} \boldsymbol{B}_{0_\sigma} \boldsymbol{v}_c + \boldsymbol{\sigma}^{\mathrm{T}} \boldsymbol{B}_{0_\sigma} \tilde{\boldsymbol{d}}_0 \\
&= \left(\boldsymbol{B}_{0_\sigma}^{\mathrm{T}} \boldsymbol{\sigma}\right)^{\mathrm{T}} \boldsymbol{B}_{0_\sigma}^{-1} \left(\boldsymbol{A}_{0_\sigma} + \frac{\partial \boldsymbol{\sigma}}{\partial t}\right) + \left(\boldsymbol{B}_{0_\sigma}^{\mathrm{T}} \boldsymbol{\sigma}\right)^{\mathrm{T}} \left(\boldsymbol{v}_c + \tilde{\boldsymbol{d}}_0\right) \\
&= \boldsymbol{\sigma}^{*^{\mathrm{T}}} \boldsymbol{B}_{0_\sigma}^{-1} \left(\boldsymbol{A}_{0_\sigma} + \frac{\partial \boldsymbol{\sigma}}{\partial t}\right) + \boldsymbol{\sigma}^{*^{\mathrm{T}}} \left(\boldsymbol{v}_c + \tilde{\boldsymbol{d}}_0\right).
\end{aligned} \tag{5.19}$$

Redefining the control law $\boldsymbol{v}_c$, initially described by Eq. (5.5), as:

$$\boldsymbol{v}_c = -\boldsymbol{B}_{0_\sigma}^{-1} \left(\boldsymbol{A}_{0_\sigma} + \frac{\partial \boldsymbol{\sigma}}{\partial t}\right) - \left(\mathbf{G}\, \mathrm{sign}(\boldsymbol{\sigma}^*) + \mathbf{K}\, h(\boldsymbol{\sigma}^*)\right), \tag{5.20}$$

and substituting the new control law, Eq. (5.20), in Eq. (5.19), one obtains $\dot{V}$ as:

$$\dot{V} = -\boldsymbol{\sigma}^{*^{\mathrm{T}}} \left(\mathbf{G}\, \mathrm{sign}(\boldsymbol{\sigma}^*) + \mathbf{K}\, h(\boldsymbol{\sigma}^*)\right) + \boldsymbol{\sigma}^{*^{\mathrm{T}}} \tilde{\boldsymbol{d}}_0. \tag{5.21}$$

As Eq. (5.21) is similar to Eq. (5.9), the same conclusions on the stability analysis are valid, considering Eqs. (5.10) and (5.11). Moreover, the sliding mode occurs in the manifold $\boldsymbol{\sigma}^*(\tilde{z}, t) = \mathbf{0}$. The transformation, in Eqs. (5.12) and (5.13), is nonsingular, and therefore, the manifolds $\boldsymbol{\sigma}(\tilde{z}, t) = \mathbf{0}$ and $\boldsymbol{\sigma}^*(\tilde{z}, t) = \mathbf{0}$ coincide, and the sliding mode takes place in the manifold $\boldsymbol{\sigma}(\tilde{z}, t) = \mathbf{0}$, which was selected to design the sliding motion with the desired properties.

5.2.3 Controller Synthesis

The synthesis of kinematic controllers for the DWMR using the SMC technique is treated in this section, starting from the premise that the dynamic control (Sect. 2.2) fulfills its objectives. The objective of kinematic control is to determine the linear and angular velocities necessary for the actual DWMR to track a given trajectory, described by posture coordinates. For this, a virtual reference DWMR is used, which determines the trajectory, and must be followed by the actual DWMR.

For the controller's synthesis are considered the kinematic behavior of posture of the actual DWMR with C = P, Eqs. (1.18), (1.19), (1.21) and (1.24), and the kinematic behavior of posture of the reference DWMR, Eqs. (2.16) and (2.17), in treatment of the trajectory tracking problem, as can be seen in Fig. 2.6. Moreover, the posture error vector $\boldsymbol{q}_{\mathrm{e}}$ expressed in frame $\mathrm{CX_cY_c}$ (Figs. 1.1 and 2.6) described by Eq. (2.18) as well as the posture error dynamics of the closed-loop system $\dot{\boldsymbol{q}}_{\mathrm{e}}$ described in Sect. 2.4 and given by the Eq. (2.26) are used.

It is emphasized that the effect of the uncertainties and disturbances affecting the system is being considered Eq. (2.26) through the auxiliary velocity tracking errors $\boldsymbol{v}_{\mathrm{e}}$. Considering the invariance principle for SMC, described by Utkin et al. [14], and assuming that the $\boldsymbol{v}_{\mathrm{e}}$ (uncertainties and disturbances) is matched by the control signal $\boldsymbol{v}_{\mathrm{c}}$, then one can conclude that if the system, Eq. (2.26), is enforced to a sliding motion under some desired constraints, it will be ideally invariant to $\boldsymbol{v}_{\mathrm{e}}$ (uncertainties and disturbances).

To satisfy the SMC theory as well as to develop the control synthesis, the posture error dynamics of the closed-loop system, Eq. (2.26), can be rewritten as Eq. (2.30) described in Sect. 2.5, i.e.,

$$\begin{aligned}
\dot{\boldsymbol{q}}_{\mathrm{e}} &= \boldsymbol{A}_0\left(\boldsymbol{q}_{\mathrm{e}}, t\right) + \boldsymbol{B}_0\left(\boldsymbol{q}_{\mathrm{e}}, t\right) \boldsymbol{v}_{\mathrm{c}}\left(\boldsymbol{q}_{\mathrm{e}}, t\right) + \boldsymbol{d}_{\mathrm{b}}(t), \\
\dot{\boldsymbol{q}}_{\mathrm{e}} &= \boldsymbol{A}_0\left(\boldsymbol{q}_{\mathrm{e}}, t\right) + \boldsymbol{B}_0\left(\boldsymbol{q}_{\mathrm{e}}, t\right) \boldsymbol{v}_{\mathrm{c}}\left(\boldsymbol{q}_{\mathrm{e}}, t\right) + \boldsymbol{B}_0\left(\boldsymbol{q}_{\mathrm{e}}, t\right) \tilde{\boldsymbol{d}}_0(t), \\
\dot{\boldsymbol{q}}_{\mathrm{e}} &= \boldsymbol{A}_0\left(\boldsymbol{q}_{\mathrm{e}}, t\right) + \boldsymbol{B}_0\left(\boldsymbol{q}_{\mathrm{e}}, t\right) \boldsymbol{v}_{\mathrm{c}}\left(\boldsymbol{q}_{\mathrm{e}}, t\right) + \boldsymbol{B}_0\left(\boldsymbol{q}_{\mathrm{e}}, t\right) \boldsymbol{v}_{\mathrm{e}}(t).
\end{aligned} \tag{5.22}$$

Equation (5.22) is represented in matrix form as:

$$\begin{bmatrix} \dot{x}_{\mathrm{e}} \\ \dot{y}_{\mathrm{e}} \\ \dot{\theta}_{\mathrm{e}} \end{bmatrix} = \begin{bmatrix} \upsilon_{\mathrm{r}} \cos\left(\theta_{\mathrm{e}}\right) \\ \upsilon_{\mathrm{r}} \sin\left(\theta_{\mathrm{e}}\right) \\ \omega_{\mathrm{r}} \end{bmatrix} + \begin{bmatrix} -1 & y_{\mathrm{e}} \\ 0 & -x_{\mathrm{e}} \\ 0 & -1 \end{bmatrix} \begin{bmatrix} \upsilon_{\mathrm{c}} \\ \omega_{\mathrm{c}} \end{bmatrix} + \begin{bmatrix} -1 & y_{\mathrm{e}} \\ 0 & -x_{\mathrm{e}} \\ 0 & -1 \end{bmatrix} \begin{bmatrix} \upsilon_{\mathrm{e}} \\ \omega_{\mathrm{e}} \end{bmatrix}. \tag{5.23}$$

Thus, the first step to the SMC design [10] is the selection of sliding surfaces such that if the trajectory of states of the system is confined to such surfaces, then the system exhibits the desired behavior. In other words, this selection aims at obtaining a desired sliding motion restricted to the manifolds $\boldsymbol{\sigma}\left(\boldsymbol{q}_{\mathrm{e}}, t\right) = \boldsymbol{0}$, $\dot{\boldsymbol{\sigma}}\left(\boldsymbol{q}_{\mathrm{e}}, t\right) = \boldsymbol{0}$, $\boldsymbol{\sigma}^*\left(\boldsymbol{q}_{\mathrm{e}}, t\right) = \boldsymbol{0}$ and $\dot{\boldsymbol{\sigma}}^*\left(\boldsymbol{q}_{\mathrm{e}}, t\right) = \boldsymbol{0}$. However, the choice of sliding surfaces is a difficult problem due to posture error $\boldsymbol{q}_{\mathrm{e}}$ represent a nonlinear multi-input system [8].

To choose $\boldsymbol{\sigma}$, it is considered that the DWMR is a sub-actuated system, in which the error dynamics consists of a state vector $\boldsymbol{q}_{\mathrm{e}} \in \mathbb{R}^3$, and must be controlled by the control inputs $\boldsymbol{v}_{\mathrm{c}} \in \mathbb{R}^2$. As the sliding surface synthesis has the aim of converging posture error $\boldsymbol{q}_{\mathrm{e}}$ to zero, the sliding surface $\boldsymbol{\sigma}$ must have the same dimension as the control velocities $\boldsymbol{v}_{\mathrm{c}}$, so that their elements σ_1 and σ_2 tends to zero and are associated with υ_{c} e ω_{c} respectively. Observing the error dynamics of the closed-loop system, Eq. (2.26), it is clear that the linear velocity υ influences only the dynamics of the state variable x_{e}, while the angular velocity ω acts on all state variables. Because of these facts, the choice of $\boldsymbol{\sigma}$ is defined as:

$$\begin{aligned}\boldsymbol{\sigma}(\boldsymbol{q}_{\mathrm{e}},t) &= \boldsymbol{\Lambda}^{\mathrm{T}}\boldsymbol{q}_{\mathrm{e}}\\ \begin{bmatrix}\sigma_1\\ \sigma_2\end{bmatrix} &= \begin{bmatrix}\lambda_1 & 0 & 0\\ 0 & \lambda_2 & \lambda_3\end{bmatrix}\begin{bmatrix}x_{\mathrm{e}}\\ y_{\mathrm{e}}\\ \theta_{\mathrm{e}}\end{bmatrix}\\ &= \begin{bmatrix}\lambda_1 x_{\mathrm{e}}\\ \lambda_2 y_{\mathrm{e}} + \lambda_3\theta_{\mathrm{e}}\end{bmatrix},\end{aligned} \tag{5.24}$$

where λ_1, λ_2, and λ_3 are positive constants chosen by the designer to achieve the desired dynamic behavior in the sliding mode. Consequently, $\boldsymbol{\sigma}^*$ is given as:

$$\begin{aligned}\boldsymbol{\sigma}^*(\boldsymbol{q}_{\mathrm{e}},t) &= \left(\frac{\partial\boldsymbol{\sigma}}{\partial\boldsymbol{q}_{\mathrm{e}}}\boldsymbol{B}_0\right)^{\mathrm{T}}\boldsymbol{\sigma}(\boldsymbol{q}_{\mathrm{e}},t)\\ \begin{bmatrix}\sigma_1^*\\ \sigma_2^*\end{bmatrix} &= \left(\begin{bmatrix}\lambda_1 & 0 & 0\\ 0 & \lambda_2 & \lambda_3\end{bmatrix}\begin{bmatrix}-1 & y_{\mathrm{e}}\\ 0 & -x_{\mathrm{e}}\\ 0 & -1\end{bmatrix}\right)^{\mathrm{T}}\begin{bmatrix}x_{\mathrm{e}}\\ y_{\mathrm{e}}\\ \theta_{\mathrm{e}}\end{bmatrix}\\ &= \begin{bmatrix}-\lambda_1^2 x_{\mathrm{e}}\\ \lambda_1^2 x_{\mathrm{e}} y_{\mathrm{e}} - (\lambda_2 x_{\mathrm{e}} + \lambda_3)(\lambda_2 y_{\mathrm{e}} + \lambda_3\theta_{\mathrm{e}})\end{bmatrix}.\end{aligned} \tag{5.25}$$

Obtaining the desired sliding motion is determined by the equivalent control method [14] for the dynamic system of Eqs. (5.22) and (5.23), considering that $\boldsymbol{\sigma}(\boldsymbol{q}_{\mathrm{e}},t) = \mathbf{0}$, $\dot{\boldsymbol{\sigma}}(\boldsymbol{q}_{\mathrm{e}},t) = \mathbf{0}$, $\boldsymbol{\sigma}^*(\boldsymbol{q}_{\mathrm{e}},t) = \mathbf{0}$ e $\dot{\boldsymbol{\sigma}}^*(\boldsymbol{q}_{\mathrm{e}},t) = \mathbf{0}$. The equivalent control $\boldsymbol{v}_{\mathrm{c_{eq}}}$ is computed, similar to Eq. (5.14), as:

$$\begin{aligned}\dot{\boldsymbol{\sigma}}(\boldsymbol{q}_{\mathrm{e}},t) &= \frac{\partial\boldsymbol{\sigma}}{\partial\boldsymbol{q}_{\mathrm{e}}}\dot{\boldsymbol{q}}_{\mathrm{e}} + \frac{\partial\boldsymbol{\sigma}}{\partial t} = \mathbf{0}\\ &= \frac{\partial\boldsymbol{\sigma}}{\partial\boldsymbol{q}_{\mathrm{e}}}(\boldsymbol{A}_0 + \boldsymbol{B}_0\boldsymbol{v}_{\mathrm{c_{eq}}} + \boldsymbol{B}_0\boldsymbol{v}_{\mathrm{e}}) + \frac{\partial\boldsymbol{\sigma}}{\partial t} = \mathbf{0},\end{aligned}$$

isolating $\boldsymbol{v}_{\mathrm{c_{eq}}}$, becomes:

$$\boldsymbol{v}_{\mathrm{c_{eq}}} = -\left(\frac{\partial\boldsymbol{\sigma}}{\partial\boldsymbol{q}_{\mathrm{e}}}\boldsymbol{B}_0\right)^{-1}\left(\frac{\partial\boldsymbol{\sigma}}{\partial\boldsymbol{q}_{\mathrm{e}}}(\boldsymbol{A}_0 + \boldsymbol{B}_0\boldsymbol{v}_{\mathrm{e}})\right) = \boldsymbol{v}^*_{\mathrm{c_{eq}}}, \tag{5.26}$$

where

$$\frac{\partial\boldsymbol{\sigma}}{\partial t} = \mathbf{0}, \qquad \frac{\partial\boldsymbol{\sigma}}{\partial\boldsymbol{q}_{\mathrm{e}}} = \boldsymbol{\Lambda}^{\mathrm{T}}. \tag{5.27}$$

After designing the sliding surfaces, Eqs. (5.24) and (5.25), for $\boldsymbol{\sigma}(\boldsymbol{q}_{\mathrm{e}},t) = \mathbf{0}$ and $\boldsymbol{\sigma}^*(\boldsymbol{q}_{\mathrm{e}},t) = \mathbf{0}$, one has that:

$$x_{\mathrm{e}} = 0, \qquad \theta_{\mathrm{e}} = -\frac{\lambda_2}{\lambda_3}y_{\mathrm{e}}, \tag{5.28}$$

and for $\dot{\sigma} = 0$ and $\dot{\sigma}^* = 0$, results in:

$$\dot{x}_e = 0 \quad \dot{\theta}_e = -\frac{\lambda_2}{\lambda_3}\dot{y}_e. \tag{5.29}$$

Substituting Eq. (5.26) into Eq. (5.22) and considering Eqs. (5.28) and (5.29), one obtains the desired sliding motion as [10]:

$$\dot{\theta}_e = -\frac{\lambda_2}{\lambda_3}\upsilon_r \sin(\theta_e), \tag{5.30}$$

which is identical to Eq. (5.29) with

$$\dot{y}_e = \upsilon_r \sin(\theta_e). \tag{5.31}$$

As long as $\lambda_1 > 0$, $\lambda_2 > 0$, $\lambda_3 > 0$, $\theta_e \in (-\pi, \ \pi)$ and $\upsilon_r > 0$ the surfaces are asymptotically stable in the sliding mode. To perform the stability analysis of the desired sliding motion, the Lyapunov function candidate is defined as:

$$V_{dsm} = \frac{1}{2}\theta_e^T\theta_e. \tag{5.32}$$

Differentiating Eq. (5.32) and substituting Eq. (5.30), $\dot{V}_{dsm}$ yields to:

$$\dot{V}_{dsm} = \theta_e\dot{\theta}_e = -\frac{\lambda_2}{\lambda_3}\upsilon_r\theta_e \sin(\theta_e) \leq 0, \tag{5.33}$$

one has $\dot{V}_{dsm} \leq 0$ for $\theta_e \in (-\pi, \ \pi)$ and $\upsilon_r > 0$, and $\dot{V}_{dsm} = 0$ if $\theta_e = 0$ and $\theta_e = -\frac{\lambda_2}{\lambda_3}y_e$. When x_e converge to $-\frac{\lambda_2}{\lambda_3}y_e$, the system state y_e will also converge to zero. Therefore, the developed sliding surfaces can have the states $x_e = 0$ and $\theta_e = -\frac{\lambda_2}{\lambda_3}y_e$ in the sliding mode, which are asymptotically stable.

Once the selection of the sliding surfaces has been established, as well as the stability of the desired sliding motion, the second step for the SMC design [10] is to determine an SMC law, which allows $\boldsymbol{\sigma}(\boldsymbol{q}_e, t) \to 0$ and $\boldsymbol{\sigma}^*(\boldsymbol{q}_e, t) \to 0$ when $x_e \to 0$ and $\theta_e \to -\frac{\lambda_2}{\lambda_3}y_e$. Finally, the aim of DWMR posture can be performed by $y_e \to 0$ and $\theta_e \to 0$.

The determination of this control law is realized through the stability analysis using the same Lyapunov function candidate given by Eq. (5.8), in which it can be seen that by substituting the sliding surfaces, Eq. (5.24), $V = 0$ for $\boldsymbol{q}_e = 0$ and $V > 0$ for $\boldsymbol{q}_e \neq 0$, i.e., V is positive definite. Differentiating Eq. (5.8) and substituting Eq. (5.26), $\dot{V}$ gives:

$$\dot{V} = \boldsymbol{\sigma}^T\dot{\boldsymbol{\sigma}} = \boldsymbol{\sigma}^T\frac{\partial\boldsymbol{\sigma}}{\partial\boldsymbol{q}_e}\boldsymbol{A}_0 + \boldsymbol{\sigma}^T\frac{\partial\boldsymbol{\sigma}}{\partial\boldsymbol{q}_e}\boldsymbol{B}_0\boldsymbol{v}_c + \boldsymbol{\sigma}^T\frac{\partial\boldsymbol{\sigma}}{\partial\boldsymbol{q}_e}\boldsymbol{B}_0\boldsymbol{v}_e. \tag{5.34}$$

Using the new sliding surfaces, Eq. (5.25), as well as the appropriate non-singular transformation, Eq. (5.13), and doing algebraic manipulations, one obtains:

$$\dot{V} = \boldsymbol{\sigma}^{*^{\mathrm{T}}} \boldsymbol{B}_{0_\sigma}^{-1} \boldsymbol{A}_{0_\sigma} + \boldsymbol{\sigma}^{*^{\mathrm{T}}} (\boldsymbol{v}_{\mathrm{c}} + \boldsymbol{v}_{\mathrm{e}}), \tag{5.35}$$

which is similar to Eq. (5.19). Therefore, the control law is defined as:

$$\boldsymbol{v}_{\mathrm{c}} = -\boldsymbol{B}_{0_\sigma}^{-1} \boldsymbol{A}_{0_\sigma} - (\mathbf{G}\ \mathrm{sign}(\boldsymbol{\sigma}^*) + \mathbf{K}\, h(\boldsymbol{\sigma}^*)), \tag{5.36}$$

with

$$\boldsymbol{A}_{0_\sigma} = \frac{\partial \boldsymbol{\sigma}}{\partial \tilde{\boldsymbol{z}}} \boldsymbol{A}_0 = \begin{bmatrix} \lambda_1 \upsilon_{\mathrm{r}} \cos(\theta_{\mathrm{e}}) \\ \lambda_2 \upsilon_{\mathrm{r}} \sin(\theta_{\mathrm{e}}) + \lambda_3 \omega_{\mathrm{r}} \end{bmatrix}, \tag{5.37}$$

$$\boldsymbol{B}_{0_\sigma} = \frac{\partial \boldsymbol{\sigma}}{\partial \tilde{\boldsymbol{z}}} \boldsymbol{B}_0 = \begin{bmatrix} -\lambda_1 & \lambda_1 y_{\mathrm{e}} \\ 0 & -\lambda_2 x_{\mathrm{e}} - \lambda_3 \end{bmatrix}, \tag{5.38}$$

$$\boldsymbol{B}_{0_\sigma}^{-1} = \begin{bmatrix} -\frac{1}{\lambda_1} & -\frac{y_{\mathrm{e}}}{\lambda_2 x_{\mathrm{e}} + \lambda_3} \\ 0 & -\frac{1}{\lambda_2 x_{\mathrm{e}} + \lambda_3} \end{bmatrix}, \tag{5.39}$$

and so that $\boldsymbol{B}_{0_\sigma}$ is always nonsingular, the following condition must be satisfied:

$$\lambda_2 = \lambda_3 \kappa, \quad \text{with} \quad 0 \leq \kappa \leq \frac{1}{|x_{\mathrm{e}}| + 1}. \tag{5.40}$$

Substituting Eq. (5.36) into $\dot{V}$, Eq. (5.35), results in:

$$\dot{V} = -\boldsymbol{\sigma}^{*^{\mathrm{T}}} (\mathbf{G}\ \mathrm{sign}(\boldsymbol{\sigma}^*) + \mathbf{K}\, h(\boldsymbol{\sigma}^*)) + \boldsymbol{\sigma}^{*^{\mathrm{T}}} \boldsymbol{v}_{\mathrm{e}}. \tag{5.41}$$

As in Eqs. (5.10) and (5.11), it is considered that $\lambda_{\min}\{\mathbf{G}\}$ is the minimum singular value of $\mathbf{G}$ and as $|\boldsymbol{\sigma}^*| = \boldsymbol{\sigma}^{*^{\mathrm{T}}}\ \mathrm{sign}(\boldsymbol{\sigma}^*)$

$$\dot{V} = -\boldsymbol{\sigma}^{*^{\mathrm{T}}} \mathbf{K}\, h(\boldsymbol{\sigma}^*) - \lambda_{\min}\{\mathbf{G}\}\, |\boldsymbol{\sigma}^*|_1 + \boldsymbol{\sigma}^{*^{\mathrm{T}}} \boldsymbol{v}_{\mathrm{e}},$$

as $|\boldsymbol{\sigma}^*|_1 >= |\boldsymbol{\sigma}^*|$ and considering $\overline{\psi}_{\max} >= |\boldsymbol{v}_{\mathrm{e}}|$ is a known scalar that represents the maximum effect of the uncertainties and disturbances. So $\overline{\psi}_{\max} |\boldsymbol{\sigma}^*|_1 >= |\boldsymbol{\sigma}^{*^{\mathrm{T}}} \boldsymbol{v}_{\mathrm{e}}|$ and Eq. (5.41) can be rewritten as:

$$\dot{V} \leq -\boldsymbol{\sigma}^{*^{\mathrm{T}}} \mathbf{K}\, h(\boldsymbol{\sigma}^*) - (\lambda_{\min}\{\mathbf{G}\} - \overline{\psi}_{\max}) |\boldsymbol{\sigma}^*|_1 \leq 0, \tag{5.42}$$

as long as $h(\boldsymbol{\sigma}^*) \neq \mathbf{0}$ and $h(\boldsymbol{\sigma}^*)$ has the same signal of $\boldsymbol{\sigma}^*$ as well as $\lambda_{\min}\{\mathbf{G}\} > \overline{\psi}_{\max}$. For the case of $h(\boldsymbol{\sigma}^*) = \mathbf{0}$, $\dot{V}$ becomes:

$$\dot{V} \leq -(\lambda_{\min}\{\mathbf{G}\} - \overline{\psi}_{\max}) |\boldsymbol{\sigma}^*|_1 \leq 0, \tag{5.43}$$

since the condition $\lambda_{\min}\{\mathbf{G}\} > \overline{\psi}_{\max}$ be satisfied. Thus, $\dot{V} < 0$ for $\boldsymbol{q}_{\mathrm{e}} \neq \mathbf{0}$. Hence, $\boldsymbol{q}_{\mathrm{e}} = \mathbf{0}$ is an asymptotically stable equilibrium point. Recalling that as Eq. (5.41) is similar to Eq. (5.21), the same conclusions on the stability analysis are valid.

5.2.4 Controller Variants

Despite the simplicity of the design, robustness to uncertainties and disturbances, the chattering phenomenon is an undesirable feature in the practical application of SMC. The chattering phenomenon can cause excitation of unmodeled dynamics, which in turn can cause high-frequency and low-amplitude oscillations, imposing unnecessary wear on actuators and even promoting system instability. Switching imperfections can also cause high-frequency oscillations, which due to the noninstantaneous switching, can cause time delays due to the sampling and/or execution time required for the control calculus.

As the discontinuous part of the SMC is responsible for the compensation of the uncertainties and disturbances, for its use is required the prior knowledge of the bounds of the uncertainties and disturbances, otherwise, high values of gains are applied, which causes still more the increase of the chattering phenomenon.

One solution to chattering phenomenon attenuation is to deal with the reaching phase of the sliding surface since the chattering phenomenon is caused by a nonideal reaching. This approach establishes the characteristics of this phase by the use of a reaching model. The result is called the reaching law method [7, 11, 12].

Given the reaching law presented by Eq. (5.2), four variants are presented:

- **Constant rate reaching**

$$\dot{\boldsymbol{\sigma}}_{\mathrm{A}}(\boldsymbol{q}_{\mathrm{e}}, t) = -\mathbf{Q}\,\mathrm{sign}(\boldsymbol{\sigma}), \quad \mathbf{G} = \mathbf{Q}, \quad h(\boldsymbol{\sigma}) = \mathbf{0}. \tag{5.44}$$

  This variant forces the sliding surfaces $\boldsymbol{\sigma}$ to reach the switching manifolds at the constant rate $|\dot{\boldsymbol{\sigma}}| = -\lambda_{\min}\{\mathbf{G}\}$. It is a simple reaching law, but if the value of $\lambda_{\min}\{\mathbf{G}\}$ is too small, the reaching time will be very long. In contrast, if the value of $\lambda_{\min}\{\mathbf{G}\}$ is too large, it will cause a severe phenomenon.

- **Constant plus proportional rate reaching**

$$\dot{\boldsymbol{\sigma}}_{\mathrm{B}}(\boldsymbol{q}_{\mathrm{e}}, t) = -\mathbf{Q}\,\mathrm{sign}(\boldsymbol{\sigma}) - \mathbf{K}\boldsymbol{\sigma}, \quad \mathbf{G} = \mathbf{Q}, \quad h(\boldsymbol{\sigma}) = \boldsymbol{\sigma}. \tag{5.45}$$

  With the addition of the proportional rate term $-\mathbf{K}\boldsymbol{\sigma}$, state vector $\boldsymbol{q}_{\mathrm{e}}$ is forced to approach the fastest sliding manifolds when $\boldsymbol{\sigma}$ is large. It is to emphasize that the reaching time for $\boldsymbol{q}_{\mathrm{e}}$ to move from an initial state $\boldsymbol{q}_{\mathrm{e}}(0)$ to the switching manifolds is finite.

- **Power rate reaching**

$$\dot{\boldsymbol{\sigma}}_C(\boldsymbol{q}_e, t) = -\mathbf{Q}\,|\boldsymbol{\sigma}|^{\alpha}\,\mathrm{sign}(\boldsymbol{\sigma}), \quad \mathbf{G} = \mathbf{Q}\,|\boldsymbol{\sigma}|^{\alpha}, \quad h(\boldsymbol{\sigma}) = \mathbf{0}. \tag{5.46}$$

This variant increases the reaching velocity when the state vector $\boldsymbol{q}_e$ is far away from the switching manifolds but reduces the rate when the state vector $\boldsymbol{q}_e$ is near the manifolds. The result is a fast reaching and low chattering phenomenon in the reaching mode. It is to emphasize that this variant also gives a finite reaching time.

- **Speed control rate reaching**

$$\dot{\boldsymbol{\sigma}}_D(\boldsymbol{q}_e, t) = -\mathbf{Q}\,e^{\alpha|\boldsymbol{\sigma}|}\,\mathrm{sign}(\boldsymbol{\sigma}), \quad \mathbf{G} = \mathbf{Q}\,e^{\alpha|\boldsymbol{\sigma}|}, \quad h(\boldsymbol{\sigma}) = \mathbf{0}. \tag{5.47}$$

Other variants for the reaching law, Eq. (5.2), can be found in the literature, such as double power rate reaching [15]; inverse sine hyperbolic rate reaching [1]; double hyperbolic rate reaching, which uses hyperbolic tangent function plus inverse sine hyperbolic [13].

Substituting the values of $\mathbf{G}$, and $h(\boldsymbol{\sigma})$ presented in the four reaching laws, Eqs. (5.44)–(5.47), in the control law, described by Eq. (5.36), results in:

- **FOSMC-A**

$$\begin{aligned} \boldsymbol{v}_{c_A} &= -\boldsymbol{B}_{0_\sigma}^{-1}\boldsymbol{A}_{0_\sigma} + \boldsymbol{v}_{c_A}^*, \quad \boldsymbol{v}_{c_A}^* = -\mathbf{Q}\,\mathrm{sign}(\boldsymbol{\sigma}^*), \\ &\mathbf{G} = \mathbf{Q}, \quad h(\boldsymbol{\sigma}^*) = \mathbf{0}. \end{aligned} \tag{5.48}$$

- **FOSMC-B**

$$\begin{aligned} \boldsymbol{v}_{c_B} &= -\boldsymbol{B}_{0_\sigma}^{-1}\boldsymbol{A}_{0_\sigma} + \boldsymbol{v}_{c_B}^*, \quad \boldsymbol{v}_{c_B}^* = -\mathbf{Q}\,\mathrm{sign}(\boldsymbol{\sigma}^*) - \mathbf{K}\,h(\boldsymbol{\sigma}^*), \\ &\mathbf{G} = \mathbf{Q}, \quad h(\boldsymbol{\sigma}^*) = \boldsymbol{\sigma}^*. \end{aligned} \tag{5.49}$$

- **FOSMC-C**

$$\begin{aligned} \boldsymbol{v}_{c_C} &= -\boldsymbol{B}_{0_\sigma}^{-1}\boldsymbol{A}_{0_\sigma} + \boldsymbol{v}_{c_C}^*, \quad \boldsymbol{v}_{c_C}^* = -\mathbf{Q}\,|\boldsymbol{\sigma}^*|^{\alpha}\,\mathrm{sign}(\boldsymbol{\sigma}^*), \\ &\mathbf{G} = \mathbf{Q}\,|\boldsymbol{\sigma}^*|^{\alpha}, \quad h(\boldsymbol{\sigma}^*) = \mathbf{0}. \end{aligned} \tag{5.50}$$

- **FOSMC-D**

$$\begin{aligned} \boldsymbol{v}_{c_D} &= -\boldsymbol{B}_{0_\sigma}^{-1}\boldsymbol{A}_{0_\sigma} + \boldsymbol{v}_{c_D}^*, \quad \boldsymbol{v}_{c_D}^* = -\mathbf{Q}\,e^{\alpha|\boldsymbol{\sigma}^*|}\,\mathrm{sign}(\boldsymbol{\sigma}^*), \\ &\mathbf{G} = \mathbf{Q}\,e^{\alpha|\boldsymbol{\sigma}^*|}, \quad h(\boldsymbol{\sigma}^*) = \mathbf{0}. \end{aligned} \tag{5.51}$$

where $\boldsymbol{v}_{c_A}^*$, $\boldsymbol{v}_{c_B}^*$, $\boldsymbol{v}_{c_C}^*$ and $\boldsymbol{v}_{c_D}^*$ are the control portions that perform the compensation of the uncertainties and disturbances.

These variants may attenuate the chattering phenomenon since by changing the control structure, the control effort can be decreased, generating smoother signals [11, 12]. Although the technique may attenuate the chattering phenomenon, it is not able to eliminate it, and it can also affect the control robustness.

5.3 Simulations Using Matlab and/or MobileSim Simulator

To demonstrate the effectiveness of the control variants FOSMC-A Eq. (5.48), FOSMC-B Eq. (5.49), FOSMC-C Eq. (5.50) and FOSMC-D Eq. (5.51) as well as the occurrence of the chattering phenomenon, results are obtained and analyzed through simulations and practical experiments.

The gains chosen for the control variants are given in Table 5.1.

Recalling that the eight-shape reference trajectory provides greater complexity in the execution, considering acceleration and deceleration, in addition to generating an initial posture error due to the initial conditions of this trajectory and the DWMR being different. More information about the reference trajectory can be seen in Sect. 3.1.2.

5.3.1 Ideal Scenario

The results of RMS of the errors in Table 5.2 show that all control variants are capable of tracking the trajectory with errors very close to zero.

Figures 5.2, 5.3, and 5.4 visually confirm the results of Table 5.2. Figure 5.2 shows that using the FOSMCs, the DWMR gets to the trajectory at a similar point, and keeps on it for the remaining time, i.e., the DWMR satisfactorily follows the reference trajectory. Nevertheless, for all the FOSMCs, the DWMR tracks the trajectory with posture tracking and orientation errors (xy_e, and θ_e) that tend to converge to zero (Figs. 5.3 and 5.4).

Figure 5.5 shows the linear and angular velocity control of the FOSMCs. The chattering phenomenon can be observed on the control velocities of the FOSMCs.

Table 5.1 FOSMC gains

Controller	λ_1	λ_2	λ_3	Q_1	Q_2	K_1	K_2	α
FOSMC-A	1.5	6.0	1.0	0.1	0.3	–	-	–
FOSMC-B	1.5	6.0	1.0	0.1	0.3	0.1	0.1	–
FOSMC-C	1.5	6.0	1.0	0.1	0.3	–	–	0.2
FOSMC-D	1.5	6.0	1.0	0.1	0.3	–	–	0.2

Table 5.2 RMS of the errors—results in the ideal scenario

Controller	x_e (m)	y_e (m)	θ_e (rad)
FOSMC-A	0.0138	0.0468	0.0505
FOSMC-B	0.0122	0.0369	0.0452
FOSMC-C	0.0154	0.0465	0.0471
FOSMC-D	0.0137	0.0396	0.0452

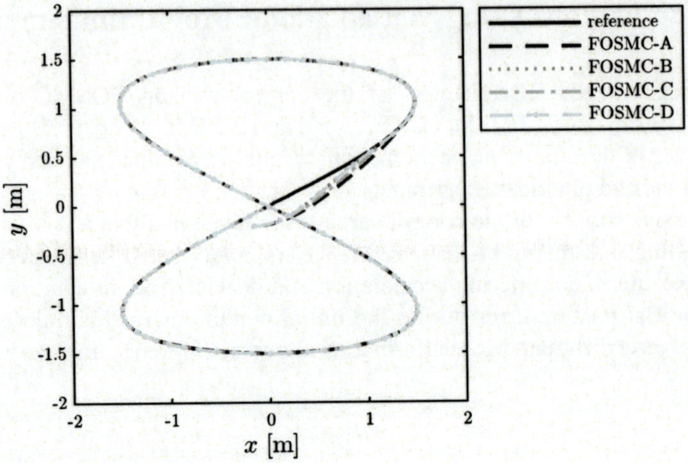

Fig. 5.2 Trajectory tracking in the ideal scenario

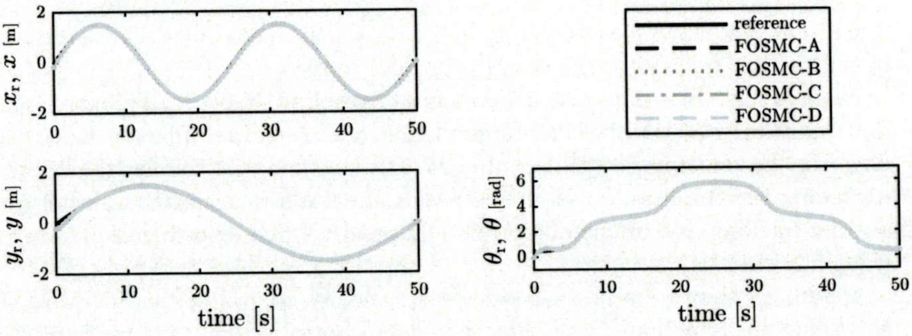

Fig. 5.3 Trackings (x, y, and θ) in the ideal scenario

In Fig. 5.6, the linear and angular velocities of the DWMR present the chattering phenomenon by using FOSMCs as well as those velocities tend towards the reference velocities.

As already previously treated in Sect. 2.4 under the robustness considerations, it is known that the kinematic controllers, here the FOSMCs, contain a function to correct the posture tracking and orientation errors, whereas the dynamic controller, using PD Eq. (2.12), aims to correct the auxiliary velocity tracking errors ($\boldsymbol{v}_\mathrm{e}$). With the integration of the FOSMCs with PD, the "perfect velocity tracking" is not maintained. Thus, these auxiliary velocity tracking errors can be viewed as uncertainties and disturbances to the kinematic model Eq. (2.24), and are compensated by the control velocities $\boldsymbol{v}_\mathrm{c}^*$ of the FOSMCs, thus ensuring that the posture tracking and orientation errors tend to converge to zero (Fig. 5.4), and producing satisfactory reference trajectory tracking (Fig. 5.2). This could be verified in Fig. 5.7, whose behaviors of the auxiliary velocity tracking errors and control velocities $\boldsymbol{v}_\mathrm{c}^*$ have opposed magnitudes

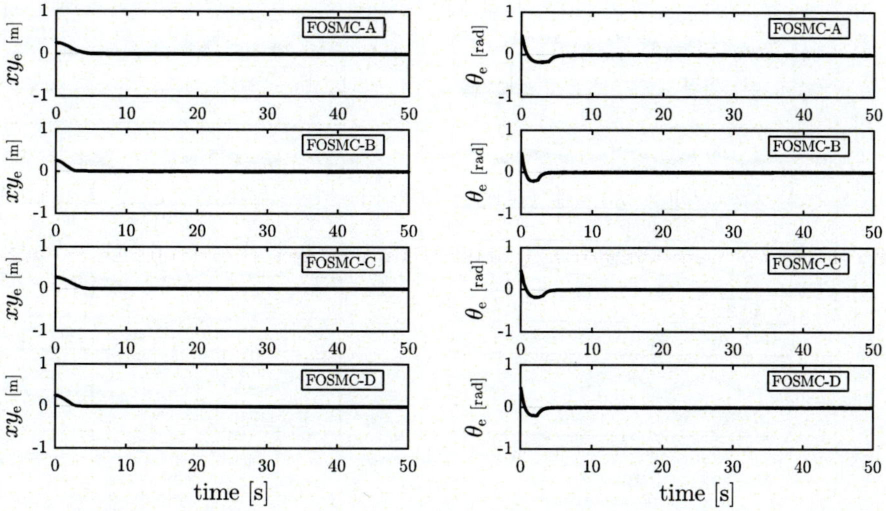

Fig. 5.4 Posture tracking errors, with $xy_e = \sqrt{x_e^2 + y_e^2}$, in the ideal scenario

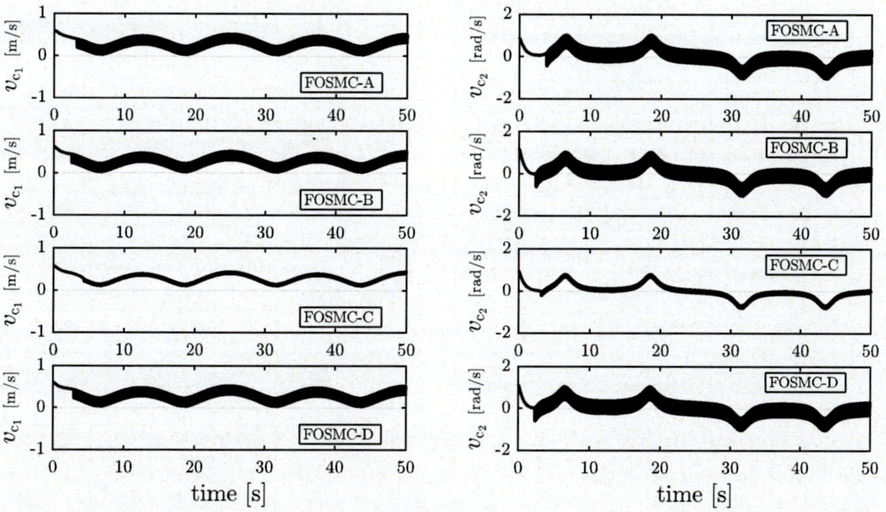

Fig. 5.5 Control velocities in the ideal scenario

in absolute terms, i.e., the difference between auxiliary velocity tracking errors and control velocities $\boldsymbol{v}_c^*$ tend to converge to zero. However, the chattering phenomenon in the compensation due to the use of FOSMCs makes such verification difficult, as can be observed in Fig. 5.7. In short, FOSMCs need prior knowledge of the bounds of uncertainties and disturbances to define or estimate a value for G to compensate for them, however, causing the chattering phenomenon.

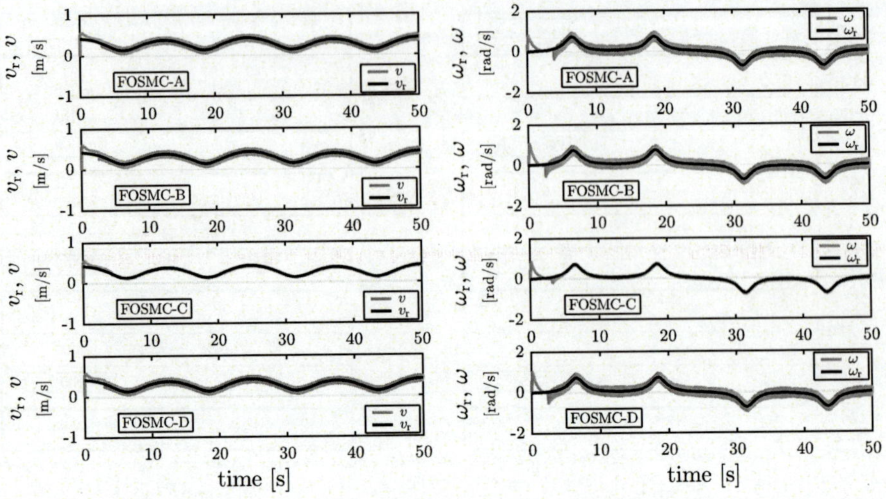

Fig. 5.6 Velocities of the DWMR and reference velocities in the ideal scenario

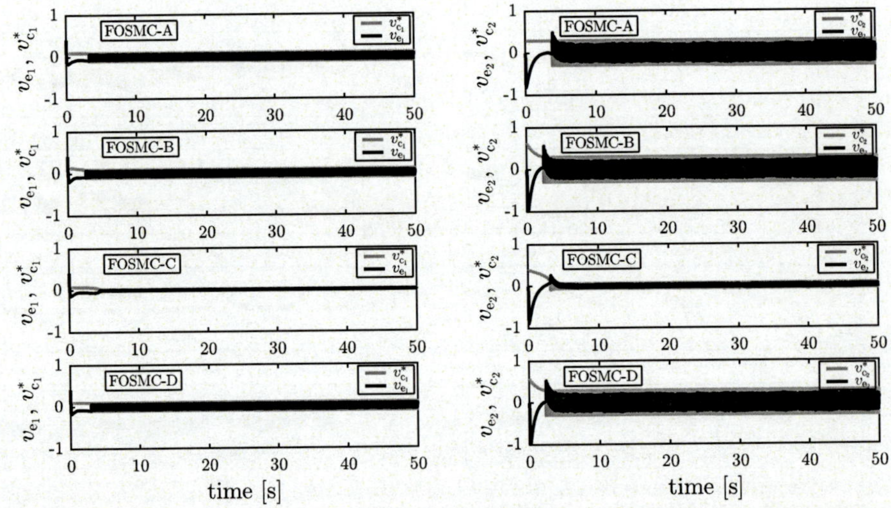

Fig. 5.7 Auxiliary velocity tracking errors and control velocities v_c^* in the ideal scenario

Both the sliding surfaces and new sliding surfaces, as well as their derivatives, tend to converge to zero, as can be observed in Figs. 5.8, 5.9, 5.10 and, 5.11 the very significant mitigation of the chattering phenomenon.

The simulation results show satisfactory tracking using FOSMCs at the cost of the chattering phenomenon. This chattering phenomenon is considered a large drawback due to the reduction of the useful life of the DWMR's actuators. It is important to recall that, in this scenario, uncertainties, and disturbances e.g., modeling imprecisions;

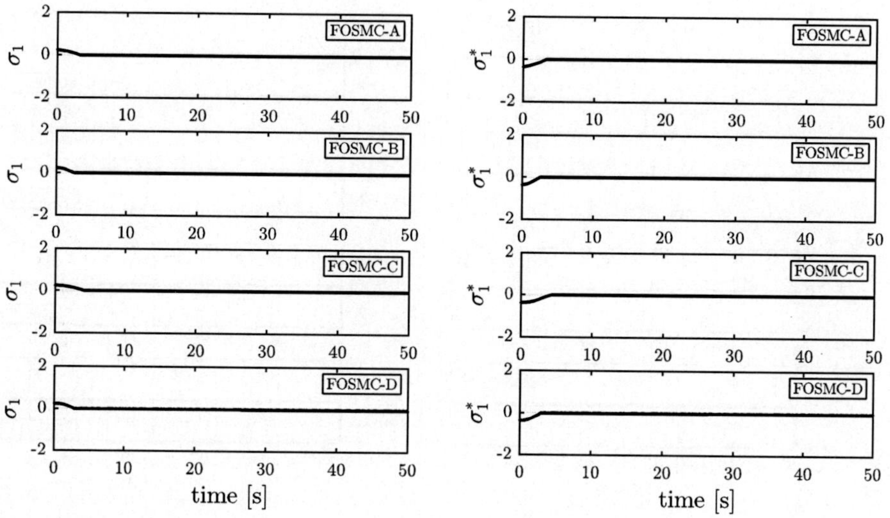

Fig. 5.8 Sliding surface σ_1 and new sliding surface σ_1^* in the ideal scenario

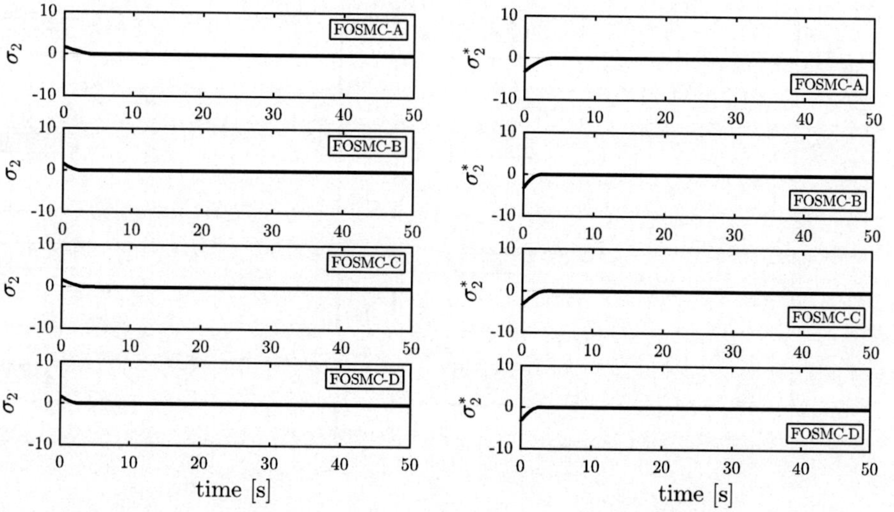

Fig. 5.9 Sliding surface σ_2 and new sliding surface σ_2^* in the ideal scenario

surface friction; external disturbances; bounded unknown disturbances; unmodeled and unstructured dynamics of actuators, sensors, and other electronic devices; integration errors; noises; operational, sampling, and physical limitations) of the DWMR are not considered. Therefore, DWMR can reproduce velocities with the chattering phenomenon, which may not be true when considering a real DWMR.

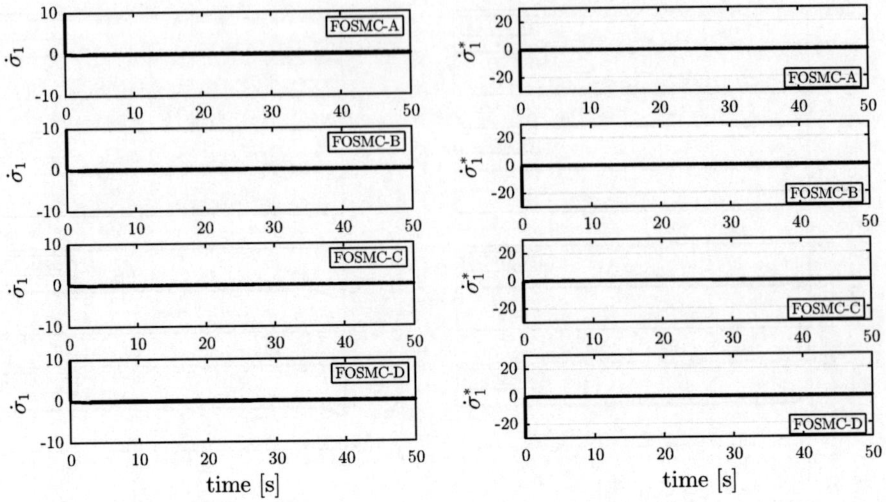

Fig. 5.10 Derivatives of sliding surface $\dot{\sigma}_1$ and derivatives of new sliding surface $\dot{\sigma}_1^*$ in the ideal scenario

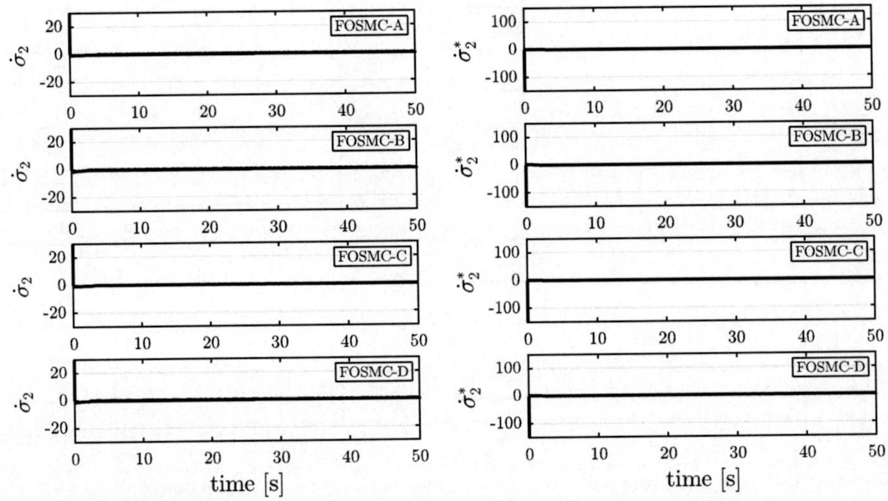

Fig. 5.11 Derivatives of sliding surface $\dot{\sigma}_2$ and derivatives of new sliding surface $\dot{\sigma}_2^*$ in the ideal scenario

5.3.2 Realistic Scenario

In this scenario, FOSMCs have been simulated with the main difference that the MobileSim simulator considers the uncertainties and disturbances (e.g., modeling imprecisions; surface friction; external disturbances; bounded unknown disturbances; unmodeled and unstructured dynamics of actuators, sensors, and other electronic devices; integration errors; noises; operational, sampling and physical limitations) of the Powerbot DWMR.

The results in Table 5.3, containing RMS of the errors, show that the FOSMCs can track the trajectories with a minimal error even with the influence of those uncertainties and disturbances.

The reference trajectory and the realized trajectory by the DWMR using the FOSMCs is shown in Fig. 5.12, in which is possible to verify that the DWMR gets to the trajectory at a similar point and remains near the trajectory during the remaining time, i.e., the DWMR tracks the reference trajectory, as can also be observed in Fig. 5.13.

Figure 5.14 shows the posture tracking and orientation errors of the FOSMCs, in which these errors vary slightly. This slight variation is more visible in the orientation error. In short, this variation of the orientation error means that instead of the DWMR tracks the trajectory in terms of the desired values of orientation, the orientation of the DWMR is varying around these values. This oscillatory behavior may be occurring due to the influence of uncertainties and disturbances.

Figure 5.15 presents the control velocities, in which the FOSMCs present the chattering phenomenon, both in linear and angular velocities. These same behaviors

Table 5.3 RMS of the errors—results in the realistic scenario

Controller	x_e (m)	y_e (m)	θ_e (rad)
FOSMC-A	0.0200	0.0449	0.0482
FOSMC-B	0.0239	0.0378	0.0433
FOSMC-C	0.0298	0.0480	0.0453
FOSMC-D	0.0281	0.0414	0.0453

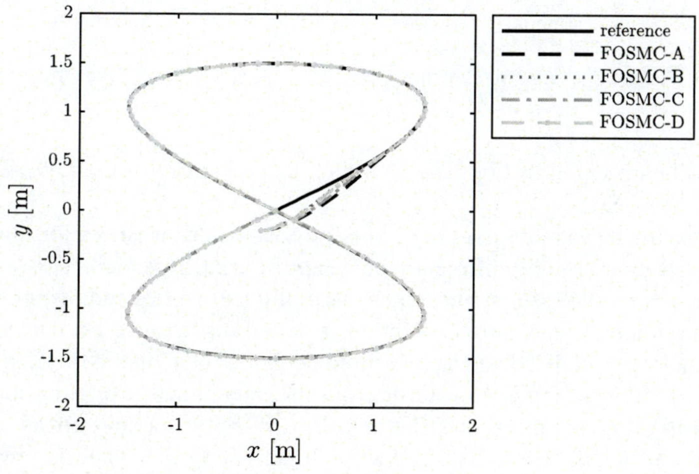

Fig. 5.12 Trajectory tracking in the realistic scenario

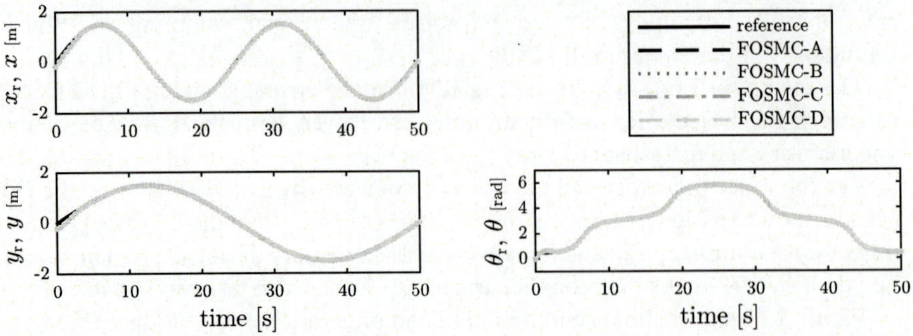

Fig. 5.13 Trackings (x, y, and θ) in the realistic scenario

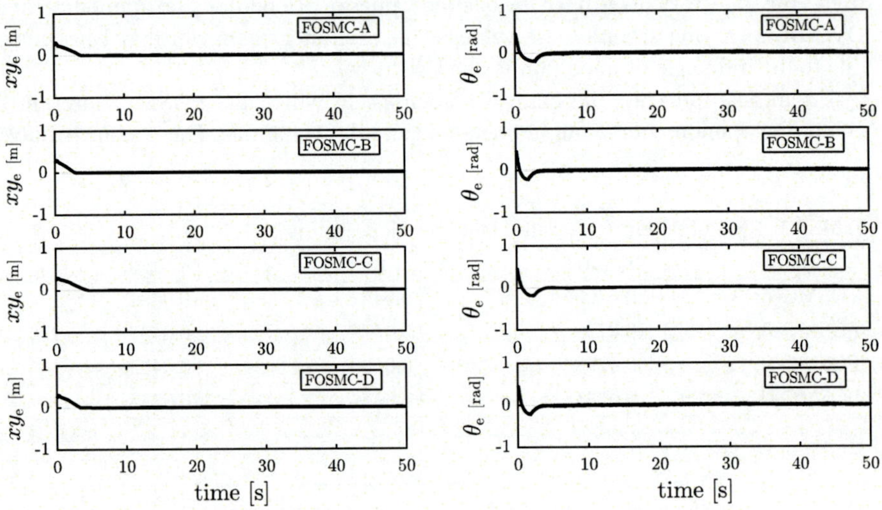

Fig. 5.14 Posture tracking errors, with $xy_e = \sqrt{x_e^2 + y_e^2}$, in the realistic scenario

can also be observed in Fig. 5.16, in which the velocities of the DWMR track the reference velocities.

The auxiliary velocity tracking errors $\boldsymbol{v}_e$, seen as disturbances for the kinematic model and given by the difference between the control velocities $\boldsymbol{v}_c$ and the real velocities of the DWMR $\boldsymbol{v}$, happen because of uncertainties and disturbances (e.g., modeling imprecisions; surface friction; external disturbances; bounded unknown disturbances; unmodeled and unstructured dynamics of actuators, sensors, and other electronic devices; integration errors; noises; operational, sampling and physical limitations). Figure 5.17 shows the auxiliary velocity tracking errors and control velocities $\boldsymbol{v}_c^*$ for the use of the FOSMCs, highlighting the chattering phenomenon caused by those control velocities, $\boldsymbol{v}_{c_A}^*$ (Eq. 5.48), $\boldsymbol{v}_{c_B}^*$ (Eq. 5.49), $\boldsymbol{v}_{c_C}^*$ (Eq. 5.50), and $\boldsymbol{v}_{c_D}^*$ (Eq. 5.51, that perform the compensations.

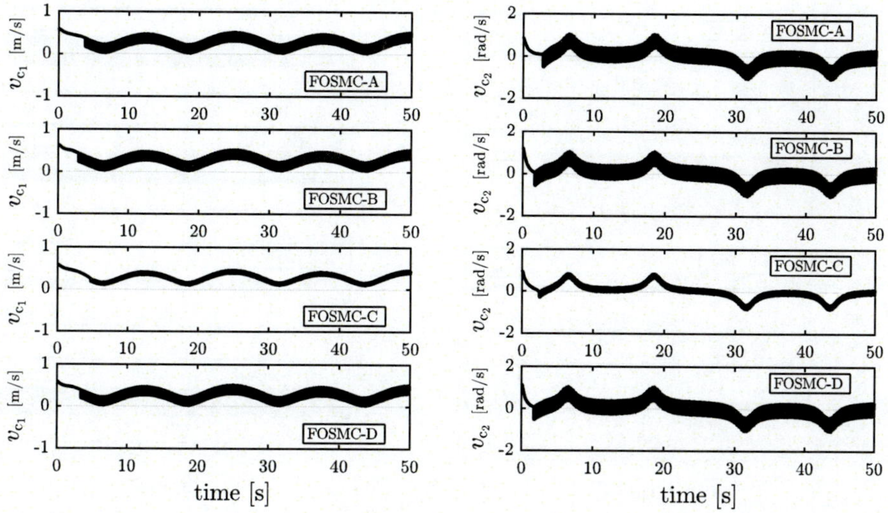

Fig. 5.15 Control velocities in the realistic scenario

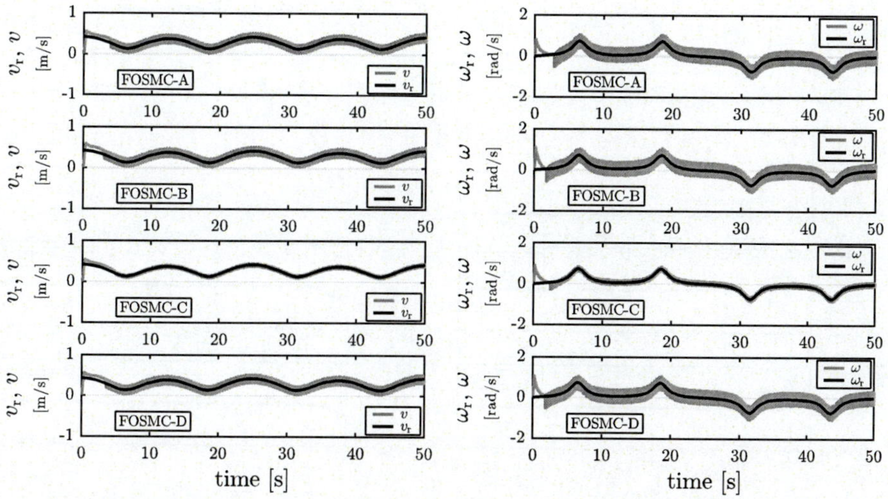

Fig. 5.16 Velocities of the DWMR and reference velocities in the realistic scenario

For the use of the FOSMCs, the sliding surfaces and new sliding surfaces present behaviors alternating near zero, as can be seen in Figs. 5.18 and 5.19. The chattering phenomenon is more visible in the derivatives of these sliding surfaces as can be verified in Figs. 5.20 and 5.21.

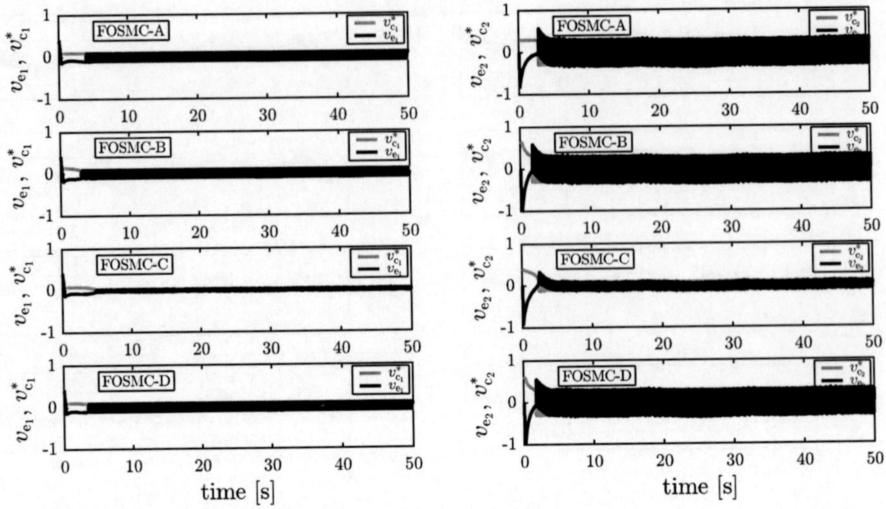

Fig. 5.17 Auxiliary velocity tracking errors and control velocities $\boldsymbol{v}_c^*$ in the realistic scenario

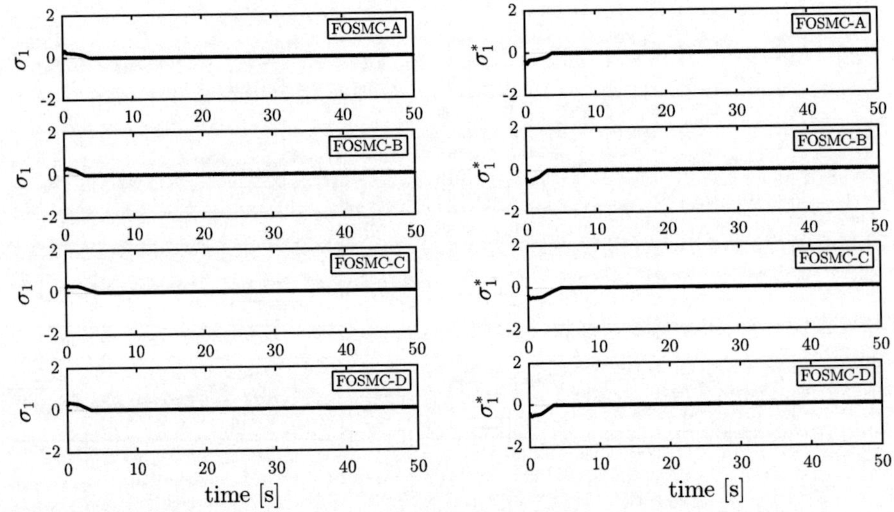

Fig. 5.18 Sliding surface σ_1 and new sliding surface σ_2^* in the realistic scenario

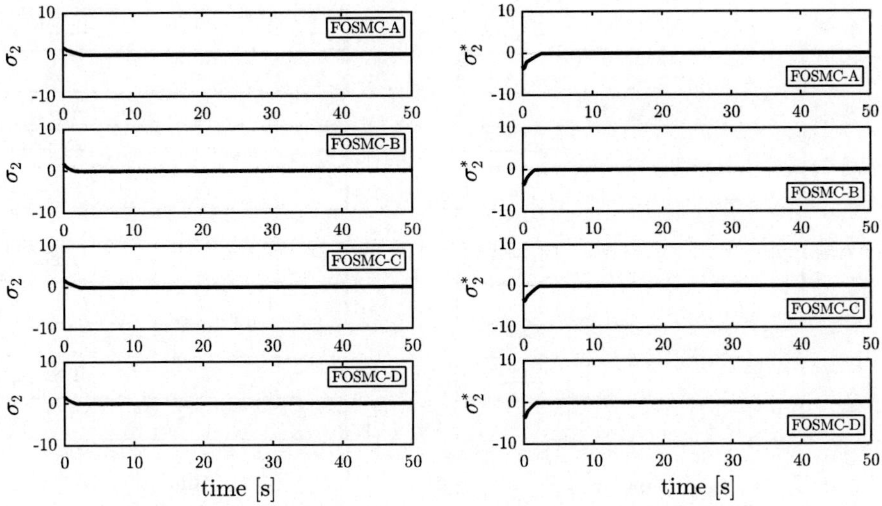

Fig. 5.19 Sliding surface σ_2 and new sliding surface σ_2^* in the realistic scenario

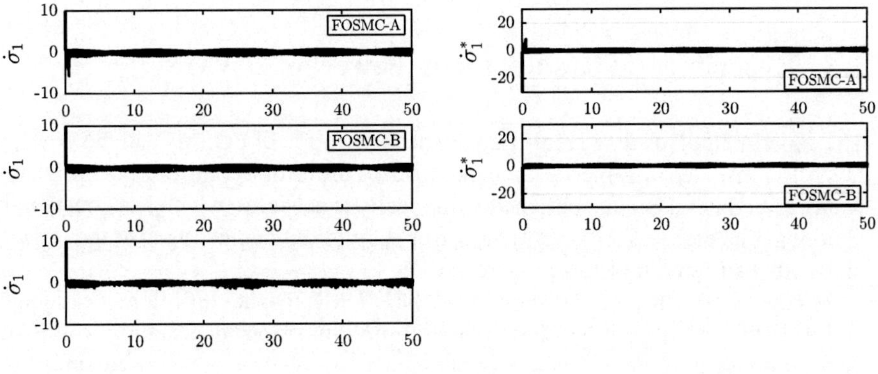

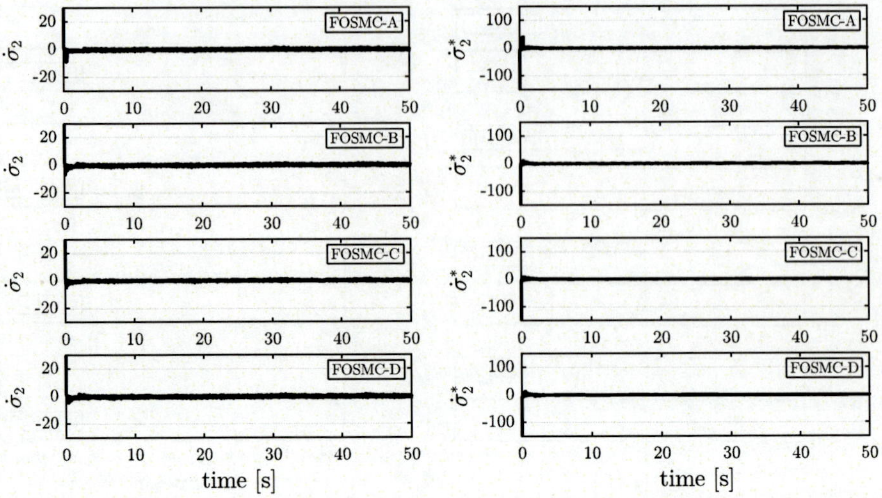

Fig. 5.21 Derivatives of sliding surface $\dot{\sigma}_2$ and derivatives of new sliding surface $\dot{\sigma}_2^*$ in the realistic scenario

5.4 Experimental Results Using Powerbot DWMR

The experimental results in real-time by the application of FOSMCs in the Powerbot DWMR, presented in Table 5.4 about the RMS of the errors, confirm what initially has been verified in the results obtained in the realistic scenario, which is a considerably significant increase of the orientation error θ_e, without neglecting also the errors x_e and y_e that suffered less abrupt increases.

For the FOSMCs, Fig. 5.22 shows how the DWMR reaches the reference trajectory and tries to remain on it for the sequential time, however, due mainly to the large orientation error, is visible a slight performance degradation as a consequence of the incidence of uncertainties and disturbances as well as the chattering phenomenon. This can also be seen in the posture trackings shown in Fig. 5.23.

In Fig. 5.24, by using FOSMCs, it is observed how the posture tracking errors xy_e have behaviors that alternate very close to zero while the orientation errors θ_e show a considerably significant variation around zero during the experiment. The

Table 5.4 RMS of the errors—results in the experimental scenario

Controller	x_e (m)	y_e (m)	θ_e (rad)
FOSMC-A	0.0651	0.0740	0.1280
FOSMC-B	0.0673	0.0657	0.1560
FOSMC-C	0.0681	0.0702	0.1000
FOSMC-D	0.0710	0.0638	0.1430

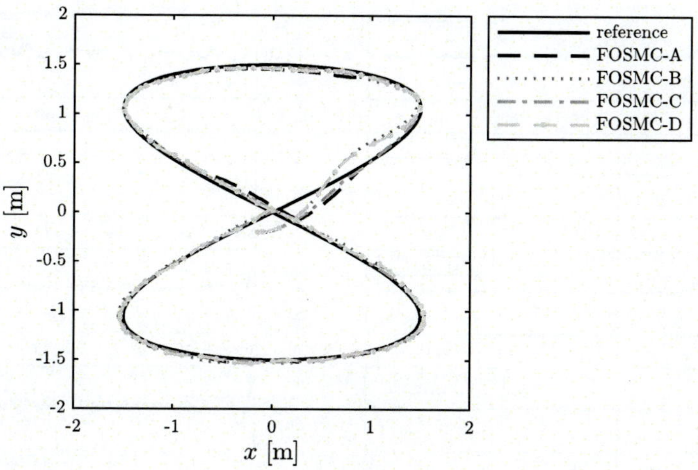

Fig. 5.22 Trajectory tracking in the experimental scenario

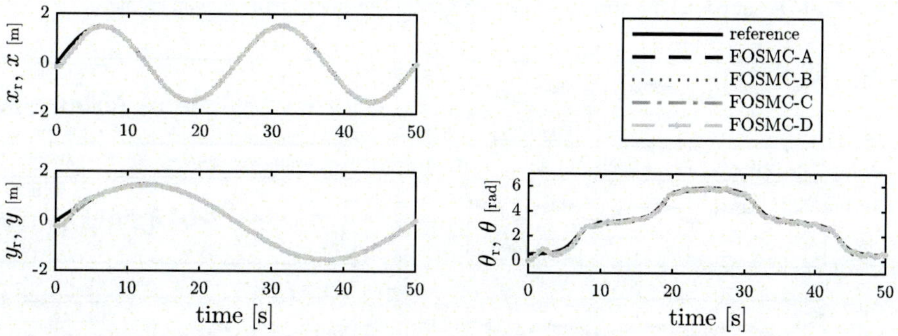

Fig. 5.23 Trackings (x, y, and θ) in the experimental scenario

variation of the orientation error is more perceptible than that of Fig. 5.14 obtained in the realistic scenario. This variation in the orientation error means that instead of the DWMR tracks the trajectory in terms of the desired values of orientation, the orientation of the DWMR is varying around these values. Moreover, the possible reasons for the occurrence of these variations are uncertainties and disturbances (e.g., modeling imprecisions; surface friction; external disturbances; bounded unknown disturbances; unmodeled and unstructured dynamics of actuators, sensors, and other electronic devices; integration errors; noises; operational, sampling and physical limitations) of the DWMR and switching imperfections of the discontinuous portion of the FOSMCs that cause the chattering phenomenon.

Unlike the ideal and realistic scenarios, Figs. 5.25 and 5.26 show how the control velocities can not switch instantaneously as well as the velocities of the DWMR, thus producing significant auxiliary velocity tracking errors. Because of this noninstantaneous switching, there is the occurrence of switching imperfections that cause

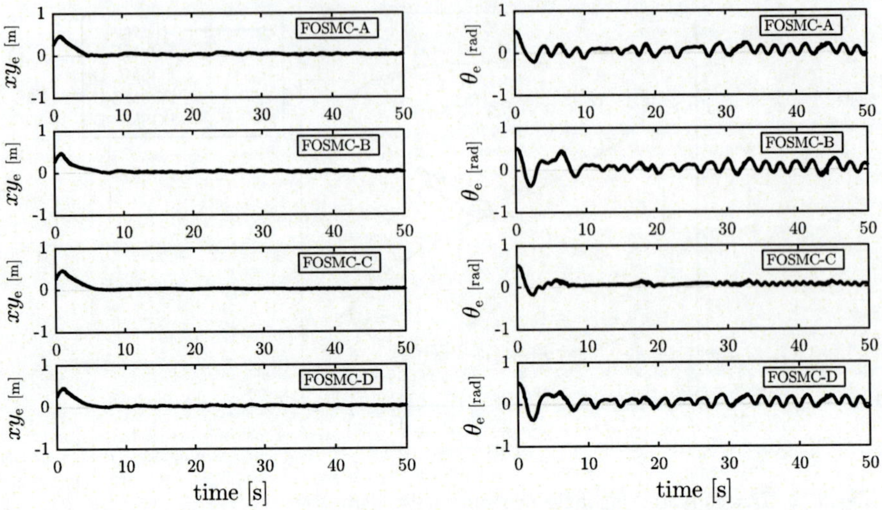

Fig. 5.24 Posture tracking errors, with $xy_e = \sqrt{x_e^2 + y_e^2}$, in the experimental scenario

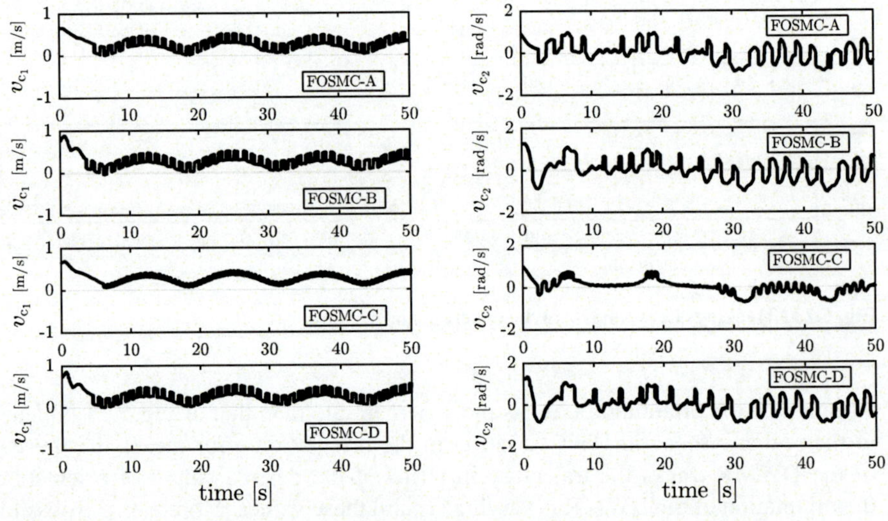

Fig. 5.25 Control velocities in the experimental scenario

high-frequency oscillations and time delays due to the sampling and/or execution times required for the control calculus. As a consequence, the compensations do not provide the control efforts necessary or sufficient with equal or larger magnitudes than the auxiliary velocity tracking errors, as can be seen in Fig. 5.27, by using the FOSMCs.

Observing very carefully the Figs. 5.28 and 5.29, it can be verified that, when using FOSMCs, the sliding surfaces σ (Eq. 5.24) and the new sliding surfaces σ^*

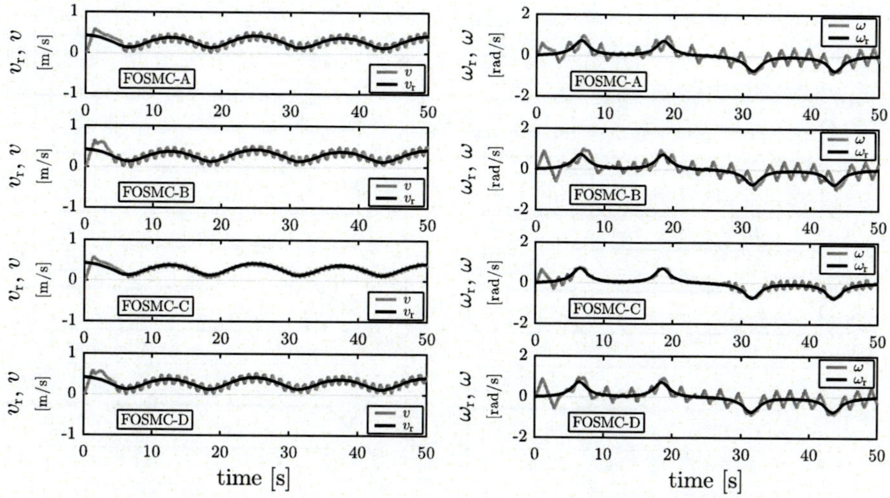

Fig. 5.26 Velocities of the DWMR and reference velocities in the experimental scenario

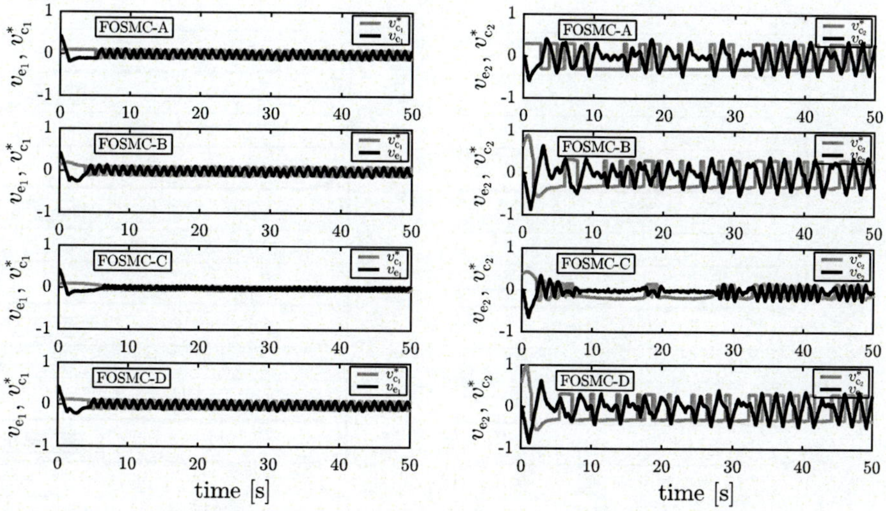

Fig. 5.27 Auxiliary velocity tracking errors and control velocities $\boldsymbol{v}^*_c$ in the experimental scenario

(Eq. 5.25) present behaviors varying around zero. This variation is more expressive on the sliding surface σ_2 and new sliding surface σ_2^* due to the switching functions $sgn(\sigma_2)$ and $sgn(\sigma_2^*)$ (Eqs. 5.48, 5.49, 5.50, and 5.51), since the behaviors of the sliding surface σ_2 and new sliding surface σ_2^* are highly affected by orientation error θ_e (Fig. 5.24). As θ_e fluctuates significantly from negative to positive around zero at all times, then σ_2 and σ_2^* are affected by the change of θ_e. This characteristic might be a cause for the chattering phenomenon. Conversely, the chattering phenomenon

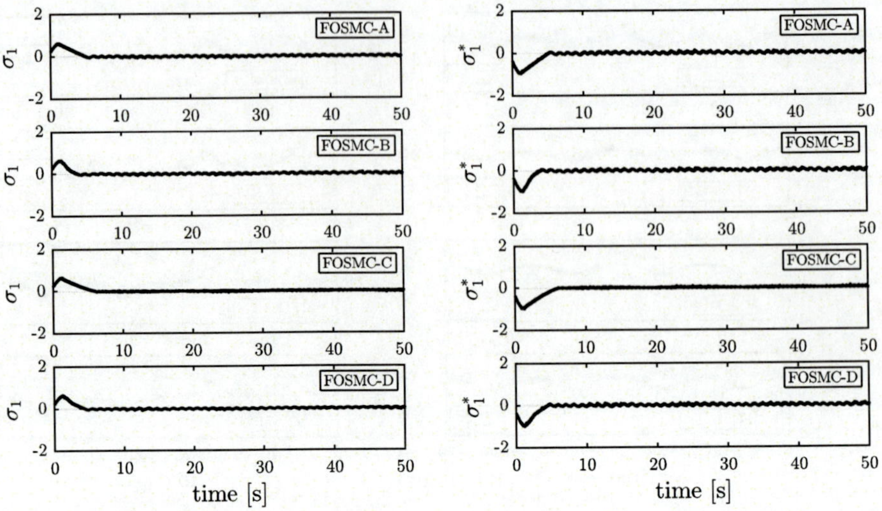

Fig. 5.28 Sliding surface σ_1 and new sliding surface σ_1^* in the experimental scenario

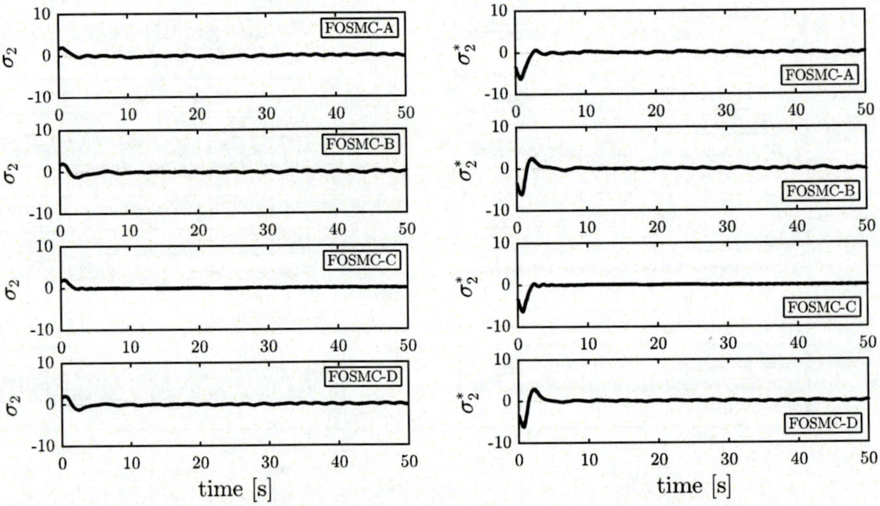

Fig. 5.29 Sliding surface σ_2 and new sliding surface σ_2^* in the experimental scenario

behavior of θ_e produced by uncertainties and disturbances may lead to the oscillation of θ_e in σ_2 and σ_2^*. Consequently, the derivatives of the sliding surfaces $\dot{\boldsymbol{\sigma}}$ and new sliding surfaces $\dot{\boldsymbol{\sigma}}^*$ have also oscillatory behaviors around zero, as can be verified in Figs. 5.30 and 5.31.

Finally, it must be emphasized that the practical implementation of the FOSMCs was carried out in the Powerbot DWMR to investigate the chattering phenomenon, which causes performance degradation, as well as to verify the occurrence of the

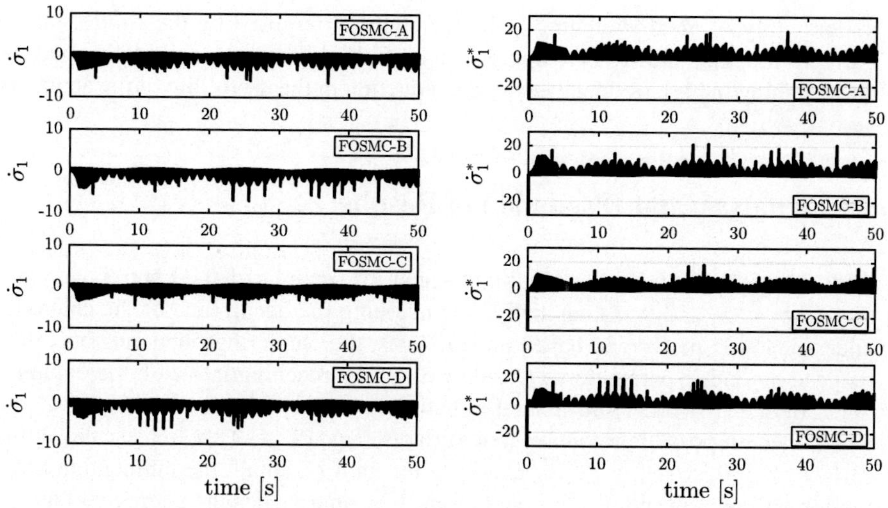

Fig. 5.30 Derivatives of sliding surface $\dot{\sigma}_1$ and derivatives of new sliding surface $\dot{\sigma}_1^*$ in the experimental scenario

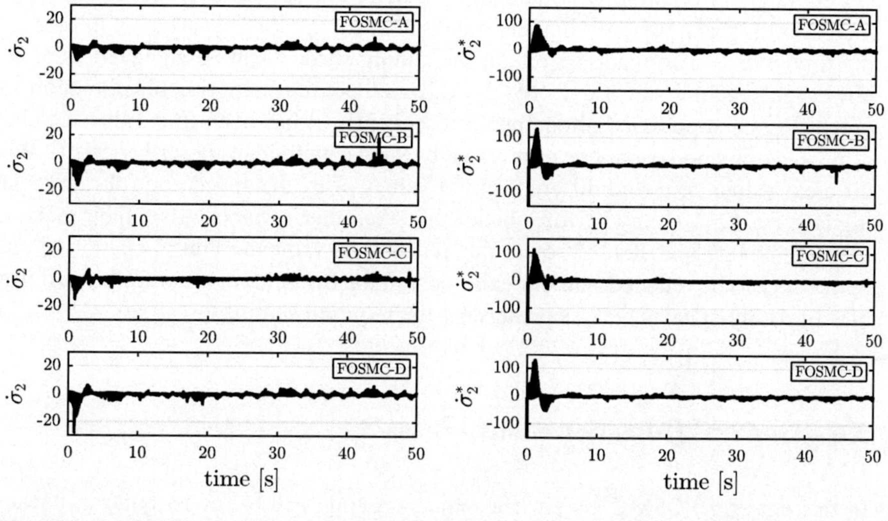

Fig. 5.31 Derivatives of sliding surface $\dot{\sigma}_2$ and derivatives of new sliding surface $\dot{\sigma}_2^*$ in the experimental scenario

impossibility of ideal switching in high frequency required by the control signal. Furthermore, the FOSMCs are not recommended for long-term daily tasks, because the switching in high frequency causes a reduction in the useful life of the actuators.

5.5 Analysis and Discussion of Results

With the results obtained in the three scenarios, mainly of the control velocities (Figs. 5.5, 5.15, and 5.25), an analysis concerning the occurrence of the chattering phenomenon is performed based on [7]. Thus, the chattering phenomenon can be reduced by tuning parameters Q, and/or K, in the reaching law with the dynamics given by the first-order differential equation, which is described in Eq. (5.2) and which made it possible to obtain four variants, Eqs. (5.44)–(5.47). Near the sliding surface, $\boldsymbol{\sigma} \approx 0$, so $|\dot{\boldsymbol{\sigma}}| \approx Q$. By choosing the gain Q, small, the momentum of the motion will be reduced as the system trajectory approaches the sliding surface. As a result, the amplitude of the chattering phenomenon will be reduced. However, Q, can not be chosen equal to zero because the reaching time would become infinite; the system fails to be a sliding mode control system. A large value for K in Eq. (5.45), increases the reaching rate when the state is not near the sliding surface. Finite reaching time with a zero reaching rate at the sliding surface can be achieved using the power rate reaching law, Eq. (5.46). By this method, the chattering phenomenon can practically be suppressed altogether. Subsequently, it has been found that the chattering phenomenon cannot be eliminated by such a method, as described in [9]. With different values of α and different initial values of $\boldsymbol{\sigma}$, the speed control rate reaching law, Eq. (5.47), may perform better than the other variants. Also, their reaching time can be less than that of power rate reaching law, the amount of chattering phenomenon can be reduced without extra cost in the evaluation of control or suffering bias in steady-state as well as preserving the required stability.

5.6 General Considerations

In this chapter, FOSMC's were chosen, as an approach to robust control, because the invariance principle applies to them and the following attractive characteristics: fast response, good performance in transient regime, and robustness concerning the uncertainties and disturbances attainable by the control inputs when in sliding mode, i.e., robustness to matched disturbances [14]. However, like all control strategies, there are drawbacks, among which can be cited: the chattering phenomenon and the need to know the limits of the disturbances to apply the appropriate gain to compensate them, thus saving the actuators of unnecessary efforts. These drawbacks are also included as objects of research in the next chapters.

Also, there is a technical difficulty in the design of robust controllers for nonholonomic systems with the uncertainties and disturbances in the kinematic model due inherently to the fact that the number of control inputs is always less than the state variables in the dynamics of nonholonomic systems, i.e., as these systems are under-actuated due to nonholonomic restrictions on movement, for FOSMC designs, it is critical to select sliding surfaces so that tracking errors (position and orientation) are reduced or converge to zero with the use of two control inputs [8]. The difficulty in selecting these surfaces comes up against the lack of methodologies. Thus, for each situation, it is necessary to choose the sliding surfaces properly and, subsequently, to analyze the behavior of the reduced dynamics of the system using the equivalent control method [14] to meet the desired performance and robustness requirements.

Therefore, the idea of these FOSMC designs is to enable development from other researches with the extension to the second-order sliding mode control, higher-order sliding mode control, integral sliding mode control, terminal sliding mode control, etc.

In summary, the chattering phenomenon is a hindrance to the widespread use of FOSMC's in many practical control systems. Mitigation of the chattering phenomenon remains an important and challenging problem, which are addressed and discussed in later chapters.

References

1. Asad, M., Ashraf, M., Iqbal, S., Bhatti, A.: Chattering and stability analysis of the sliding mode control using inverse hyperbolic function. Int. J. Control Autom. Syst. **15**, 2608–2618 (2017)
2. Begnini, M., Bertol, D., Martins, N.:. Practical implementation of a simple and effective robust adaptive fuzzy variable structure trajectory tracking control for differential wheeled mobile robots. Int. J. Innov. Comput. Inf. Control: IJICIC **13**, 341–364 (2017)
3. Begnini, M., Bertol, D., Martins, N.: A robust adaptive fuzzy variable structure tracking control for the wheeled mobile robot: Simulation and experimental results. Control Eng. Pract. **64**, 27–43 (2017)
4. Begnini, M., Bertol, D., Martins, N.: Design of an adaptive fuzzy variable structure compensator for the nonholonomic mobile robot in trajectory tracking task. Control Cybernet. **47**, 239–275 (2018)
5. Begnini, M., Bertol, D., Martins, N.: Practical implementation of an effective robust adaptive fuzzy variable structure tracking control for a wheeled mobile robot. J. Intell. Fuzzy Syst. **35**, 1087–1101 (2018)
6. DeCarlo, R.A., Zak, S.H., Matthews, G.P.: Variable structure control of nonlinear multivariable systems: a tutorial. Proc. IEEE **76**(3), 212–232 (1988)
7. Hung, J.Y., Gao, W., Hung, J.C.: Variable structure control: a survey. IEEE Trans. Indus. Electron. **40**(1), 2–22 (1993)
8. Lee, J.H., Lin, C., Lim, H., Lee, J.M.: Sliding mode control for trajectory tracking of mobile robot in the RFID sensor space. Int. J. Control Autom. Syst. **7**(3), 429–435 (2009)
9. Loh, A., Yeung, L.: Chattering reduction in sliding mode control: An improvement for nonlinear systems. WSEAS Trans. Circ. Syst. **10**, 2090–2098 (2004)
10. Martins, N.A., Elyoussef, E.S., Bertol, D.W., De Pieri, E.R., Moreno, U.F., Castelan, E.B.: Nonholonomic mobile robot with kinematic disturbances in the trajectory tracking: a variable structure controller. Learn. Nonlinear Models **8**(1), 23–40 (2010)

11. Solea, R., Cernega, D.: Modeling and performance evaluation for trajectory tracking control of a wheeled mobile robot. In: Proceedings of the 1st Workshop on Energy, Transport, and Environment Control Applications (ETECA'2009), pages 1–7 (2009)
12. Solea, R., Cernega, D.: Sliding mode control for trajectory tracking problem—performance evaluation. In: Proceedings of the 19th International Conference on Artificial Neural Networks (ICANN'2009), pages 865–874. Springer (2009)
13. Tao, L., Chen, Q., Nan, Y., Wu, C.: Double hyperbolic reaching law with chattering-free and fast convergence. IEEE Access **6**, 27717–27725 (2018)
14. Utkin, V.I., Guldner, J., Shi, J.: Sliding Mode Control in Electro-mechanical Systems, 2nd edn. CRC Press, Boca Raton (2009)
15. Yang, Rongni, Zhao, Y-X, Wu, T., Yan, M.: A Double Power Reaching Law of Sliding Mode Control Based on Neural Network. Hindawi Publishing Corporation, pp. 1–9 (09) (2013)

Chapter 6
Approximated Robust Control: First-Order Quasi-sliding Mode Control Techniques

6.1 Introduction

SMCs guarantee system insensitivity concerning uncertainties and disturbances, and cause reduction of the system order. Moreover, they are computationally efficient and may be applied to a wide range of various, possibly nonlinear, and time-varying systems [3]. However, often they also exhibit a serious drawback which essentially hinders their practical applications. This drawback, the high-frequency oscillations which inevitably appear in any real system whose input is supposed to switch infinitely fast, is usually called the chattering phenomenon. This is a highly undesirable phenomenon because it causes serious wear on the actuator components. In other words, the discontinuous portion of control of the FOSMCs is responsible for the compensation of the uncertainties and disturbances, requiring prior knowledge of bounds of the uncertainties and disturbances, otherwise high values of gains are applied, causing the increase of the chattering phenomenon. Therefore, continuous approximation methods to attenuate the chattering phenomenon have been proposed in substitution to the discontinuous control term of the SMCs.

In short, the controller variants considered in Chap. 5 show that, while making it possible to prescribe arbitrarily the rate of convergence of the error, results in a control signal with chattering phenomenon whose direct application to the system may be impractical. Thus, in this chapter, a first-order quasi-sliding mode control (FOQSMC) with basis on the stability analysis using Lyapunov theory is presented as well as the effectiveness of the controller variants is verified through simulations and experiments. It is important to emphasize that FOQSMCs are intended to mitigate the chattering phenomenon, but that the application of unnecessary control efforts

---

Electronic supplementary material The online version of this chapter (https://doi.org/10.1007/978-3-030-77912-2_6) contains supplementary material, which is available to authorized users.

N. A. Martins and D. W. Bertol, *Wheeled Mobile Robot Control*, Studies in Systems, Decision and Control 380, https://doi.org/10.1007/978-3-030-77912-2_6

on the actuators, because the high gains of **G**, may persist as a consequence of the need for knowledge of bounds of the uncertainties and disturbances.

6.2 Control Design

The controller variants are given by Eqs. (5.48)–(5.51) are discontinuous and the synthesis of such variants gives rise to the chattering phenomenon of trajectories about the surface $\boldsymbol{\sigma}^* = \mathbf{0}$. The discontinuous nature of these variants is due to the sign function, i.e., $\mathrm{sign}(\boldsymbol{\sigma}^*)$.

From the definition of the discontinuous function, similar to Eq. (5.3), i.e.,

$$\mathrm{sign}(\boldsymbol{\sigma}) = \boldsymbol{\sigma} \oslash |\boldsymbol{\sigma}|_\circ = \left[\mathrm{sign}(\sigma_1) \ \ldots \ \mathrm{sign}(\sigma_\mathrm{n})\right]^\mathrm{T}, \tag{6.1}$$

where $\oslash$ represents the Hadamard (element-wise) division and $|\cdot|_\circ$ is the element-wise module of a vector. So the ith element of $\mathrm{sign}(\boldsymbol{\sigma}^*)$ is defined as:

$$\mathrm{sign}(\sigma_i^*) = \frac{\sigma_i^*}{|\sigma_i^*|} = \begin{cases} +1 \text{ if } \sigma_i^* > 0 \\ 0 \text{ if } \sigma_i^* = 0 \\ -1 \text{ if } \sigma_i^* < 0 \end{cases},$$

and in the control law, Eq. (5.36), the discontinuous portion of the control results in:

$$\mathrm{g_i}\,\mathrm{sign}(\sigma_i^*) = \begin{cases} \mathrm{g}_i \text{ if } \boldsymbol{\sigma}_i^* > 0 \\ 0 \text{ if } \boldsymbol{\sigma}_i^* = 0 \\ -\mathrm{g}_i \text{ if } \sigma_i^* < 0 \end{cases}, \tag{6.2}$$

where $\mathrm{G} = \mathrm{diag}\left(\left[\mathrm{g}_1\ \mathrm{g}_2 \ldots \mathrm{g}_\mathrm{n}\right]\right)$ is the control amplitude.

The idea then is to attenuate the chattering phenomenon problem for the controller variants to perform properly through a good approximation to the inherent properties of ideal sliding motion. There are several approximation methods available to convert the discontinuous control into suitable continuous control law to perform the chattering phenomenon attenuation, including [3]:

- The most popular of them uses the linear saturation function

$$\mathrm{sign}(\sigma_i^*) \approx \mathrm{sat}(\sigma_i^*) = \begin{cases} -1 \text{ if } \sigma_i^* < -\rho_i \\ \frac{\sigma_i^*}{\rho_i} \text{ if } |\sigma_i^*| \le \rho_i \\ 1 \text{ if } \sigma_i^* > \rho_i \end{cases}. \tag{6.3}$$

With this modification the term becomes continuous and the switching variable does not converge to zero but to the closed interval $[-\rho_i,\ \rho_i]$, where ρ_i is the boundary layer thickness. Consequently, the system representative point after the reaching phase termination belongs to a layer around the switching hyperplane and therefore this strategy is called the boundary layer controller. In the work of [6] the approximation is treated with the use of a nonlinear saturation function;

- Introduction of other nonlinear approximations of the discontinuous control term, such as:

  – Fractional continuous approximation or proper continuous function

$$\mathrm{sign}(\boldsymbol{\sigma}^*) \approx \boldsymbol{\sigma}^* \oslash |\boldsymbol{\sigma}^*|_\circ + \boldsymbol{\epsilon}. \tag{6.4}$$

  – Logistic Sigmoid and Shifted Sigmoid functions

$$\begin{aligned}\mathrm{sign}(\boldsymbol{\sigma}^*) &\approx \mathbf{1} \oslash \left(\mathbf{1} + e^{\circ(-\boldsymbol{\epsilon}\circ\boldsymbol{\sigma}^*)}\right),\\ \mathrm{sign}(\boldsymbol{\sigma}^*) &\approx \left(\mathbf{1} - e^{\circ(-\boldsymbol{\epsilon}\circ\boldsymbol{\sigma}^*)}\right) \oslash \left(\mathbf{1} + e^{\circ(-\boldsymbol{\epsilon}\circ\boldsymbol{\sigma}^*)}\right).\end{aligned} \tag{6.5}$$

  where $\mathbf{1}$ stands for a vector of all terms equal to 1 and $\cdot^\circ$ stands for an element-wise power.

  – Hyperbolic tangent function

$$\mathrm{sign}(\boldsymbol{\sigma}^*) \approx \circ\tanh(\boldsymbol{\epsilon} \circ \boldsymbol{\sigma}^*), \tag{6.6}$$

  where $\circ\tanh$ stands for an element-wise hyperbolic tangent function and $\boldsymbol{\epsilon} = [\epsilon_1\ \epsilon_2\ \ldots\ \epsilon_n]$ in the Eqs. (6.4)–(6.6) is a small positive constant selected by the designers, thus leading to an exponentially stable robust control system;

- Substituting the boundary layer with a sliding sector;
- Using dynamic SMCs;
- Using adaptive SMCs;
- Using soft computing SMCs, such as fuzzy logic or/and neural networks;
- Using second (or higher) order SMCs.

Some of these methods are similar and both smooth the discontinuous control laws using the approximation functions. The fractional continuous approximated method or proper continuous function is adopted in the development of the first order quasi-sliding mode control in this chapter.

6.2.1 Approximated Controller Variants

Substituting the discontinuous term $\mathrm{sign}(\boldsymbol{\sigma}^*)$ of the four control variants, Eqs. (5.48)–(5.51), by the proper continuous function, Eq. (6.4), results in:

- FOQSMC-A ⟶ Constant rate reaching

$$\begin{aligned}\boldsymbol{v}_{\mathrm{cQA}} &= -\boldsymbol{B}_{0\sigma}^{-1}\boldsymbol{A}_{0\sigma} + \boldsymbol{v}_{\mathrm{cA}}^*,\\ \boldsymbol{v}_{\mathrm{cA}}^* &= -\mathbf{Q}\left(\boldsymbol{\sigma}^* \oslash (|\boldsymbol{\sigma}^*|_\circ + \boldsymbol{\epsilon})\right),\\ \mathbf{G} = \mathbf{Q}, &\qquad h(\boldsymbol{\sigma}^*) = \mathbf{0}.\end{aligned} \tag{6.7}$$

- FOQSMC-B ⟶ Constant plus proportional rate reaching

$$\begin{aligned} \boldsymbol{v}_{\mathrm{cQB}} &= -\boldsymbol{B}_{0\sigma}^{-1}\boldsymbol{A}_{0\sigma} + \boldsymbol{v}_{\mathrm{cB}}^{*}, \\ \boldsymbol{v}_{\mathrm{cB}}^{*} &= -\mathbf{Q}\left(\boldsymbol{\sigma}^{*} \oslash \left(|\boldsymbol{\sigma}^{*}|_{\circ} + \boldsymbol{\epsilon}\right)\right) - \mathbf{K}h(\boldsymbol{\sigma}^{*}), \\ \mathbf{G} &= \mathbf{Q}, \qquad h(\boldsymbol{\sigma}^{*}) = \boldsymbol{\sigma}^{*}. \end{aligned} \tag{6.8}$$

- FOQSMC-C ⟶ Power rate reaching

$$\begin{aligned} \boldsymbol{v}_{\mathrm{cQC}} &= -\boldsymbol{B}_{0\sigma}^{-1}\boldsymbol{A}_{0\sigma} + \boldsymbol{v}_{\mathrm{cC}}^{*}, \\ \boldsymbol{v}_{\mathrm{cC}}^{*} &= -\mathbf{Q}|\boldsymbol{\sigma}^{*}|^{\alpha}\left(\boldsymbol{\sigma}^{*} \oslash \left(|\boldsymbol{\sigma}^{*}|_{\circ} + \boldsymbol{\epsilon}\right)\right), \\ \mathbf{G} &= \mathbf{Q}|\boldsymbol{\sigma}^{*}|^{\alpha}, \qquad h(\boldsymbol{\sigma}^{*}) = \mathbf{0}. \end{aligned} \tag{6.9}$$

- FOQSMC-D ⟶ Speed control rate reaching

$$\begin{aligned} \boldsymbol{v}_{\mathrm{cQD}} &= -\boldsymbol{B}_{0\sigma}^{-1}\boldsymbol{A}_{0\sigma} + \boldsymbol{v}_{\mathrm{cD}}^{*}, \\ \boldsymbol{v}_{\mathrm{cD}}^{*} &= -\mathbf{Q}e^{(\alpha|\boldsymbol{\sigma}^{*}|)}\left(\boldsymbol{\sigma}^{*} \oslash \left(|\boldsymbol{\sigma}^{*}|_{\circ} + \boldsymbol{\epsilon}\right)\right), \\ \mathbf{G} &= \mathbf{Q}e^{(\alpha|\boldsymbol{\sigma}^{*}|)}, \qquad h(\boldsymbol{\sigma}^{*}) = \mathbf{0}. \end{aligned} \tag{6.10}$$

where $\boldsymbol{v}_{\mathrm{cA}}^{*}$, $\boldsymbol{v}_{\mathrm{cB}}^{*}$, $\boldsymbol{v}_{\mathrm{cC}}^{*}$, and $\boldsymbol{v}_{\mathrm{cD}}^{*}$ are the control portions that perform the compensation of the uncertainties and disturbances.

It is emphasized that these controller variants are quite effective in chattering phenomenon attenuation. However, the substitution of the discontinuous term by an approximation method causes the control to lose the invariance property to uncertainties and disturbances.

6.2.2 *Stability Analysis*

Stability analysis is similar to that described in Sects. 5.2.2 and 5.2.3. Therefore, to prove the stability of the closed-loop system are used the same Lyapunov function candidate, the sliding surfaces, given by Eqs. (5.8) and (5.24) respectively, thus ensuring that V is positive definite.

By differentiation of the Eq. (5.8) and substitution of the Eq. (5.26), as well as by using the new sliding surfaces, the appropriate nonsingular transformation, Eqs. (5.25) and (5.13) respectively, and doing the proper mathematical manipulations, resulting in Eq. (5.35), which is similar to Eq. (5.19).

With the substitution of the discontinuous term $\mathrm{sign}(\boldsymbol{\sigma}^{*})$ by the proper continuous function, Eq. (6.4), the control law defined by Eq. (5.36) is rewritten as:

$$\boldsymbol{v}_{\mathrm{c}} = -\boldsymbol{B}_{0\sigma}^{-1}\boldsymbol{A}_{0\sigma} - \Big[\mathbf{G}\big(\boldsymbol{\sigma}^{*} \oslash \big(|\boldsymbol{\sigma}^{*}|_{\circ} + \boldsymbol{\epsilon}\big)\big) + \mathbf{K}h(\boldsymbol{\sigma}^{*})\Big], \tag{6.11}$$

with $\boldsymbol{A}_{0\sigma}$, $\boldsymbol{B}_{0\sigma}$, and $\boldsymbol{B}_{0\sigma}^{-1}$ defined as in Eqs. (5.37)–(5.39), remembering that $\boldsymbol{B}_{0\sigma}$ should always be nonsingular by the satisfied condition given in Eq. (5.40).

Substituting Eq. (6.11) into $\dot{V}$, Eq. (5.35), results in:

$$\dot{V} = -\boldsymbol{\sigma}^{*^T}\Big[\mathbf{G}\big(\boldsymbol{\sigma}^* \oslash (|\boldsymbol{\sigma}^*|_\circ + \boldsymbol{\epsilon})\big) + \mathbf{K}h(\boldsymbol{\sigma}^*)\Big] + \boldsymbol{\sigma}^{*^T}\boldsymbol{v}_e. \tag{6.12}$$

Since the proper continuous function, Eq. (6.4), is nonlinear, analysis of the resulting motion is extremely complex. To reduce this complexity and retain a good approximation, the function is expanded in Taylor series yields to

$$\begin{aligned}
-\mathbf{G}\Big(\boldsymbol{\sigma}^* \oslash (|\boldsymbol{\sigma}^*|_\circ + \boldsymbol{\epsilon})\Big) &= -\mathbf{G}\Big(|\boldsymbol{\sigma}^*|_\circ + \boldsymbol{\epsilon}\Big)^{\circ(-1)} \circ \boldsymbol{\sigma}^* \\
&= -\Big(\mathbf{G} \oslash \boldsymbol{\epsilon}\Big)\Big((|\boldsymbol{\sigma}^*|_\circ \oslash \boldsymbol{\epsilon}) + \mathbf{1}\Big)^{\circ(-1)} \circ \boldsymbol{\sigma}^* \\
&= -\Big(\mathbf{G} \oslash \boldsymbol{\epsilon}\Big)\Big(\mathbf{1} - |\boldsymbol{\sigma}^*|_\circ \oslash \boldsymbol{\epsilon} + (|\boldsymbol{\sigma}^*|_\circ)^{\circ 2}\boldsymbol{\epsilon}^{\circ 2}\dots\Big) \circ \boldsymbol{\sigma}^*.
\end{aligned} \tag{6.13}$$

In the region $\boldsymbol{\sigma}^*_\epsilon$, the neighborhood of the manifolds of $\boldsymbol{\sigma}^*$ defined by

$$\boldsymbol{\sigma}^*_\epsilon = \Big\{\boldsymbol{q}_e : |\boldsymbol{\sigma}^*|_\circ \circ\!< \boldsymbol{\epsilon}\Big\}, \tag{6.14}$$

where $\circ<$ stands for an element-wise minor operation. So, from Eq. (6.14) a first-order approximation gives

$$-\mathbf{G}\Big(|\boldsymbol{\sigma}^*|_\circ + \boldsymbol{\epsilon}\Big)^{\circ(-1)} \circ \boldsymbol{\sigma}^* \approx \mathbf{G}_{\sigma^*} \circ \boldsymbol{\sigma}^*, \tag{6.15}$$

where $\mathbf{G}_{\sigma^*} = \mathbf{G} \oslash \boldsymbol{\epsilon}$. In this region, the system is effectively a high gain feedback system and possesses the properties of certain classes of disturbance rejection and parameter invariance [1, 4, 9].

Using $\mathbf{G}_{\sigma^*}$ of the Eq. (6.15) and considering that $\overline{\psi}_{\max} >= |\boldsymbol{v}_e|$ is a known scalar that represents the maximum effect of the uncertainties and disturbances, just like in the Eqs. (5.10) and (5.11), the Eq. (6.12) can be rewritten as:

$$\dot{V} \leq -\boldsymbol{\sigma}^{*^T}\mathbf{K}h(\boldsymbol{\sigma}^*) - (\mathbf{G}_{\sigma^*} - \overline{\psi}_{\max})|\boldsymbol{\sigma}^*|_1 \leq 0, \tag{6.16}$$

as long as $h(\boldsymbol{\sigma}^*) \neq \mathbf{0}$ and $h(\boldsymbol{\sigma}^*)$ has the same signal of $\boldsymbol{\sigma}^*$ as well as $\mathbf{G}_{\sigma^*} > \overline{\psi}_{\max}$. For the case of $h(\boldsymbol{\sigma}^*) = \mathbf{0}$, $\dot{V}$ becomes:

$$\dot{V} \leq -(\mathbf{G}_{\sigma^*} - \overline{\psi}_{\max})|\boldsymbol{\sigma}^*|_1 \leq 0, \tag{6.17}$$

since the condition $\mathbf{G}_{\sigma^*} > \overline{\psi}_{\max}$ be satisfied. Thus, $\dot{V} < 0$ for $\boldsymbol{q}_e \neq 0$. Hence, $\boldsymbol{q}_e = 0$ is an asymptotically stable equilibrium point. Recalling that as Eq. (6.12) is similar to Eqs. (5.21) and (5.41), the same conclusions on the stability analysis are valid.

6.3 Simulations Using Matlab and/or MobileSim Simulator

To verify the performance, the FOQSMCs are implemented in Matlab/Simulink software and evaluated for the trajectory tracking control problem through the three scenarios, whose reference is an eight-shape trajectory described in Sect. 3.1.2, and the gains used in the FOQSMCs are presented in Table 6.1.

6.3.1 Ideal Scenario

In this scenario, simulations are performed only in Matlab/Simulink and uncertainties and disturbances are disregarded. The simulation results are presented in Table 6.2 in terms of RMS of the errors, from which it is verified that the DWMR can track the trajectory with minimum errors using FOQSMCs.

Figures 6.1 and 6.2 show that using the FOQSMCs, the DWMR satisfactorily follows the reference trajectory as well as posture trackings generated by DWMR also follow the reference posture trackings. Consequently, the DWMR tracks the trajectory with posture tracking and orientational errors (xy_e and θ_e) that converge to zero, as can be seen in Fig. 6.3.

Figure 6.21 shows how the DWMR reaches the reference trajectory and tries to remain on it for the rest of the experiment, although it is possible to observe a certain performance degradation in the trajectory tracking due to mainly intrinsic uncertainties and disturbances of the DWMR. This observation can also be verified in the posture trackings shown in Fig. 6.22.

The linear and angular control velocities provided by the FOQSMCs as well as the linear and angular velocities of the DWMR are shown in Figs. 6.4 and 6.5

Table 6.1 FOQSMC gains

Controller	λ_1	λ_2	λ_3	Q_1	Q_2	ϵ	K_1	K_2	α
FOQSMC-A	1.5	6.0	1.0	0.1	0.3	0.1	–	–	–
FOQSMC-B	1.5	6.0	1.0	0.1	0.3	0.1	0.1	0.1	–
FOQSMC-C	1.5	6.0	1.0	0.1	0.3	0.1	–	–	0.2
FOQSMC-D	1.5	6.0	1.0	0.1	0.3	0.1	–	–	0.2

Table 6.2 RMS of the errors—Results in the ideal scenario

Controller	x_e (m)	y_e (m)	θ_e (rad)
FOQSMC-A	0.0144	0.0506	0.0484
FOQSMC-B	0.0141	0.0389	0.0442
FOQSMC-C	0.0208	0.0500	0.0468
FOQSMC-D	0.0178	0.0426	0.0444

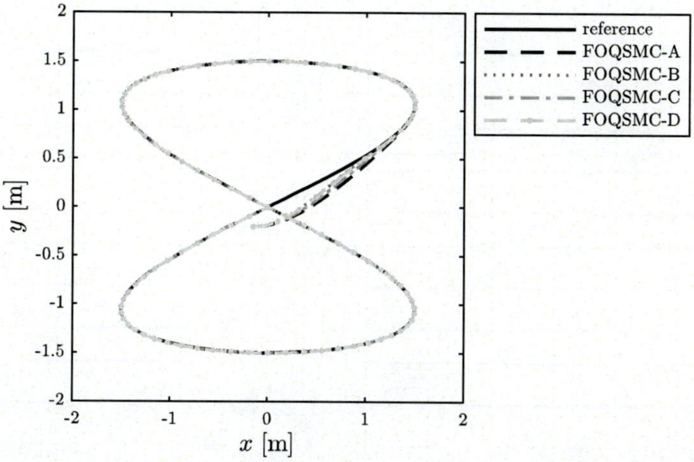

Fig. 6.1 Trajectory tracking in the ideal scenario

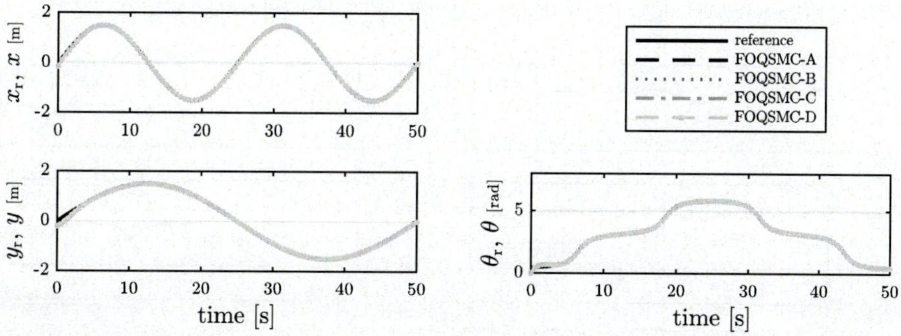

Fig. 6.2 Tracking (x, y, and θ) in the ideal scenario

respectively. These velocities present smooth behaviors with a very well-attenuated chattering phenomenon. In Fig. 6.5, it is observed that the linear and angular velocities of the DWMR also tend towards the reference velocities.

The control portions $\boldsymbol{v}_c^*$ of the FOQSMCs (Eqs. 6.7–6.10) and the auxiliary velocity tracking errors are shown in Fig. 6.6. The difference between auxiliary velocity tracking errors and control portions of compensations tends to converge to zero. Moreover, the well-attenuated chattering phenomenon in the compensations due to the use of FOQSMCs can also be verified in Fig. 6.6.

The sliding surfaces, new sliding surfaces, and their derivatives tend to converge to zero, as can be observed in Figs. 6.7, 6.8, 6.9, and 6.10, with a very well-mitigated chattering phenomenon.

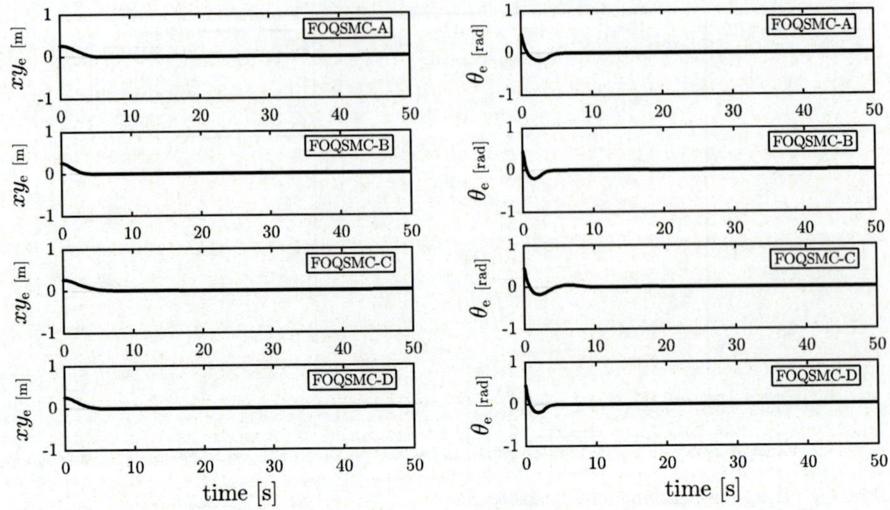

Fig. 6.3 Posture tracking errors, with $xy_e = \sqrt{x_e^2 + y_e^2}$, in the ideal scenario

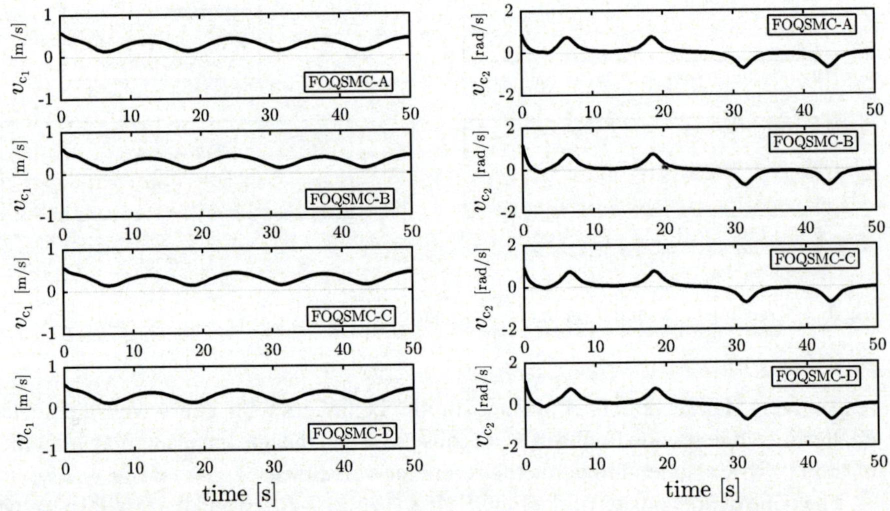

Fig. 6.4 Control velocities in the ideal scenario

6.3.2 Realistic Scenario

The results in Table 6.3, containing the RMS of the errors, show that the FOQSMCs can track the trajectories with admissible errors even with uncertainties and disturbances provided by the MobileSim simulator.

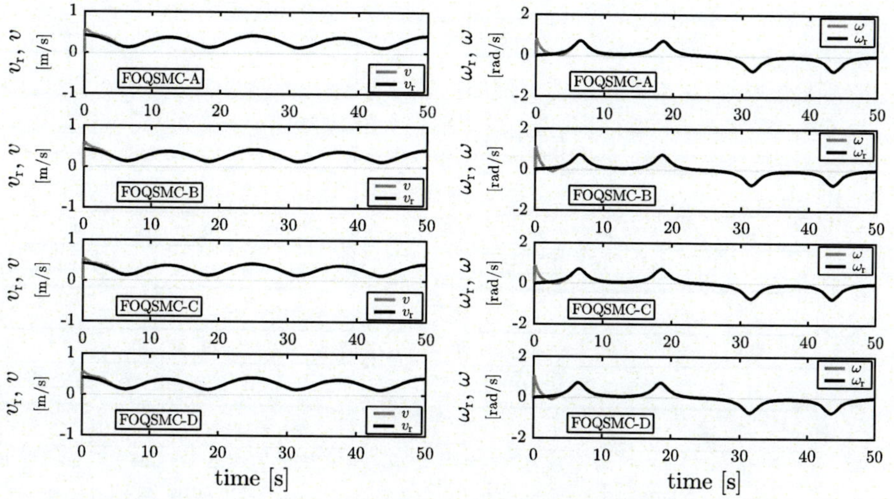

Fig. 6.5 Velocities of the DWMR and reference velocities in the ideal scenario

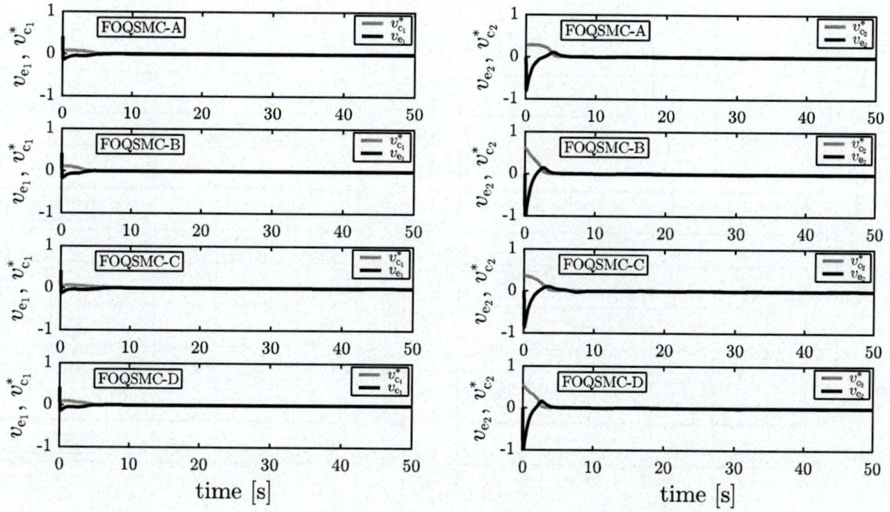

Fig. 6.6 Auxiliary velocity tracking errors and control velocities $\boldsymbol{v}^*_c$ in the ideal scenario

The trajectory performed by DWMR, using the FOSQMCs, shows that it adequately tracks the reference trajectory as can be verified in Fig. 6.11 and also by posture trackings in Fig. 6.12.

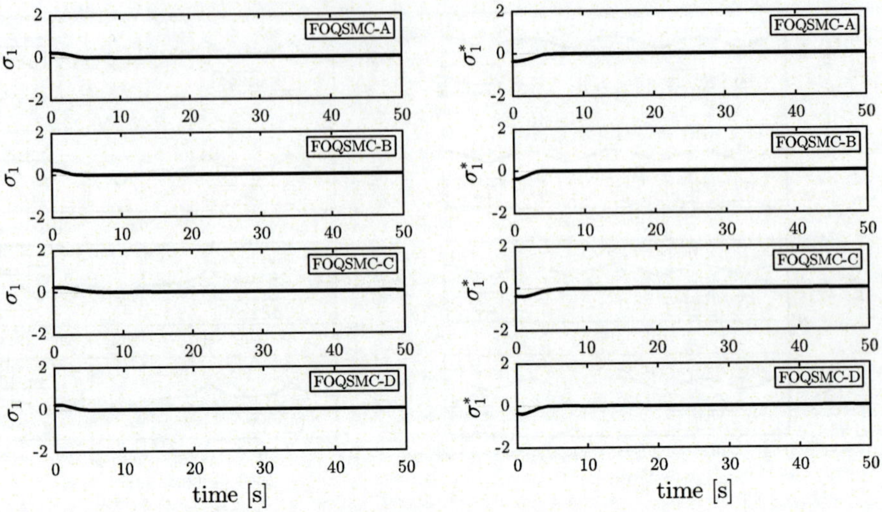

Fig. 6.7 Sliding surface σ_1 and new sliding surface σ_1^* in the ideal scenario

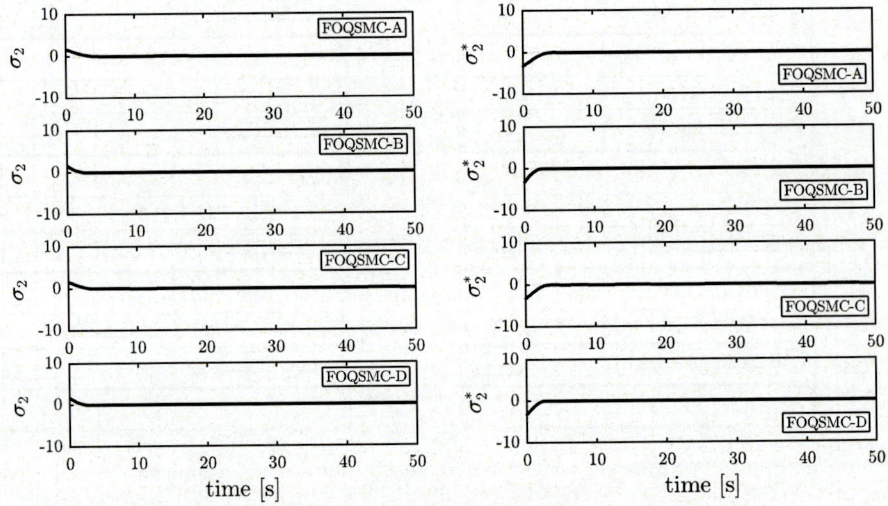

Fig. 6.8 Sliding surface σ_2 and new sliding surface σ_2^* in the ideal scenario

Figure 6.13 shows that posture tracking and orientational errors suffer a slight variation, mainly when the DWMR makes a curve, but tend to converge to zero.

In Fig. 6.14, the control velocities present smooth control signals, almost without any chattering phenomenon. This behavior can also be observed in Fig. 6.15, in which velocities of the DWMR track the reference trajectory, with the chattering phenomenon being well attenuated.

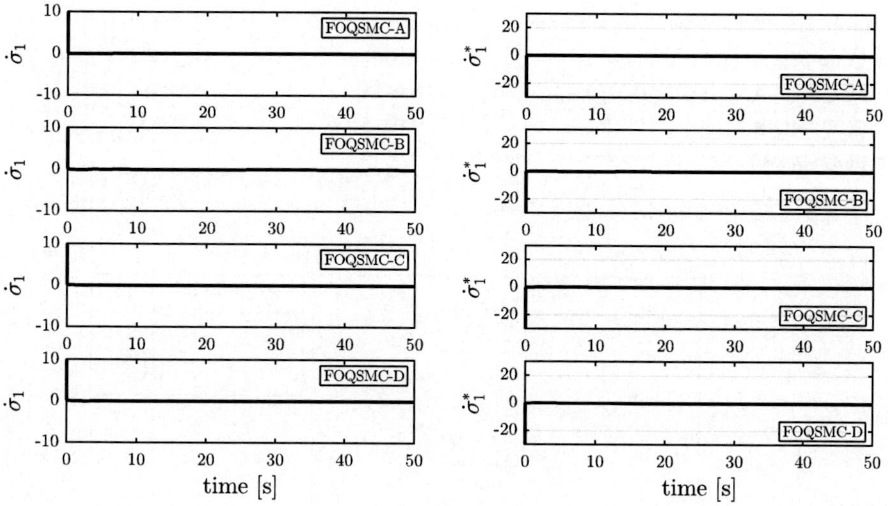

Fig. 6.9 Derivatives of sliding surface σ_1 and derivatives of new sliding surface σ_1^* in the ideal scenario

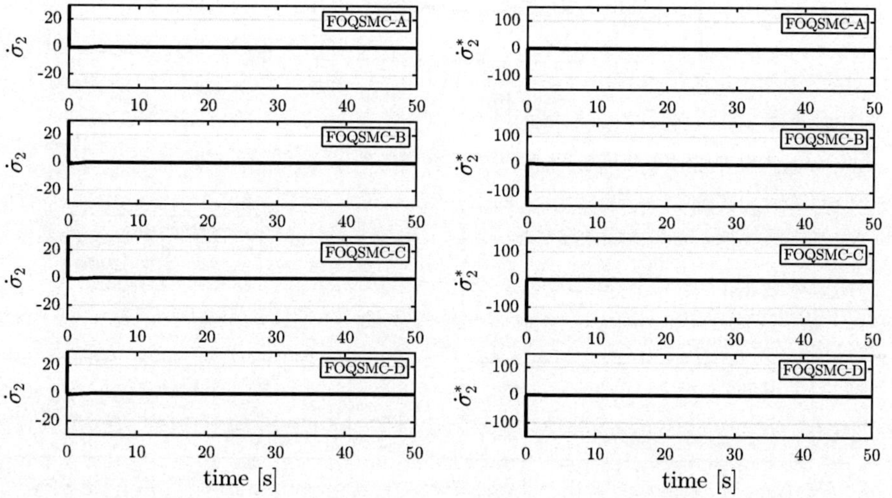

Fig. 6.10 Derivatives of sliding surface σ_2 and derivatives of new sliding surface σ_2^* in the ideal scenario

Table 6.3 RMS of the errors—Results in the realistic scenario

Controller	x_e (m)	y_e (m)	θ_e (rad)
FOQSMC-A	0.0253	0.0498	0.0459
FOQSMC-B	0.0262	0.0395	0.0431
FOQSMC-C	0.0343	0.0515	0.0476
FOQSMC-D	0.0323	0.0442	0.0432

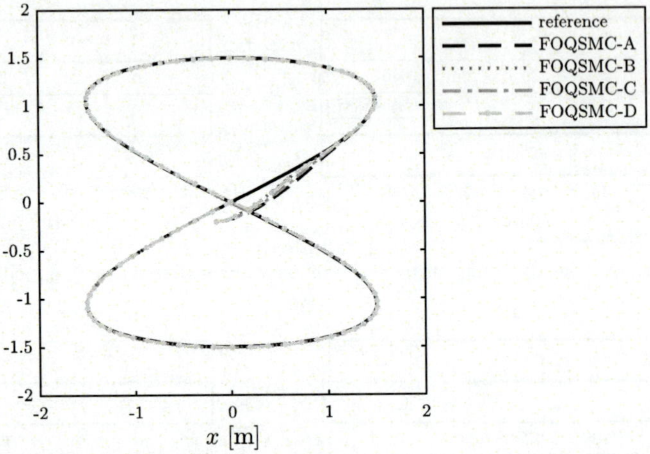

Fig. 6.11 Trajectory tracking in the realistic scenario

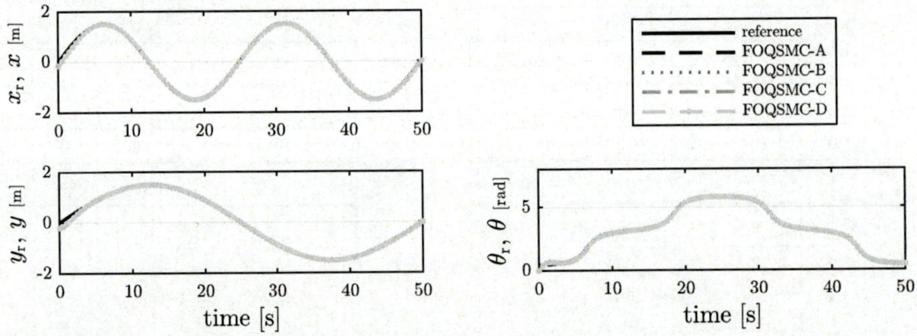

Fig. 6.12 Tracking (x, y, and θ) in the realistic scenario

Figure 6.16 showing the auxiliary velocity tracking errors $\boldsymbol{v}_e^*$ and control portions of compensations $\boldsymbol{v}_c^*$ of the FOQSMCs (Eqs. 6.7–6.10) also present behaviors with well-minimized chattering phenomenon as well as the opposite magnitudes of absolute values so that the difference between them tends to cancel. Recalling that the auxiliary velocity tracking errors $\boldsymbol{v}_e$, seen as disturbances for the kinematic model

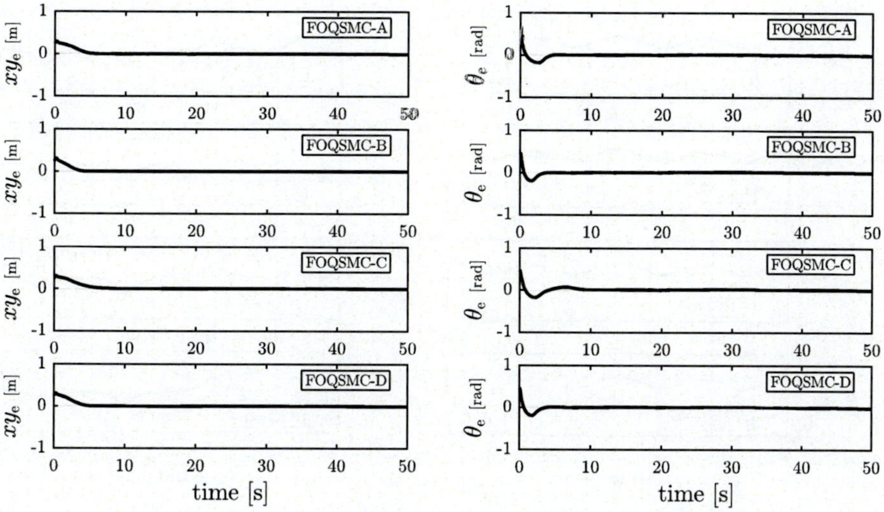

Fig. 6.13 Posture tracking errors, with $xy_e = \sqrt{x_e^2 + y_e^2}$, in the realistic scenario

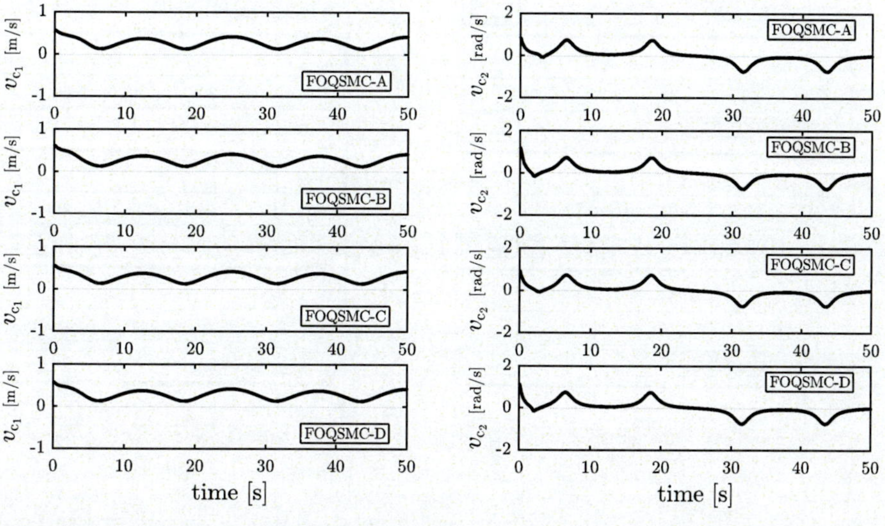

Fig. 6.14 Control velocities in the realistic scenario

and given by the difference between the control velocities $\boldsymbol{v}_c$ and the real velocities of the DWMR $\boldsymbol{v}$, happen mainly because of uncertainties and disturbances.

Concerning the sliding surfaces and new sliding surfaces, it can be said that they are slightly oscillating close to zero, as can be seen in Figs. 6.17 and 6.18, featuring a significantly reduced chattering phenomenon. However, their derivatives, shown in Figs. 6.19 and 6.20, present the chattering phenomenon with a considerable magnitude oscillating around to zero.

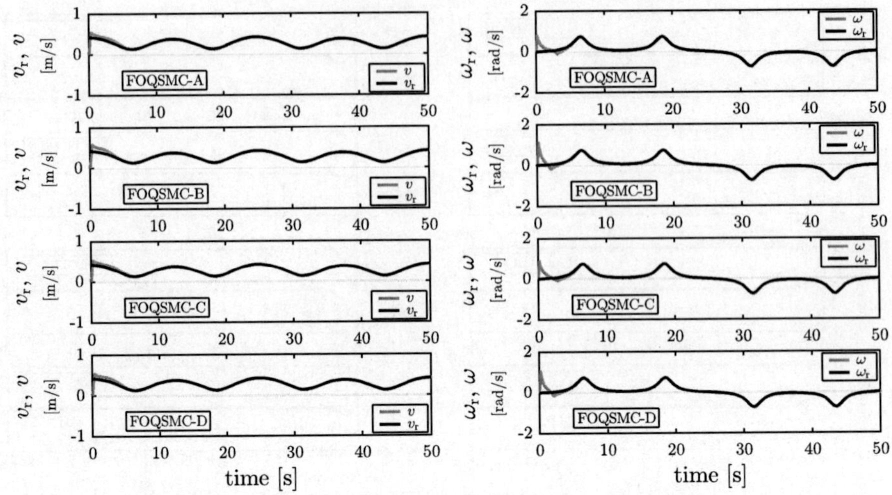

Fig. 6.15 Velocities of the DWMR and reference velocities in the realistic scenario

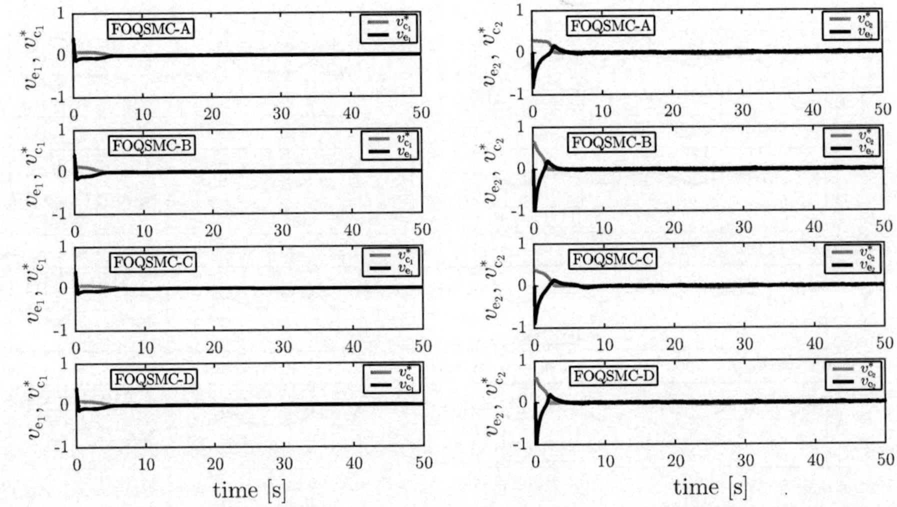

Fig. 6.16 Auxiliary velocity tracking errors and control velocities $\boldsymbol{v}_c^*$ in the realistic scenario

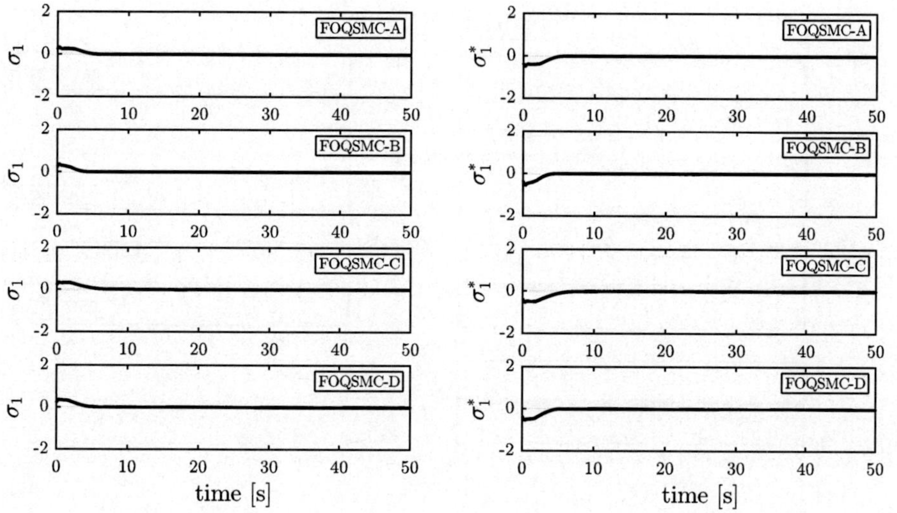

Fig. 6.17 Sliding surface σ_1 and new sliding surface σ_1^* in the realistic scenario

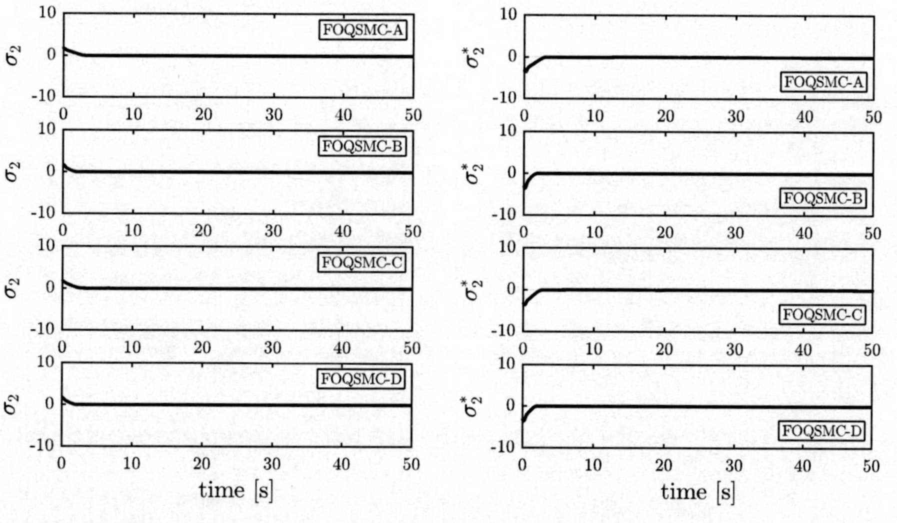

Fig. 6.18 Sliding surface σ_2 and new sliding surface σ_2^* in the realistic scenario

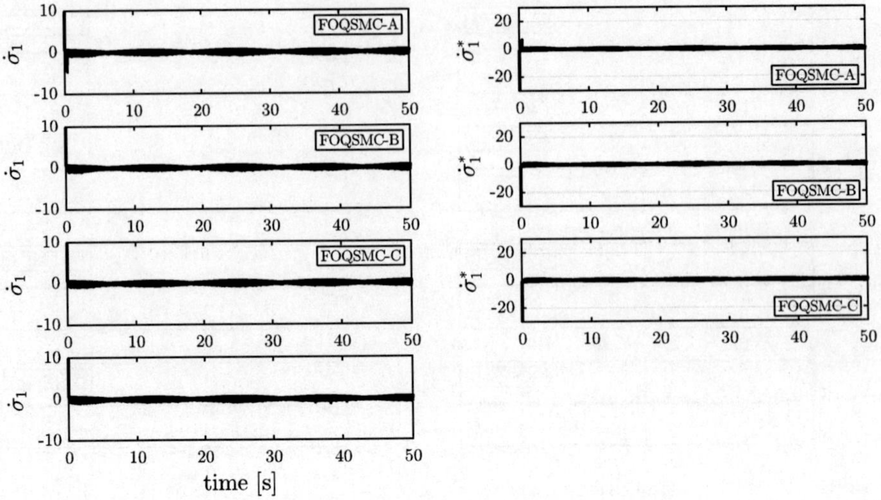
FOQSMC-A
FOQSMC-B
FOQSMC-C
FOQSMC-A
FOQSMC-B
FOQSMC-C
$\dot{\sigma}_1$
$\dot{\sigma}_1^*$
time [s]

6.4 Experimental Results Using PowerBot DWMR

In this scenario, the FOQSMCs are also tested on the Powerbot DWMR for the eight-shape trajectory. In the experimental results in real-time, presenting the values of RMS of the errors in Table 6.4, it is confirmed what the FOQSMCs can track the trajectory with acceptable errors.

In Fig. 6.23, by using FOQSMCs, it is observed how the posture tracking and orientation errors tend to oscillate very near to zero and behaving so for the remainder of the experiment. The oscillation of the orientation error has a behavior similar to that obtained in simulator MobileSim (realistic scenario), being more perceptible when the DWMR makes a curve.

The control velocities generated by the FOQSMCs have smooth control signals, as shown in Fig. 6.24. Regarding the velocities of DWMR, it can be observed in Fig. 6.25 that these velocities track the reference velocities. From the visualization of these figures, it appears that both control velocities and velocities of the DWMR present a slight chattering phenomenon.

Figure 6.26 shows how the FOQSMCs compensate for the uncertainties and disturbances, whose auxiliary velocity tracking errors and control portions of

Table 6.4 RMS of the errors—Results in the experimental scenario

Controller	x_e (m)	y_e (m)	θ_e (rad)
FOQSMC-A	0.0644	0.0725	0.1020
FOQSMC-B	0.0649	0.0615	0.1180
FOQSMC-C	0.0703	0.0757	0.1180
FOQSMC-D	0.0709	0.0622	0.1170

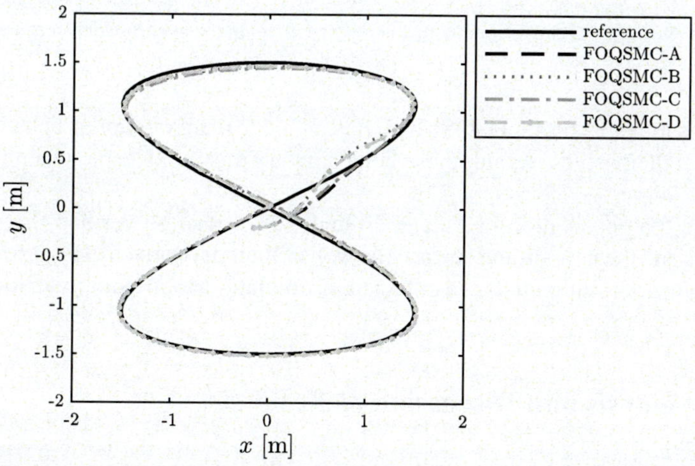

Fig. 6.21 Trajectory tracking in the experimental scenario

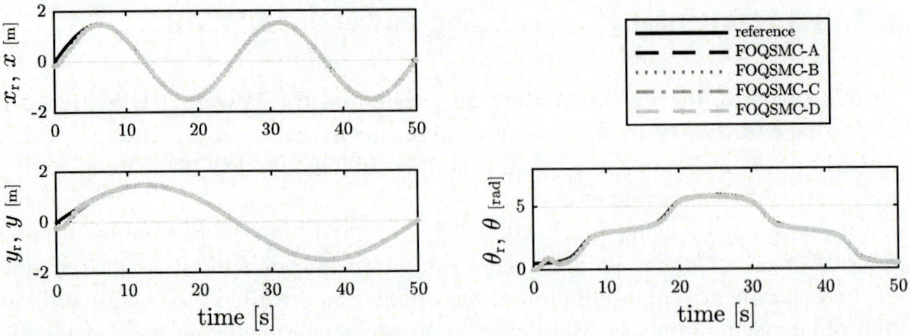

Fig. 6.22 Tracking (x, y and, θ) in the experimental scenario

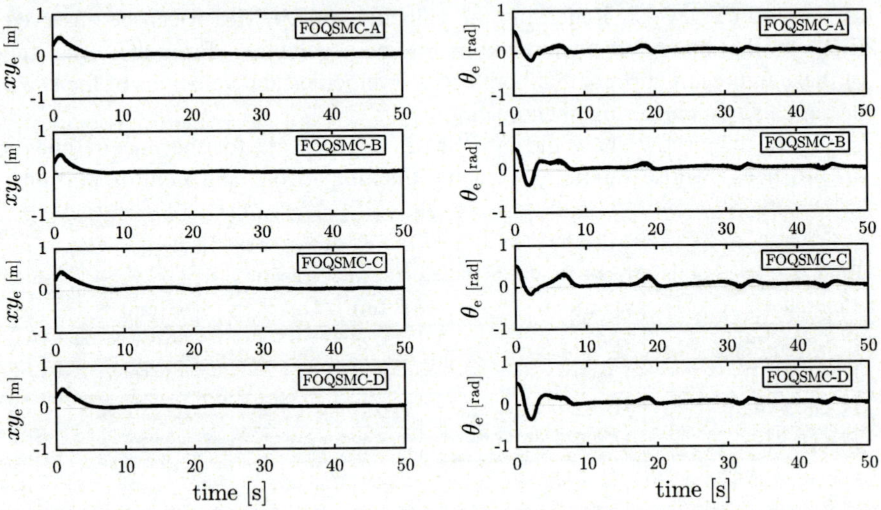

Fig. 6.23 Posture tracking errors, with $xy_e = \sqrt{x_e^2 + y_e^2}$, in the experimental scenario

compensations $\boldsymbol{v}_c^*$ of the FOQSMCs (Eqs. 6.7–6.10) are of similar behaviors, trying to cancel themselves. Again, these behaviors are most noticeable when the DWMR makes a curve.

From the observation of Figs. 6.27 and 6.28, it can be verified that the sliding surfaces and the new sliding surfaces as well as their derivatives in Figs. 6.29 and 6.30 have behaviors similar to those obtained in simulator MobileSim (realistic scenario).

6.5 Analysis and Discussion of Results

As with FOSMCs, FOSQSMCs also need prior knowledge of the bounds of uncertainties and disturbances to define the gain values $\mathbf{G}$ and/or $\mathbf{K}$ (Eqs. 6.7–6.10) that

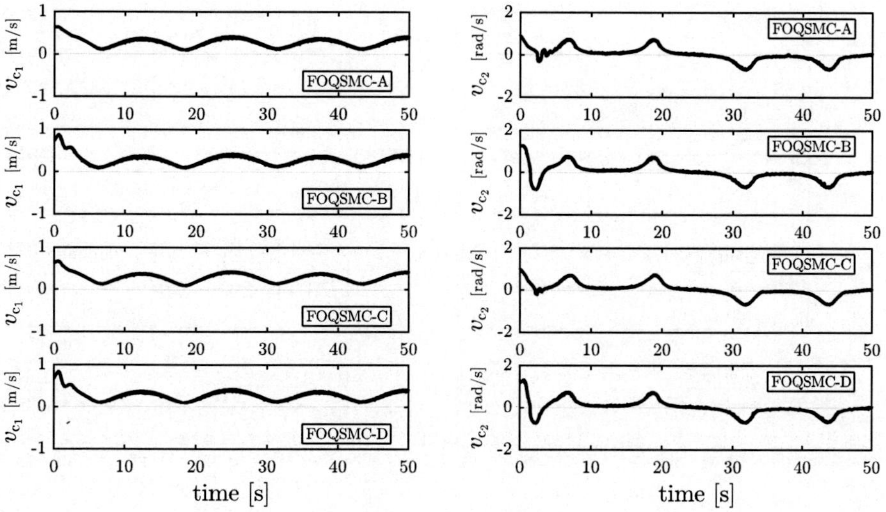

Fig. 6.24 Control velocities in the experimental scenario

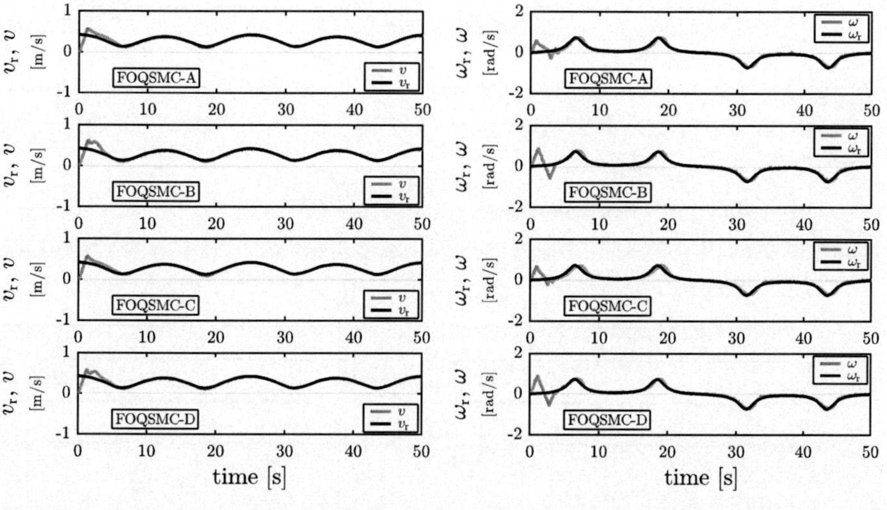

Fig. 6.25 Velocities of the DWMR and reference velocities in the experimental scenario

achieves satisfactory tracking performance but maybe applying unnecessary control efforts on the DWMR actuators.

Recalling that the FOSMCs also have the drawback of significant chattering phenomenon at their velocities mainly, being that the occurrence and amplitude of this phenomenon are due to the discontinuous function $\mathbf{sign}(\sigma^*)$ and the value defined for the gain $\mathbf{G}$, respectively.

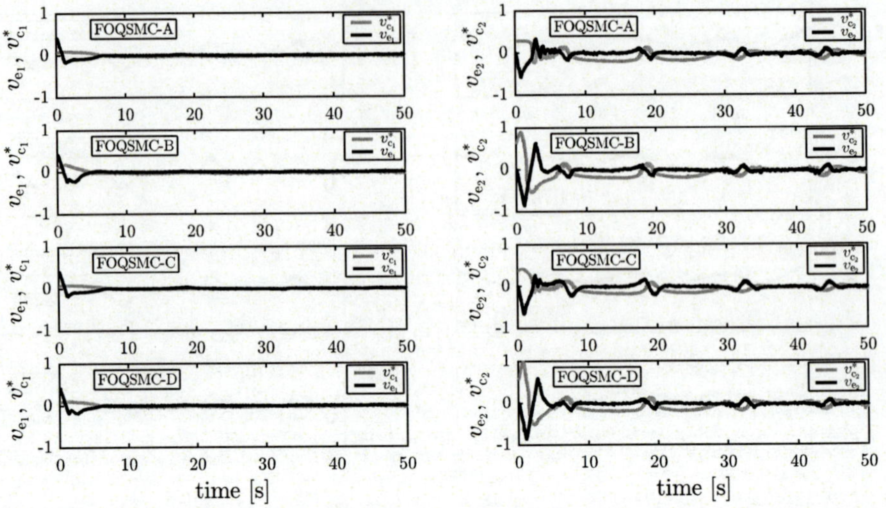

Fig. 6.26 Auxiliary velocity tracking errors and control velocities $\boldsymbol{v}_{\mathrm{c}}^{*}$ in the experimental scenario

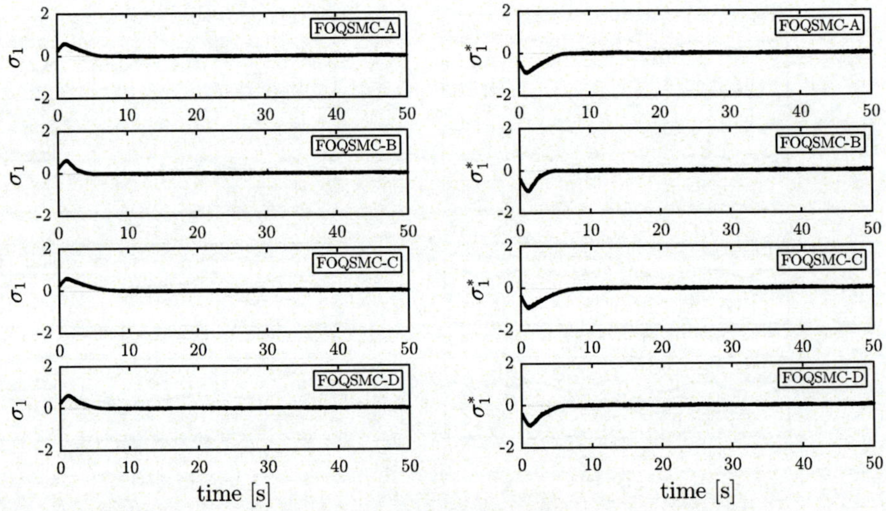

Fig. 6.27 Sliding surface σ_1 and new sliding surface σ_1^* in the experimental scenario

In the case of the FOQSMCs, the amplitude and occurrence of the chattering phenomenon depend respectively on the gain defined for **G** and the parameter ϵ of the fractional continuous approximation or proper continuous function in the Eqs. (6.7)–(6.10), being ϵ a small positive constant selected by the designers, whose choice must have a tradeoff between tracking performance and occurrence of the chattering phenomenon. If the choice of a value of ϵ is close to zero, the tracking performance response may be satisfactory, but with a significantly considerable incidence of

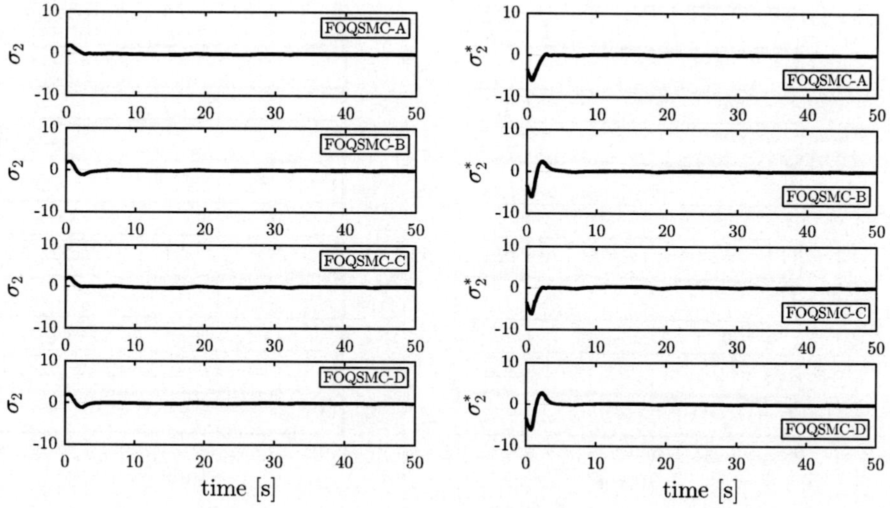

Fig. 6.28 Sliding surface σ_2 and new sliding surface σ_2^* in the experimental scenario

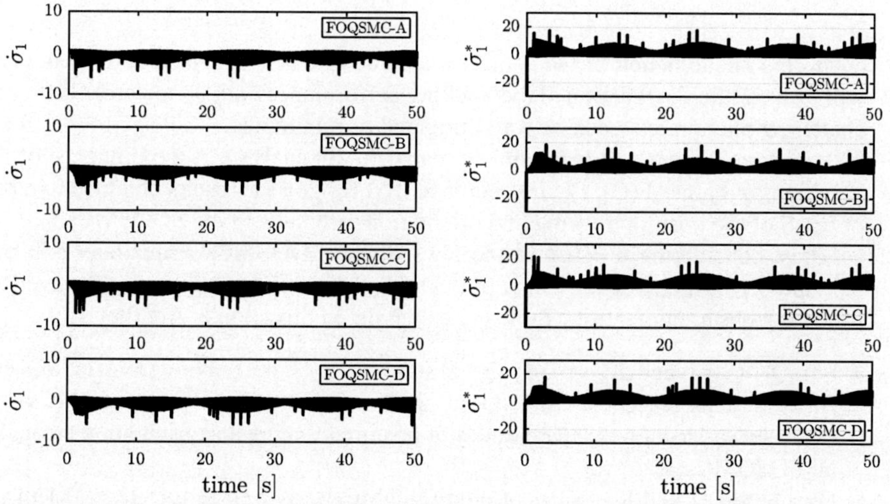

Fig. 6.29 Derivatives of sliding surface σ_1 and derivatives of new sliding surface σ_1^* in the experimental scenario

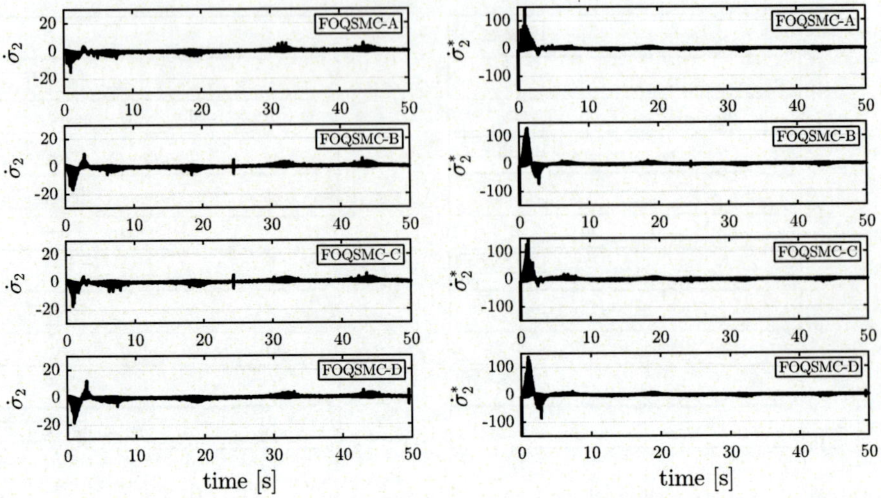

Fig. 6.30 Derivatives of sliding surface σ_2 and derivatives of new sliding surface σ_2^* in the experimental scenario

chattering phenomenon. If the choice of a value of ϵ is a little distant to zero, there will be a certain degradation in the tracking performance and, on the other hand, the chattering phenomenon can be well mitigated or minimized.

Based on Table 6.5, it is possible to carry out an analysis of the influence of the variations of gains **G** (Q_1, Q_2, α) and/or **K** (K_1, K_2), on the results obtained in terms of RMS of the posture tracking and orientation errors using the FOQSMCs.

Thus, considering $\epsilon = 0.1$ in the laws of the FOQSMCs (Equations 6.7, 6.8, 6.9, and 6.10), it can be seen how critical the choice of these gains is, however the following observations from Table 6.5 can assist in this choice. Are they:

- For $Q_1 \ll 0.1$ and $K_1 \ll 0.1$, $Q_2 \ll 0.3$ and $K_2 \ll 0.1$, both from FOSQMC-B, as well as for $Q_1 << 0.1$, $Q_2 \ll 0.3$, both from the other FOSQMCs—an increase was seen in the magnitudes of posture tracking and orientation errors for all FOSQMCs;
- For $Q_1 \gg 0.1$ and $K_1 \gg 0.1$ of the FOSQMC-B, as well as for $Q_1 \gg 0.1$ of the other FOSQMCs—there is a decrease in the magnitudes of posture tracking errors, but an increase in the magnitude of the orientation error;
- For $Q_2 \gg 0.3$ and $K_2 \gg 0.1$ of the FOSQMC-B, as well as for $Q_2 \gg 0.3$ of the other FOSQMCs—there is an increase in the magnitude of the posture tracking error in the x-direction, but a decrease in the magnitudes of the posture tracking error in the y-direction and orientation error;
- For $\alpha \ll 0.2$—it is seen that, for the FOSQMC-C, the magnitude of the posture tracking error in x-direction increases and the posture tracking error in the y-direction and orientation error decrease, whereas, for the FOSQMC-D, the opposite behavior of this occurs, i.e., the magnitude of the posture tracking error in x-

Table 6.5 Errors for FOQSMC laws—Simulation results in the realistic scenario using $\epsilon = 0.1$ and different gain values

Law	Gains					Errors		
						RMS		
	Q_1	Q_2	K_1	K_2	α	x_e (m)	y_e (m)	θ_e (rad)
FOQSMC-A	0.01	0.30	–	–	–	0.1229	0.1171	0.2212
	0.10					0.0281	0.0517	0.0466
	1.00					0.0275	0.0428	0.0537
	0.10	0.03	–	–	–	0.0706	0.1146	0.0443
		0.30				0.0281	0.0517	0.0466
		3.00				0.0391	0.0375	0.0416
FOQSMC-B	0.01	0.30	0.10	0.10	–	0.0459	0.0494	0.0696
	0.10					0.0276	0.0386	0.0533
	1.00					0.0193	0.0307	0.0482
	0.10	0.03	0.10	0.10	–	0.0419	0.0690	0.0527
		0.30				0.0276	0.0386	0.0533
		3.00				0.0311	0.0314	0.0476
	0.10	0.30	0.01	0.10	–	0.0352	0.0422	0.0520
			0.10			0.0276	0.0386	0.0533
			1.00			0.0182	0.0318	0.0501
	0.10	0.30	0.10	0.01	–	0.0269	0.0454	0.0528
				0.10		0.0276	0.0386	0.0533
				1.00		0.0318	0.0317	0.0500
FOQSMC-C	0.01	0.30	–	–	0.20	0.1356	0.1234	0.2400
	0.10					0.0367	0.0505	0.0604
	1.00					0.0235	0.0389	0.0552
	0.10	0.03	–	–	0.20	0.0769	0.1235	0.0446
		0.30				0.0367	0.0505	0.0604
		3.00				0.0443	0.0374	0.056
	0.10	0.30	–	–	0.02	0.0306	0.0506	0.0538
					0.20	0.0367	0.0505	0.0604
					2.00	0.1010	0.0947	0.1752
FOQSMC-D	0.01	0.30	–	–	0.20	0.1189	0.1074	0.2112
	0.10					0.0338	0.0424	0.0609
	1.00					0.0204	0.0324	0.0527
	0.10	0.03	–	–	0.20	0.0720	0.1123	0.0449
		0.30				0.0338	0.0424	0.0609
		3.00				0.0382	0.0351	0.0487
	0.10	0.30	–	–	0.02	0.0296	0.0493	0.0527
					0.20	0.0338	0.0424	0.0609
					2.00	0.0264	0.0289	0.0556

direction decreases and the posture tracking error in the y-direction and orientation error increase;
- For $\alpha \gg 0.2$—it is observed that, for FOSQMC-C, the magnitudes of posture tracking and orientation errors increase, however, for FOSQMC-D, these magnitudes of errors decrease.

In short, it is important to note that the choice of these gains can or can not:

- Influence the amplitude of the chattering phenomenon;
- Adequately compensate for the incidence of uncertainties and disturbances;
- Provides the necessary control efforts for satisfactory tracking performance;
- Shortens the life of the actuators and therefore damages the DWMR.

6.6 General Considerations

In this chapter, by using the continuous approximation method of the discontinuous portion of the control, the FOSQSMC's can avoid the chattering phenomenon but the loss of the invariance principle occurs, ensures acceptable robustness, and results in a smooth control signal [8]. Based on the results obtained, it can be concluded that despite the invariance requires switching between the reaching phase and sliding phase to be ideal, which is impractical, the invariance is ideal and has little practical meaning [12].

It is emphasized that there is a diversity of approximation methods in the literature, such as in [2, 3, 5, 7, 10, 11, 13, 14], which can be used in the FOSQSMC's proposed here.

However, other problems persist concerning the FOSQSMC's, which are the necessity of a knowledge of the limits of uncertainties and disturbances in the system, and the application of a large value to the gains **G** that can cause a high control effort (large authority control), affecting the trajectory tracking and deteriorating the system performance [8].

In order not to need the knowledge of the limits of uncertainties and disturbances in the system and nor the use of trial and error method to assign the value of the gains **G** as well as to deal with the chattering phenomenon, solutions are proposed in the next chapters.

References

1. Ambrosino, G., Celentano, G., Garofalo, F.: Variable structure mode-reference adaptive-control systems. Int. J. Control **39**, 1339–1349 (1984)
2. Asad, M., Ashraf, M., Iqbal, S., Bhatti, A.: Chattering and stability analysis of the sliding mode control using inverse hyperbolic function. Int. J. Control Autom. Syst. **15**, 2608–2618 (2017)

3. Bartoszewicz, A.: Variable structure control: from principles to applications. In: International Symposium on System Theory, Automation, Robotics, Computers, Informatics, Electronics and Instrumentation, pp. 17–24 (2007)
4. Burton, J.A., Zinober, A.S.I.: Continuous approximation of variable structure control. Int. J. Syst. Sci. **17**(6), 875–885 (1986)
5. Cheng, N., Guan, L., Wang, L., Han, J.: Chattering reduction of sliding mode control by adopting nonlinear saturation function. Adv. Mater. Res. **143–144**, 53–61 (2011)
6. Cheng, N.B., Guan, L.W., Wang, L.P., Han, J.: Chattering reduction of sliding mode control by adopting nonlinear saturation function. In: Smart Materials and Intelligent Systems—SMIS2010, vol. 143 of Advanced Materials Research, pp 53–61. Trans Tech Publications Ltd. (2011)
7. Cui, M., Liu, W., Liu, H., Jiang, H., Wang, Z.: Extended state observer-based adaptive sliding mode control of differential-driving mobile robot with uncertainties. Nonlinear Dyn. **83**, 667–683 (2016)
8. Martins, N.A., Elyoussef, E.S., Bertol, D.W., De Pieri, E.R., Moreno, U.F., Castelan, E.B.: Nonholonomic mobile robot with kinematic disturbances in the trajectory tracking: a variable structure controller. Learn. Nonlinear Models **8**(1), 23–40 (2010)
9. Saksena, V.R., O'Reilly, J., Kokotovid, P.V.: Singular perturbations and time-scale methods in control theory: survey 1976–1983. Automatica **20**, 273–293 (1984)
10. Tahir, K., Belfedal, c., Allaoui, T., Denaï, M., Doumi, M.: A new sliding mode control strategy for variable-speed wind turbine power maximization. Int. Trans. Electr. Energ. Syst. 1–22 (2018)
11. Tao, L., Chen, Q., Nan, Y., Wu, C.: Double hyperbolic reaching law with chattering-free and fast convergence. IEEE Access **6**, 27717–27725 (2018)
12. Wang, S., Gao, W.: Robustness and invariance of variable structure systems with multiple inputs. In: Proceedings of the American Control Conference (ACC'1995), vol. 1, pp. 1035–1039. IEEE (1995)
13. Zhai, W., Ge, Y., Song, S.: Sliding mode control of manipulator based on improved reaching law. Sens. Trans. **20**(Special Issue), 69–78 (2013)
14. Zhang, J., Liu, B., Jiang, Z., Hu, H.: The application of sliding mode control with improved approaching law in manipulator control. In: Proceedings of the IEEE Conference on Industrial Electronics and Applications (ICIEA), pp. 807–812. IEEE (2018)

Chapter 7
Adaptive Robust Control: Adaptive Fuzzy Sliding Mode Control Technique

7.1 Introduction

Unfortunately, in practical implementations, due to delays, neglected dynamics, sampling frequency limitation, physical limitations of actuators, and imperfections of switching, it is not possible to switch the control from a value to another instantaneously [5, 15]. Thus, the trajectory of states varies in a vicinity around the sliding surface, instead of sliding over it. This is known as the chattering phenomenon, which occurs because the controllers in Chap. 5 will not always be able to maintain the trajectory of states slipping on the sliding surfaces, being that the chattering phenomenon is more evident the higher the value of $\mathbf{G}$ mainly. Figure 7.1 illustrates the comparison between the ideal sliding mode and the sliding mode with the chattering phenomenon, respectively [17–19].

Moreover, it is possible to verify the chattering phenomenon in simulation and experimental results of the controllers in Sects. 5.3 and 5.4 of Chap. 5, in which can be observed that the chattering phenomenon occurs in the graphs of linear and angular velocities of the DWMR, sliding surfaces, and their derivatives. Thus, such a phenomenon can be avoided or at least reduced by replacing the discontinuous function $\text{sign}(\boldsymbol{\sigma}^*)$ over a continuous approximation, as already discussed in Chap. 6.

The controllers in Chap. 6 can avoid the chattering phenomenon but the invariance principle is not verified anymore, theoretically reducing the robustness (such reduction is not significant and the robustness is ensured); however, the smooth control signal is achieved [14, 15]. This occurs because the system dynamics, in this case, is confined to a neighborhood of the sliding surfaces, and no longer over it, i.e., as a consequence of the approximation with the fractional continuous approximation or proper continuous function, the system is enforced to a neighborhood of

---

Electronic supplementary material The online version of this chapter (https://doi.org/10.1007/978-3-030-77912-2_7) contains supplementary material, which is available to authorized users.

N. A. Martins and D. W. Bertol, *Wheeled Mobile Robot Control*, Studies in Systems, Decision and Control 380, https://doi.org/10.1007/978-3-030-77912-2_7

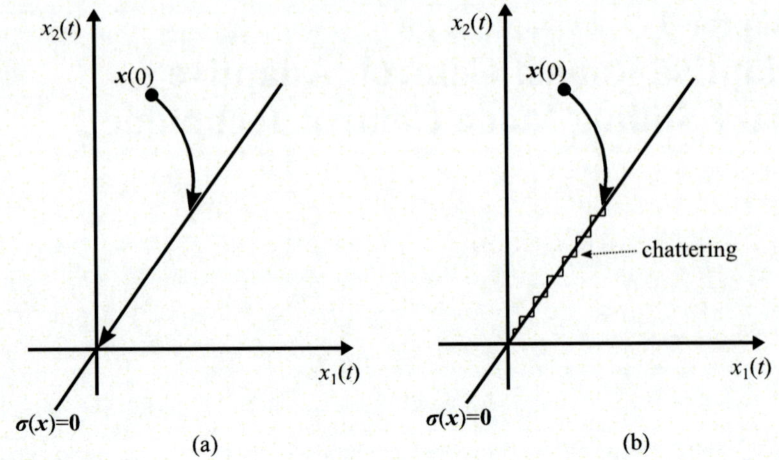

Fig. 7.1 Trajectory of states for **a** ideal sliding mode and **b** sliding mode with chattering phenomenon

the manifolds $\boldsymbol{\sigma}(\tilde{z}, t) = \mathbf{0}$ and $\boldsymbol{\sigma}^*(\tilde{z}, t) = \mathbf{0}$, resulting in a reduction of the original robustness that can be acceptable. Moreover, invariance requires switching between the reaching phase and sliding phase to be ideal, which is impractical. Therefore, the invariance is ideal and has little practical meaning [21]. Another problem related to the controllers in Chap. 6 is the necessity of knowing the bounds of uncertainties and disturbances in the system, and the application of a large value to the gains $\mathbf{G}$ that can cause a high control effort (large authority control), affecting the trajectory tracking and deteriorating the system performance [15].

To improve the results without needing to know of the bounds of uncertainties and disturbances in the system, a control solution using fuzzy systems, through an adaptive fuzzy sliding mode control (AFSMC) considering two variants of the parameter adaptation law, is described in this chapter to deal with the chattering phenomenon. The approaches of the AFSMC are made similarly to [1–4].

7.2 Background on Fuzzy Systems

A fuzzy system has four basic parts, as shown in the block diagram in Fig. 7.2. Given a set of non-fuzzy entry, from an external system, the fuzzification is responsible for mapping these entries to input fuzzy sets, evaluating the degree of membership of each entry for each fuzzy set. The rules are written in the form of "if ... then ...", describing a relationship between the input space and the output space. Then, for each rule, the inference machine maps an input set to an output fuzzy set, according to the relationship defined by the rules. Afterward, it combines the fuzzy sets from all the rules in the rule base into the output fuzzy set. Finally, the defuzzification translates fuzzy output to a real number for the system. Further details can be found in [13, 20].

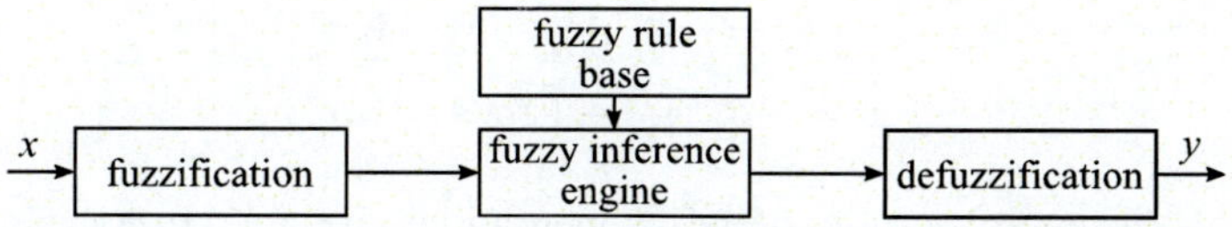

Fig. 7.2 Representation in a block diagram of a fuzzy system

All four parts can be mathematically formulated. In this paper, by choosing singleton fuzzification, center average defuzzification, the Sugeno implication in the rule base and product inference engine, the ith output of a fuzzy system with n outputs can be written as:

$$y_i = \frac{\sum_{j=1}^{\mathrm{m}} \beta_i^j \, \mu_i^j(\boldsymbol{x})}{\sum_{j=1}^{\mathrm{m}} \mu_i^j(\boldsymbol{x})} = \boldsymbol{\beta}_i^{\mathrm{T}} \boldsymbol{\phi}_i(\boldsymbol{x}), \tag{7.1}$$

where $\boldsymbol{x}$ is the input, $\boldsymbol{\beta}_i = [\beta_i^1 \ \ldots \ \beta_i^{\mathrm{m}}]^{\mathrm{T}}$ is the vector of consequences, $\boldsymbol{\phi}_i(\boldsymbol{x}) = [\phi_i^1(\boldsymbol{x}) \ \ldots \ \phi_i^{\mathrm{m}}(\boldsymbol{x})]^{\mathrm{T}}$ is the vector of rule weights in which:

$$\phi_i^j(\boldsymbol{x}) = \frac{\mu_i^j(\boldsymbol{x})}{\sum_{k=1}^{\mathrm{m}} \mu_i^k(\boldsymbol{x})},$$

with $\mu_i^j(\boldsymbol{x})$ being the membership function of the jth rule of the ith output, and m the number of rules.

7.3 Control Design

The AFSMC design is based on the first-order sliding mode control technique (FOSMC) discussed in Sect. 5.2.1 and the stability analysis of the closed-loop control system covered in Sect. 5.2.2. The AFSMC synthesis is similar to the FOSMC synthesis described in Sect. 5.2.3, emphasizing only that the discontinuous portion of the control is replaced by an adaptive fuzzy gain.

7.3.1 Controller Synthesis

The chattering phenomenon in the control presented in Eqs. (5.36) and (5.49) is caused by the constant value of $\mathbf{G}$ and the discontinuous function $\mathrm{sign}(\boldsymbol{\sigma}^*)$. Let the control gain $\mathbf{G}\,\mathrm{sign}(\boldsymbol{\sigma}^*)$ be replaced by a fuzzy gain $\hat{\boldsymbol{F}}(\boldsymbol{\sigma}^*)$, and therefore, the new control input (AFSMC) is then written as:

$$\begin{aligned} \boldsymbol{v}_{\mathrm{c}} &= -\boldsymbol{B}_{0_\sigma}^{-1}\boldsymbol{A}_{0_\sigma} - \boldsymbol{v}_{\mathrm{c}}^{*} - \mathbf{K}\boldsymbol{\sigma}^{*}, \\ \boldsymbol{v}_{\mathrm{c}} &= -\boldsymbol{B}_{0_\sigma}^{-1}\boldsymbol{A}_{0_\sigma} - \hat{\boldsymbol{F}}(\boldsymbol{\sigma}^{*}) - \mathbf{K}\boldsymbol{\sigma}^{*}, \end{aligned} \tag{7.2}$$

where $\boldsymbol{v}_{\mathrm{c}}^{*}$ is the control portions of the compensations given by a fuzzy gain defined as $\hat{\boldsymbol{F}}(\boldsymbol{\sigma}^{*}) = [\hat{f}_1(\boldsymbol{\sigma}^{*}) \ \ldots \ \hat{f}_{\mathrm{n}}(\boldsymbol{\sigma}^{*})]^{\mathrm{T}}$, and each $\hat{f}_i(\boldsymbol{\sigma}^{*})$ is estimated by an individual fuzzy system that can be written as:

$$\hat{f}_i(\boldsymbol{\sigma}^{*}) = \frac{\sum_{j=1}^{\mathrm{m}} \hat{\beta}_i^{j}\, \mu_i^{j}(\boldsymbol{\sigma}^{*})}{\sum_{j=1}^{\mathrm{m}} \mu_i^{j}(\boldsymbol{\sigma}^{*})} = \hat{\boldsymbol{\beta}}_i^{\mathrm{T}} \boldsymbol{\phi}_i(\boldsymbol{\sigma}^{*}), \tag{7.3}$$

where $\hat{\boldsymbol{\beta}}_i = [\hat{\beta}_i^1 \ \ldots \ \hat{\beta}_i^{\mathrm{m}}]^{\mathrm{T}}$ is the vector of online updated consequences, $\boldsymbol{\phi}_i(\boldsymbol{\sigma}^{*}) = [\phi_i^1(\boldsymbol{\sigma}^{*}) \ \ldots \ \phi_i^{\mathrm{m}}(\boldsymbol{\sigma}^{*})]^{\mathrm{T}}$ is the vector of rule weights. For an online update of the parameters of the output functions $\hat{f}_i(\boldsymbol{\sigma}^{*})$, $\hat{\boldsymbol{\beta}}_i$ was chosen as the parameter to be updated.

Defining $\boldsymbol{\beta}_i$, so that $f_i(\boldsymbol{\sigma}^{*}) = \boldsymbol{\beta}_i^{\mathrm{T}}\boldsymbol{\phi}_i(\boldsymbol{\sigma}^{*})$ is the optimal compensation for $\tilde{\boldsymbol{d}}_{0_i}$. According to Wang's theorem [7, 13, 20] there exists $\psi_i > 0$, satisfying

$$|\tilde{\boldsymbol{d}}_{0_i} - \boldsymbol{\beta}_i^{\mathrm{T}}\boldsymbol{\phi}_i(\boldsymbol{\sigma}^{*})| \leq \psi_i, \quad \tilde{\boldsymbol{d}}_0 = \boldsymbol{v}_{\mathrm{e}}(t), \tag{7.4}$$

where ψ_i can be as small as possible. Now, the estimation error and its derivative are defined in the following form:

$$\begin{aligned} \tilde{\boldsymbol{\beta}}_i &= \boldsymbol{\beta}_i - \hat{\boldsymbol{\beta}}_i, \\ \dot{\tilde{\boldsymbol{\beta}}}_i &= -\dot{\hat{\boldsymbol{\beta}}}_i, \end{aligned} \tag{7.5}$$

the Eq. (7.3) can be rewritten as:

$$\begin{aligned} \hat{f}_i(\boldsymbol{\sigma}^{*}) &= \boldsymbol{\beta}_i^{\mathrm{T}}\boldsymbol{\phi}_i(\boldsymbol{\sigma}^{*}) - \tilde{\boldsymbol{\beta}}_i^{\mathrm{T}}\boldsymbol{\phi}_i(\boldsymbol{\sigma}^{*}), \\ \hat{\boldsymbol{F}}(\boldsymbol{\sigma}^{*}) &= \boldsymbol{F}(\boldsymbol{\sigma}^{*}) - \tilde{\boldsymbol{F}}(\boldsymbol{\sigma}^{*}), \end{aligned} \tag{7.6}$$

and the adaptation law of the AFSMC can be chosen as:

$$\dot{\hat{\boldsymbol{\beta}}}_i = \sigma_i^{*}\boldsymbol{\phi}_i(\boldsymbol{\sigma}^{*}). \tag{7.7}$$

7.3.2 *Stability Analysis*

The stability analysis of the AFSMC can be done using the Lyapunov function candidate as:

$$V = \frac{1}{2}\left(\boldsymbol{\sigma}^{\mathrm{T}}\boldsymbol{\sigma} + \sum_{i=1}^{\mathrm{n}} \tilde{\boldsymbol{\beta}}_i^{\mathrm{T}} \tilde{\boldsymbol{\beta}}_i\right), \tag{7.8}$$

where $\tilde{\boldsymbol{\beta}}_i^{\mathrm{T}} \tilde{\boldsymbol{\beta}}_i > 0$, therefore V is a positive definite.

Differentiating Eq. (7.8), using the definition of Eq. (7.5) and considering the Eqs. (5.14) and (5.26), one obtains:

$$\dot{V} = \boldsymbol{\sigma}^{\mathrm{T}}\dot{\boldsymbol{\sigma}} + \sum_{i=1}^{\mathrm{n}} \tilde{\boldsymbol{\beta}}_i^{\mathrm{T}} \dot{\tilde{\boldsymbol{\beta}}}_i,$$

$$\dot{V} = \boldsymbol{\sigma}^{\mathrm{T}}\frac{\partial \boldsymbol{\sigma}}{\partial \boldsymbol{q}_{\mathrm{e}}}\boldsymbol{A}_0 + \boldsymbol{\sigma}^{\mathrm{T}}\frac{\partial \boldsymbol{\sigma}}{\partial \boldsymbol{q}_{\mathrm{e}}}\boldsymbol{B}_0\boldsymbol{v}_{\mathrm{c}} + \boldsymbol{\sigma}^{\mathrm{T}}\frac{\partial \boldsymbol{\sigma}}{\partial \boldsymbol{q}_{\mathrm{e}}}\boldsymbol{d}_{\mathrm{b}} - \sum_{i=1}^{\mathrm{n}} \tilde{\boldsymbol{\beta}}_i^{\mathrm{T}} \dot{\hat{\boldsymbol{\beta}}}_i,$$

$$\dot{V} = \boldsymbol{\sigma}^{\mathrm{T}}\frac{\partial \boldsymbol{\sigma}}{\partial \boldsymbol{q}_{\mathrm{e}}}\boldsymbol{A}_0 + \boldsymbol{\sigma}^{\mathrm{T}}\frac{\partial \boldsymbol{\sigma}}{\partial \boldsymbol{q}_{\mathrm{e}}}\boldsymbol{B}_0\boldsymbol{v}_{\mathrm{c}} + \boldsymbol{\sigma}^{\mathrm{T}}\frac{\partial \boldsymbol{\sigma}}{\partial \boldsymbol{q}_{\mathrm{e}}}\boldsymbol{B}_0\tilde{\boldsymbol{d}}_0 - \sum_{i=1}^{\mathrm{n}} \tilde{\boldsymbol{\beta}}_i^{\mathrm{T}} \dot{\hat{\boldsymbol{\beta}}}_i,$$

considering $\mathbf{I} = \boldsymbol{B}_{0_\sigma}\boldsymbol{B}_{0_\sigma}^{-1}$,

$$\dot{V} = \boldsymbol{\sigma}^{\mathrm{T}}\boldsymbol{B}_{0_\sigma}\boldsymbol{B}_{0_\sigma}^{-1}\boldsymbol{A}_{0_\sigma} + \boldsymbol{\sigma}^{\mathrm{T}}\boldsymbol{B}_{0_\sigma}\boldsymbol{v}_{\mathrm{c}} + \boldsymbol{\sigma}^{\mathrm{T}}\boldsymbol{B}_{0_\sigma}\tilde{\boldsymbol{d}}_0 - \sum_{i=1}^{\mathrm{n}} \tilde{\boldsymbol{\beta}}_i^{\mathrm{T}} \dot{\hat{\boldsymbol{\beta}}}_i,$$

$$\dot{V} = \left(\boldsymbol{B}_{0_\sigma}^{\mathrm{T}}\boldsymbol{\sigma}\right)^{\mathrm{T}}\boldsymbol{B}_{0_\sigma}^{-1}\boldsymbol{A}_{0_\sigma} + \left(\boldsymbol{B}_{0_\sigma}^{\mathrm{T}}\boldsymbol{\sigma}\right)^{\mathrm{T}}\left(\boldsymbol{v}_{\mathrm{c}} + \tilde{\boldsymbol{d}}_0\right) - \sum_{i=1}^{\mathrm{n}} \tilde{\boldsymbol{\beta}}_i^{\mathrm{T}} \dot{\hat{\boldsymbol{\beta}}}_i,$$

$$\dot{V} = \boldsymbol{\sigma}^{*^{\mathrm{T}}}\boldsymbol{B}_{0_\sigma}^{-1}\boldsymbol{A}_{0_\sigma} + \boldsymbol{\sigma}^{*^{\mathrm{T}}}(\boldsymbol{v}_{\mathrm{c}} + \tilde{\boldsymbol{d}}_0) - \sum_{i=1}^{\mathrm{n}} \tilde{\boldsymbol{\beta}}_i^{\mathrm{T}} \dot{\hat{\boldsymbol{\beta}}}_i.$$

Using the AFSMC, Eq. (7.2), and replacing Eqs. (7.4) and (7.6), $\dot{V}$ results in:

$$\dot{V} = \boldsymbol{\sigma}^{*^{\mathrm{T}}}\boldsymbol{B}_{0_\sigma}^{-1}\boldsymbol{A}_{0_\sigma} - \boldsymbol{\sigma}^{*^{\mathrm{T}}}\boldsymbol{B}_{0_\sigma}^{-1}\boldsymbol{A}_{0_\sigma} - \boldsymbol{\sigma}^{*^{\mathrm{T}}}\hat{\boldsymbol{F}}(\boldsymbol{\sigma}^*) - \boldsymbol{\sigma}^{*^{\mathrm{T}}}\mathbf{K}\boldsymbol{\sigma}^* + \boldsymbol{\sigma}^*\tilde{\boldsymbol{d}}_0 - \sum_{i=1}^{\mathrm{n}} \tilde{\boldsymbol{\beta}}_i^{\mathrm{T}} \dot{\hat{\boldsymbol{\beta}}}_i,$$

$$\dot{V} = -\boldsymbol{\sigma}^{*^{\mathrm{T}}}\boldsymbol{F}(\boldsymbol{\sigma}^*) + \boldsymbol{\sigma}^{*^{\mathrm{T}}}\tilde{\boldsymbol{F}}(\boldsymbol{\sigma}^*) - \boldsymbol{\sigma}^{*^{\mathrm{T}}}\mathbf{K}\boldsymbol{\sigma}^* + \boldsymbol{\sigma}^{*^{\mathrm{T}}}\boldsymbol{v}_{\mathrm{e}} - \sum_{i=1}^{\mathrm{n}} \tilde{\boldsymbol{\beta}}_i^{\mathrm{T}} \dot{\hat{\boldsymbol{\beta}}}_i,$$

$$\dot{V} \leq -\boldsymbol{\sigma}^{*^{\mathrm{T}}}\mathbf{K}\boldsymbol{\sigma}^* + \sum_{i=1}^{\mathrm{n}} \sigma_i^* \left|v_{\mathrm{e}_i} - \boldsymbol{\beta}_i^{\mathrm{T}}\boldsymbol{\phi}_i(\boldsymbol{\sigma}^*) + \tilde{\boldsymbol{\beta}}_i^{\mathrm{T}}\boldsymbol{\phi}_i(\boldsymbol{\sigma}^*)\right| - \sum_{i=1}^{\mathrm{n}} \tilde{\boldsymbol{\beta}}_i^{\mathrm{T}} \dot{\hat{\boldsymbol{\beta}}}_i,$$

$$\dot{V} \leq -\boldsymbol{\sigma}^{*^{\mathrm{T}}}\mathbf{K}\boldsymbol{\sigma}^* + \sum_{i=1}^{\mathrm{n}} \sigma_i^* \left|v_{\mathrm{e}_i} - \boldsymbol{\beta}_i^{\mathrm{T}}\boldsymbol{\phi}_i(\boldsymbol{\sigma}^*)\right| + \sum_{i=1}^{\mathrm{n}} \tilde{\boldsymbol{\beta}}_i^{\mathrm{T}}\left(\sigma_i^*\boldsymbol{\phi}_i(\boldsymbol{\sigma}^*) - \dot{\hat{\boldsymbol{\beta}}}_i\right).$$

Replacing the adaptation law, Eq. (7.7), $\dot{V}$ yields in:

$$\dot{V} \leq -\boldsymbol{\sigma}^{*^{\mathrm{T}}}\mathbf{K}\boldsymbol{\sigma}^{*} + \sum_{i=1}^{\mathrm{n}} \sigma_i^{*} \left| v_{\mathrm{e}_i} - \boldsymbol{\beta}_i^{\mathrm{T}}\boldsymbol{\phi}_i(\boldsymbol{\sigma}^{*}) \right| + \sum_{i=1}^{\mathrm{n}} \tilde{\boldsymbol{\beta}}_i^{\mathrm{T}}\left(\sigma_i^{*}\boldsymbol{\phi}_i(\boldsymbol{\sigma}^{*}) - \sigma_i^{*}\boldsymbol{\phi}_i(\boldsymbol{\sigma}^{*})\right),$$
$$\dot{V} \leq -\boldsymbol{\sigma}^{*^{\mathrm{T}}}\mathbf{K}\boldsymbol{\sigma}^{*} + \sum_{i=1}^{\mathrm{n}} \sigma_i^{*} \left| v_{\mathrm{e}_i} - \boldsymbol{\beta}_i^{\mathrm{T}}\boldsymbol{\phi}_i(\boldsymbol{\sigma}^{*}) \right|. \tag{7.9}$$

Thus, the second term on the right side of Eq. (7.9) satisfies the condition of Eq. (7.4), i.e.,

$$\dot{V} \leq -\boldsymbol{\sigma}^{*^{\mathrm{T}}}\mathbf{K}\boldsymbol{\sigma}^{*} + \sum_{i=1}^{\mathrm{n}} \sigma_i^{*} \%_i. \tag{7.10}$$

The right side of Eq. (7.10) can be written as

$$\dot{V} \leq -\boldsymbol{\sigma}^{*^{\mathrm{T}}}(\mathbf{K} - \Psi)\boldsymbol{\sigma}^{*} \leq -\lambda_{\min}\{\mathbf{K}\}\left|\boldsymbol{\sigma}^{*}\right|^2 + \overline{\Psi}_{\max}\left|\boldsymbol{\sigma}^{*}\right| \leq 0, \tag{7.11}$$

where $\mathbf{K} = \mathrm{diag}[\kappa_{p_1} \ \ldots \ \kappa_{\mathrm{p}_1}]$ and $\Psi = \mathrm{diag}[\psi_1 \ \ldots \ \psi_{\mathrm{n}}]$ are positive definite matrices, $\lambda_{\min}\{\mathbf{K}\}$ is the minimum singular value of $\mathbf{K}$ and $\overline{\Psi}_{\max} > |\Psi|$ is a known scalar that represents the maximum effect of the uncertainties and disturbances.

Therefore, $\dot{V} \leq 0$ as long as the term in parenthesis in Eq. (7.12) is positive,

$$\dot{V} \leq -\left|\boldsymbol{\sigma}^{*}\right|\left(\lambda_{\min}\{\mathbf{K}\}\left|\boldsymbol{\sigma}^{*}\right| - \overline{\Psi}_{\max}\right) \leq 0, \tag{7.12}$$

either

$$\left|\boldsymbol{\sigma}^{*}\right| > \frac{\overline{\Psi}_{\max}}{\lambda_{\min}\{\mathbf{K}\}}. \tag{7.13}$$

This demonstrates that $\boldsymbol{\sigma}^{*}$ and $\boldsymbol{\sigma}$ are bounded and the continuity of all functions shows as well the boundedness of $\dot{\boldsymbol{\sigma}}^{*}$ and $\dot{\boldsymbol{\sigma}}$. The demonstration of boundedness $\hat{\boldsymbol{\beta}}$, or equivalently $\tilde{\boldsymbol{\beta}}$, is similar to that of [11, 12]. Moreover, supposing that the vector of rule weights $\boldsymbol{\phi}_i(\boldsymbol{x})$ is persistently exciting or meets the condition of the persistence of excitation (PE) [11, 12]. Then $\boldsymbol{\sigma}^{*}$ is uniformly ultimate bounded (UUB), with a practical bound given by the right-hand side of the Eq. (7.13), and the estimates of the consequences $\hat{\boldsymbol{\beta}}$ are bounded. Moreover, $\boldsymbol{\sigma}^{*}$ may be kept as small as desired by increasing the gain $\mathbf{K}$.

Another stability analysis can also be performed in a similar way to [14]. Thus, assuming from Eq. (7.4) that:

$$|\boldsymbol{v}_{e_i} - \boldsymbol{\beta}_i^{\mathrm{T}}\boldsymbol{\phi}_i(\boldsymbol{\sigma}^{*})| \leq \psi_i \leq \eta_i\,|\sigma_i^{*}|, \tag{7.14}$$

where $0 < \eta_i < 1$. Thereby, the second term on the right side of Eq. (7.9) satisfies the requirement of Eq. (7.14), resulting in:

$$\sigma_i^{*}|\boldsymbol{v}_{e_i} - \boldsymbol{\beta}_i^{\mathrm{T}}\boldsymbol{\phi}_i(\boldsymbol{\sigma}^{*})| \leq \eta_i\,|\sigma_i^{*}|^2 = \eta_i\sigma_i^{*^2} \tag{7.15}$$

consequently, using Eq. (7.15), $\dot{V}$ becomes:

$$\dot{V} \leq -\boldsymbol{\sigma}^{*^{\mathrm{T}}}\mathbf{K}\boldsymbol{\sigma}^{*} + \sum_{i=1}^{\mathrm{n}} \eta_i \sigma_i^{*^2}. \tag{7.16}$$

Defining $\Psi = \mathrm{diag}[\eta_1 \ \ldots \ \eta_{\mathrm{n}}]$ as positive definite matrix and $\overline{\psi}_{\max} > |\Psi|$, the right side of Eq. (7.16) can be written as

$$\dot{V} \leq -\boldsymbol{\sigma}^{*^{\mathrm{T}}}(\mathbf{K} - \Psi)\boldsymbol{\sigma}^{*} \leq \left|\boldsymbol{\sigma}^{*}\right|^2 (-\lambda_{\min}\{\mathbf{K}\} + \overline{\psi}_{\max}) \leq 0, \tag{7.17}$$

where $\mathbf{K} - \Psi > 0$ or $\lambda_{\min}\{\mathbf{K}\} > \overline{\psi}_{\max}$. Since these conditions must be satisfied, then $\dot{V}$ is negative definite, i.e., $\dot{V} \leq 0$. Moreover, $\dot{V} < 0$ for $\boldsymbol{\sigma} \neq 0$, $\boldsymbol{\sigma}^{*} \neq 0$ and, therefore, $\boldsymbol{q}_{\mathrm{e}} \neq 0$ while that $\dot{V} = 0$ only when $\boldsymbol{\sigma} = 0$, $\boldsymbol{\sigma}^{*} = 0$ and, consequently, $\boldsymbol{q}_{\mathrm{e}} = 0$, which is an asymptotically stable equilibrium point. Regarding the estimates of the consequences, $\hat{\boldsymbol{\beta}}$, or equivalently $\tilde{\boldsymbol{\beta}}$, these remain bounded.

7.3.3 Augmented Adaptation Law for Removing the PE Condition

In adaptive control theory, the possible unboundedness of the consequences (e.g., parameter or weight) estimates when PE fails to hold is known as parameter drift. Thus, the PE condition in adaptation law, Eq. (7.7), is meant to ensure that drift does not occur. An alternative to correcting this problem that does not require the PE condition is modifying the adaptation law using techniques from adaptive control, such as σ-modification, ϵ-modification, or dead-zone techniques [11, 12]. With an augmented adaptation law, the stability analysis relies on an extension to Lyapunov theory.

Considering the control law, Eq. (7.2), one defines the following augmented adaptation law as:

$$\dot{\hat{\boldsymbol{\beta}}}_i = \gamma \sigma_i^{*} \boldsymbol{\phi}_i\left(\boldsymbol{\sigma}^{*}\right) - \gamma \zeta \left|\boldsymbol{\sigma}^{*}\right| \hat{\boldsymbol{\beta}}_i, \tag{7.18}$$

where γ and ζ are positive constants. The term $\gamma \zeta \left|\boldsymbol{\sigma}^{*}\right| \hat{\boldsymbol{\beta}}_i$, Eq. (7.18), corresponding to ϵ-modification from the adaptive control theory, which must be added to eliminate the PE condition and to ensure bounded consequences estimates, i.e., $\hat{\boldsymbol{\beta}}_i$.

The stability proof is similar analysis described in Sect. 7.3.2 with the Lyapunov function candidate, Eq. (7.8), being modified and given as:

$$V = \frac{1}{2}\left(\boldsymbol{\sigma}^{\mathrm{T}}\boldsymbol{\sigma} + \sum_{i=1}^{\mathrm{n}} \tilde{\boldsymbol{\beta}}_i^{\mathrm{T}} \gamma^{-1} \tilde{\boldsymbol{\beta}}_i\right). \tag{7.19}$$

Differentiating Eq. (7.19), considering the Eqs. (5.14) and (5.26), replacing Eqs. (7.2), (7.4), (7.6), and (7.18), making the necessary mathematical manipulations, $\dot{V}$ is obtained as:

$$\begin{aligned}
\dot{V} &\leq -\boldsymbol{\sigma}^{*^{\mathrm{T}}}\mathbf{K}\boldsymbol{\sigma}^{*} + \sum_{i=1}^{\mathrm{n}} \sigma_i^{*}\left|v_{\mathrm{e}_i} - \boldsymbol{\beta}_i^{\mathrm{T}}\boldsymbol{\phi}_i(\boldsymbol{\sigma}^{*})\right| + \sum_{i=1}^{\mathrm{n}} \tilde{\boldsymbol{\beta}}_i^{\mathrm{T}}\,\zeta\,|\boldsymbol{\sigma}^{*}|\,\hat{\boldsymbol{\beta}}_i,\\
\dot{V} &\leq -\boldsymbol{\sigma}^{*^{\mathrm{T}}}\mathbf{K}\boldsymbol{\sigma}^{*} + \sum_{i=1}^{\mathrm{n}} \sigma_i^{*}\psi_i + \zeta\,|\boldsymbol{\sigma}^{*}| \sum_{i=1}^{\mathrm{n}} \tilde{\boldsymbol{\beta}}_i^{\mathrm{T}}\,\hat{\boldsymbol{\beta}}_i,\\
\dot{V} &\leq -\boldsymbol{\sigma}^{*^{\mathrm{T}}}\mathbf{K}\boldsymbol{\sigma}^{*} + \Psi\boldsymbol{\sigma}^{*} + \zeta\,|\boldsymbol{\sigma}^{*}|\,\mathrm{tr}\left(\tilde{\boldsymbol{\beta}}^{\mathrm{T}}\hat{\boldsymbol{\beta}}\right),
\end{aligned} \tag{7.20}$$

with $\zeta > 0$ being a positive constant, and $\mathrm{tr}(\cdot)$ being the trace function, which is the sum of the diagonal elements of a determined matrix. Using the Schwartz inequality [13], the trace function can be written as:

$$\mathrm{tr}\left(\tilde{\boldsymbol{\beta}}^{\mathrm{T}}\hat{\boldsymbol{\beta}}\right) = \mathrm{tr}\left(\tilde{\boldsymbol{\beta}}^{\mathrm{T}}\left(\boldsymbol{\beta} - \tilde{\boldsymbol{\beta}}\right)\right) = \left\langle\tilde{\boldsymbol{\beta}}, \boldsymbol{\beta}\right\rangle - \left|\tilde{\boldsymbol{\beta}}\right|^2 \leq \left|\tilde{\boldsymbol{\beta}}\right|\left|\boldsymbol{\beta}\right| - \left|\tilde{\boldsymbol{\beta}}\right|^2 \leq \left|\tilde{\boldsymbol{\beta}}\right|\beta_{\max} - \left|\tilde{\boldsymbol{\beta}}\right|^2, \tag{7.21}$$

where $\beta_{\max} \geq |\boldsymbol{\beta}|$ is a positive constant.

Replacing Eq. (7.21) into Eq. (7.20), $\dot{V}$ leads to:

$$\dot{V} \leq -\boldsymbol{\sigma}^{*^{\mathrm{T}}}\mathbf{K}\boldsymbol{\sigma}^{*} + \Psi\boldsymbol{\sigma}^{*} + \zeta\,|\boldsymbol{\sigma}^{*}|\left(\left|\tilde{\boldsymbol{\beta}}\right|\beta_{\max} - \left|\tilde{\boldsymbol{\beta}}\right|^2\right), \tag{7.22}$$

either

$$\dot{V} \leq -\lambda_{\min}\{\mathbf{K}\}\,|\boldsymbol{\sigma}^{*}|^2 + \overline{\Psi}_{\max}\,|\boldsymbol{\sigma}^{*}| + \zeta\,|\boldsymbol{\sigma}^{*}|\left(\left|\tilde{\boldsymbol{\beta}}\right|\beta_{\max} - \left|\tilde{\boldsymbol{\beta}}\right|^2\right), \tag{7.23}$$

with $\overline{\Psi}_{\max}$ being defined as in Eq. (7.11).

Making the necessary mathematical manipulations, $\dot{V}$ yields to:

$$\begin{aligned}
\dot{V} &\leq -\,|\boldsymbol{\sigma}^{*}|\left(\lambda_{\min}\{\mathbf{K}\}\,|\boldsymbol{\sigma}^{*}| - \overline{\Psi}_{\max} + \zeta\left(\left|\tilde{\boldsymbol{\beta}}\right|^2 - \left|\tilde{\boldsymbol{\beta}}\right|\beta_{\max}\right)\right),\\
\dot{V} &\leq -\,|\boldsymbol{\sigma}^{*}|\left(\lambda_{\min}\{\mathbf{K}\}\,|\boldsymbol{\sigma}^{*}| - \overline{\Psi}_{\max} + \zeta\left(\left|\tilde{\boldsymbol{\beta}}\right| - \frac{\beta_{\max}}{2}\right)^2 - \zeta\,\frac{\beta_{\max}^2}{4}\right).
\end{aligned} \tag{7.24}$$

Thus, $\dot{V}$ is guaranteed negative as long as the term in parenthesis in Eq. (7.24) is positive, either:

$$|\boldsymbol{\sigma}^{*}| > \frac{\zeta\,\dfrac{\beta_{\max}^2}{4} + \overline{\Psi}_{\max}}{\lambda_{\min}\{\mathbf{K}\}}, \tag{7.25}$$

or

$$\left|\tilde{\boldsymbol{\beta}}\right| > \frac{\beta_{\max}}{2} + \sqrt{\frac{\beta_{\max}^2}{4} + \frac{\overline{\Psi}_{\max}}{\zeta}}. \tag{7.26}$$

Therefore, $\dot{V}$ is negative definite within a particular compact set and it is negative semidefinite outside this set, which is defined by Eqs. (7.25) and (7.26). The stability of the closed-loop control system is ensured since $\dot{V}$ is guaranteed to be negative definite. According to a standard Lyapunov theory and LaSalle Theorem [11, 12], all signals $|\boldsymbol{\sigma}^*|$ and $|\tilde{\boldsymbol{\beta}}|$ are UUB.

It is important to note that PE is not needed to establish the bounds on $\tilde{\boldsymbol{\beta}}$ with the augmented adaptation law. The importance of the term ζ added to the augmented adaptation law is that it adds a quadratic term in $|\tilde{\boldsymbol{\beta}}|$ in Eq. (7.24), so that it is possible to establish that $\dot{V}$ is negative outside a compact set in the plane $(|\boldsymbol{\sigma}^*|\,,\,|\tilde{\boldsymbol{\beta}}|)$ [11, 12]. The term ζ is known in adaptive control as ϵ-modification of Narendra, whose function is to make the augmented adaptation law robust to unmodelled dynamics so that the PE condition is not needed. Furthermore, the right-hand sides of Eqs. (7.25) and (7.26) respectively may be taken as practical bounds on the $\boldsymbol{\sigma}^*$ and $\tilde{\boldsymbol{\beta}}$ in the sense that excursions beyond these bounds will be very small. Note, moreover, from the former that arbitrarily small bounds of $\boldsymbol{\sigma}^*$ may be achieved by selecting large control gains $\mathbf{K}$. This gain plays an important role in the control law (Eq. 7.2) because in the transient state the estimated consequences $\hat{\boldsymbol{\beta}}$ are not properly adjusted, which can move the states of the system away from the desired point. Therefore, this gain will prevent states from moving away from the desired point while the consequences are adjusted. On the other hand, the $\tilde{\boldsymbol{\beta}}$ is fundamentally bounded by $\beta_{\max}$, the known bound on the ideal consequences $\boldsymbol{\beta}$. The adaptation parameter ζ offers a design tradeoff between the relative eventual magnitudes of $|\boldsymbol{\sigma}^*|$ and $|\tilde{\boldsymbol{\beta}}|$; a smaller ζ yields a smaller $|\boldsymbol{\sigma}^*|$ and a larger $|\tilde{\boldsymbol{\beta}}|$, and vice versa.

7.3.4 Extraction of the Rule Base

To decide the rules for the fuzzy systems [7], V is considered as declared in Eq. (5.8). V is regarded as an indicator of the energy of $\boldsymbol{\sigma}$. The stability of the system is guaranteed by choosing a control law such that $\dot{V} \leq 0$. In the AFSMC, the fuzzy gain $\hat{\boldsymbol{F}}(\boldsymbol{\sigma}^*)$ is applied to compensate uncertainties and disturbances in the system and to reduce the energy of $\boldsymbol{\sigma}^*$.

Because of the function sign$(\boldsymbol{\sigma}^*)$, the control gain has the same signal as $\boldsymbol{\sigma}^*$. Therefore, $\hat{f}_i(\boldsymbol{\sigma}^*)$ should have the same signal as σ_i^*. Now consider $\sigma_i^*(\tilde{d}_{0_i} - \hat{f}_i(\boldsymbol{\sigma}^*))$. When $|\sigma_i^*|$ is large, it is expected that $|\hat{f}_i(\boldsymbol{\sigma}^*)|$ is larger so that $\dot{V}$ has a large negative value. This causes the energy of $\boldsymbol{\sigma}^*$ to decay fast. When $|\sigma_i^*|$ is small, $\sigma_i^*(\tilde{d}_{0_i} - \hat{f}_i(\boldsymbol{\sigma}^*))$ is also small and has little effect on the value of $\dot{V}$. Therefore, $|\hat{f}_i(\boldsymbol{\sigma}^*)|$ can be small to avoid the chattering phenomenon. When $|\sigma_i^*|$ is zero, $|\hat{f}_i(\boldsymbol{\sigma}^*)|$ is also zero. Based on this analysis, the rule base is chosen as:

- IF σ_i^* is NB, THEN $\hat{f}_i(\boldsymbol{\sigma}^*)$ is $\hat{\beta}_i^{1^T}\ \phi_i^1(\boldsymbol{\sigma}^*)$,
- IF σ_i^* is NM, THEN $\hat{f}_i(\boldsymbol{\sigma}^*)$ is $\hat{\beta}_i^{2^T}\ \phi_i^2(\boldsymbol{\sigma}^*)$,
- IF σ_i^* is NS, THEN $\hat{f}_i(\boldsymbol{\sigma}^*)$ is $\hat{\beta}_i^{3^T}\ \phi_i^3(\boldsymbol{\sigma}^*)$,
- IF σ_i^* is ZE, THEN $\hat{f}_i(\boldsymbol{\sigma}^*)$ is $\hat{\beta}_i^{4^T}\ \phi_i^4(\boldsymbol{\sigma}^*)$,
- IF σ_i^* is PS, THEN $\hat{f}_i(\boldsymbol{\sigma}^*)$ is $\hat{\beta}_i^{5^T}\ \phi_i^5(\boldsymbol{\sigma}^*)$,
- IF σ_i^* is PM, THEN $\hat{f}_i(\boldsymbol{\sigma}^*)$ is $\hat{\beta}_i^{6^T}\ \phi_i^6(\boldsymbol{\sigma}^*)$,
- IF σ_i^* is PB, THEN $\hat{f}_i(\boldsymbol{\sigma}^*)$ is $\hat{\beta}_i^{7^T}\ \phi_i^7(\boldsymbol{\sigma}^*)$,

where N stands for negative, P positive, ZE zero, S small, M medium, and B big.

The membership functions are chosen to be triangular-shaped functions as:

$$\phi_i^j(\boldsymbol{\sigma}^*) = \begin{cases} 0, & \sigma_i^* < \alpha_1^{i,j} \\ \dfrac{\sigma_i^* - \alpha_1^{i,j}}{\alpha_2^{i,j} - \alpha_1^{i,j}}, & \alpha_1^{i,j} \le \sigma_i^* \le \alpha_2^{i,j} \\ \dfrac{\alpha_3^{i,j} - \sigma_i^*}{\alpha_3^{i,j} - \alpha_2^{i,j}}, & \alpha_2^{i,j} \le \sigma_i^* \le \alpha_3^{i,j} \\ 0, & \alpha_3^{i,j} < \sigma_i^* \end{cases}, \tag{7.27}$$

where $\alpha_1^{i,j}$, and $\alpha_3^{i,j}$ are the "feet" of the triangle, and the parameter $\alpha_2^{i,j}$ locates the peak, all for the $j$rule of the ith output. The parameters of the input membership functions are predefined and given in Table 7.1 as well as can be seen in Fig. 7.3, while those of the output, $\hat{\beta}_i^j$, are updated online. Therefore, determining the adaptive characteristic of the controller.

It must be emphasized that the processing time required when using fuzzy logic control depends on the number of rules that must be evaluated. Moreover, large systems with many rules would require very powerful and fast processors to compute

Table 7.1 Parameters of the membership functions of $\boldsymbol{\sigma}^*$

		σ_1^*			σ_2^*		
	jth rule	$\alpha_1^{1,j}$	$\alpha_2^{1,j}$	$\alpha_3^{1,j}$	$\alpha_1^{2,j}$	$\alpha_2^{2,j}$	$\alpha_3^{2,j}$
NB	1	$-\infty$	−0.3	−0.2	$-\infty$	−0.5	−0.33
NM	2	−0.3	-0.2	−0.1	−0.5	−0.33	−0.16
NS	3	−0.2	−0.1	0	−0.33	−0.16	0
ZE	4	−0.1	0	0.1	−40.16	0	0.16
PS	5	0	0.1	0.2	0	0.16	0.33
PM	6	0.1	0.2	0.3	0.16	0.33	0.5
PB	7	0.2	0.3	∞	0.33	0.5	∞

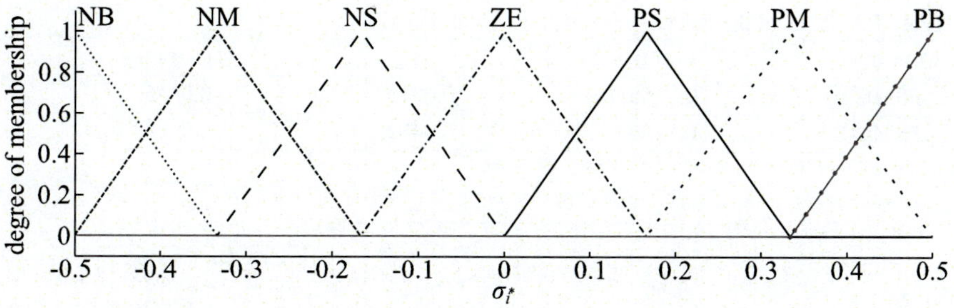

Fig. 7.3 Triangular-shaped input membership functions

in real-time. The smaller the rule base, the less computational power will be needed [16]. Thus, unlike a pure fuzzy logic controller which is encountered in the rule expanding problem, the AFSMC uses only 7 if-then rules in the rule base in respect to the sliding surfaces, as well as it uses triangular membership functions, making their structure even simpler. Therefore, it is more suitable to implement in real DWMRs compared to the works performed by [8, 10, 16, 22].

7.4 Simulations Using Matlab and/or MobileSim Simulator

Due to variants in the parameter adaptation law, two AFSMCs are evaluated for the tracking control problem of an eight-shape trajectory through simulation and experimental scenarios. They are: AFSMC-A described in Sect. 7.3.1, considering the control and adaptation laws given by Eqs. (7.2) and (7.7) respectively; and AFSMC-B of the Sect. 7.3.3, taking into account respectively the control and adaptation laws given by Eqs. (7.2) and (7.18). The gains used for this evaluation in the different scenarios are provided in Table 7.2.

7.4.1 Ideal Scenario

The results are shown in Table 7.3 after simulations of the control variants are performed in this scenario. From this table and Figs. 7.4 and 7.5, it is verified that the

Table 7.2 AFSMC gains

Controller	λ_1	λ_2	λ_3	K_1	K_2	γ	ζ
AFSMC-A	1.5	6.0	1.0	0.1	0.1	–	–
AFSMC-B	1.5	6.0	1.0	0.1	0.1	0.1	1×10^{-6}

Table 7.3 RMS of the errors—results in the ideal scenario

Controller	x_e (m)	y_e (m)	θ_e (rad)
AFSMC-A	0.0166	0.0313	0.0445
AFSMC-B	0.0236	0.0482	0.0468

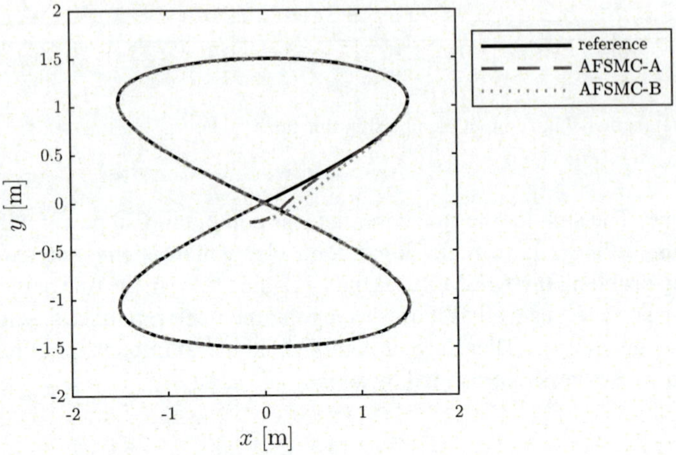

Fig. 7.4 Trajectory tracking in the ideal scenario

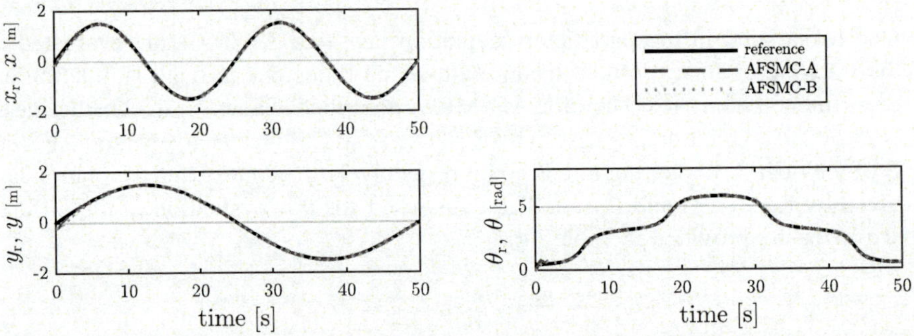

Fig. 7.5 Tracking (x, y, and θ) in the ideal scenario

AFSMC-A and AFSMC-B can follow the trajectory and the DWMR tries to track the reference posture.

For the AFSMC-A and AFSMC-B, the DWMR follows the trajectory with posture tracking and orientation errors that tend to converge to zero, as can be seen in Fig. 7.6 and Table 7.3, in which it can also be observed that the AFSMC-A presents lower errors than AFSMC-B.

Viewing Fig. 7.7 regarding linear and angular control velocities as well as the linear and angular velocities of the DWMR in Fig. 7.8, these velocities present smooth

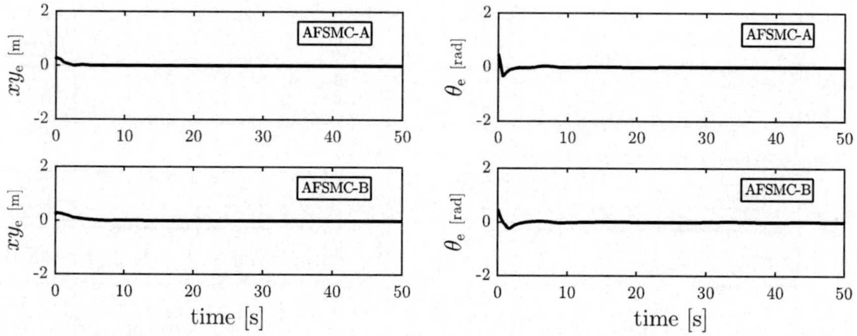

Fig. 7.6 Posture tracking errors, with $xy_e = \sqrt{x_e^2 + y_e^2}$, in the ideal scenario

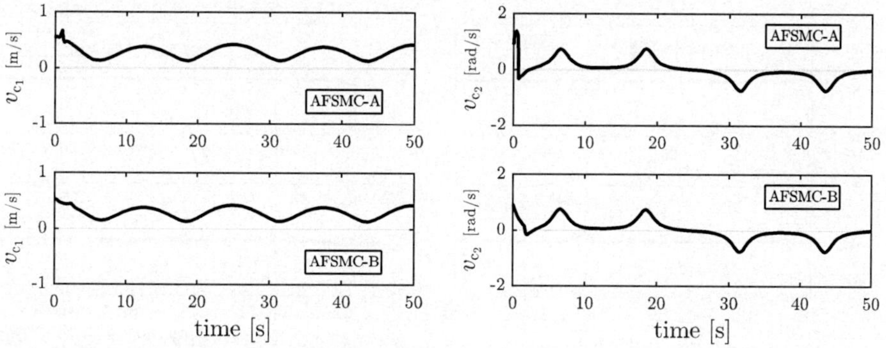

Fig. 7.7 Control velocities in the ideal scenario

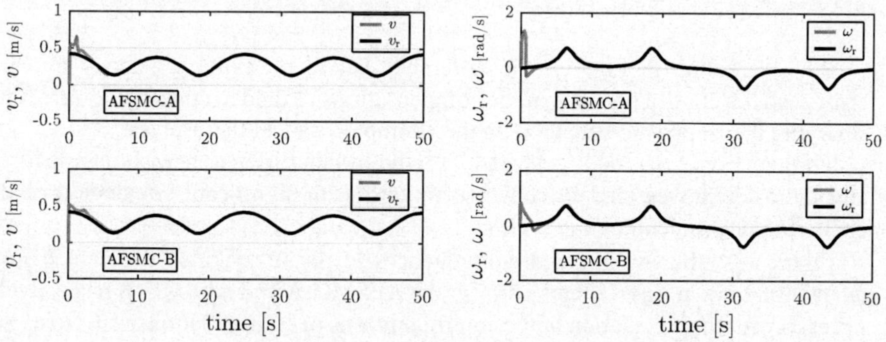

Fig. 7.8 Velocities of the DWMR and reference velocities in the ideal scenario

behaviors with chattering phenomenon significantly attenuated for both control variants. Also, it is possible to observe that the linear and angular velocities of the DWMR also tend towards the reference velocities.

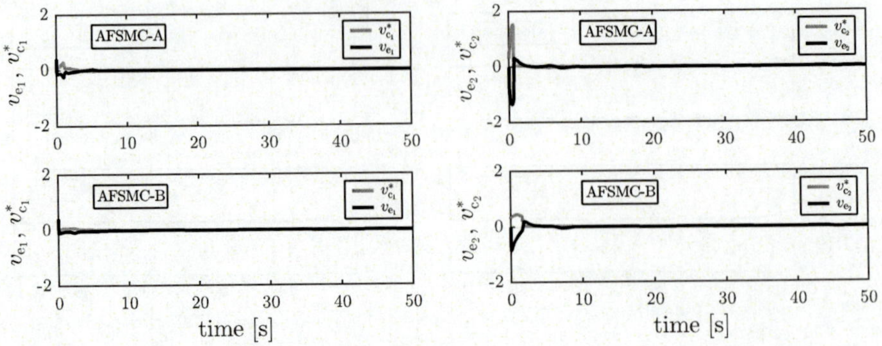

Fig. 7.9 Auxiliary velocity tracking errors and control velocities $\boldsymbol{v}_c^*$ in the ideal scenario

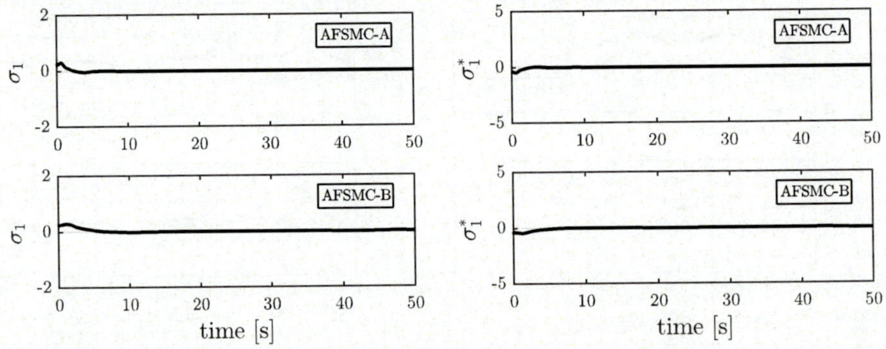

Fig. 7.10 Sliding surface σ_1 and new sliding surface σ_1^* in the ideal scenario

By observation of Fig. 7.9, the difference between auxiliary velocity tracking errors and the control portions of the compensations $\boldsymbol{v}_c^*$ tend to converge to zero, thus verifying their opposite behaviors in the attempt to cancel themselves.

Through Figs. 7.10, 7.11, 7.12, and 7.13, the sliding surfaces, new sliding surfaces, and their derivatives tend to converge to zero with significant attenuation of the chattering phenomenon.

Figure 7.14 shows estimated consequences of the fuzzy rules $\hat{\boldsymbol{\beta}}$ obtained from adaptation laws in Eqs. 7.7 and 7.18, for the AFSMC-A and AFSMC-B respectively, necessary for the calculation of the control portions of compensations $\boldsymbol{v}_c^* = \hat{\boldsymbol{F}}(\boldsymbol{\sigma}^*)$ of the control law in Eq. 7.2. The difference in the behavior of these consequences is due to the gains γ and ς (to see Table 7.2) chosen for the adaptation law of the AFSMC-B, which resulted in a degradation of tracking performance and, consequently, higher RMS of errors related to the AFSMC-A, as listed in Table 7.3.

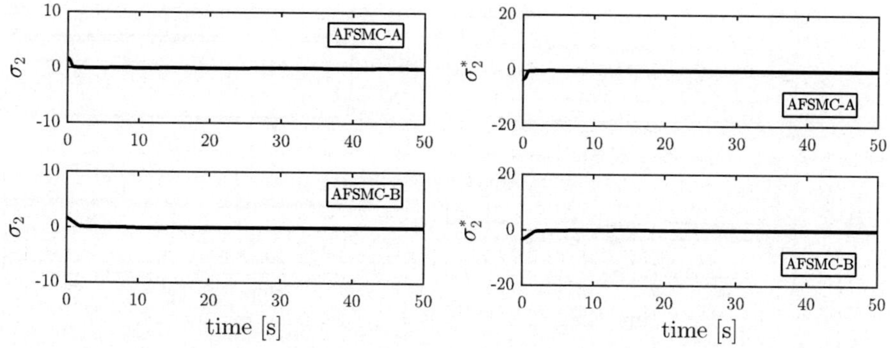

Fig. 7.11 Sliding surface σ_2 and new sliding surface σ_2^* in the ideal scenario

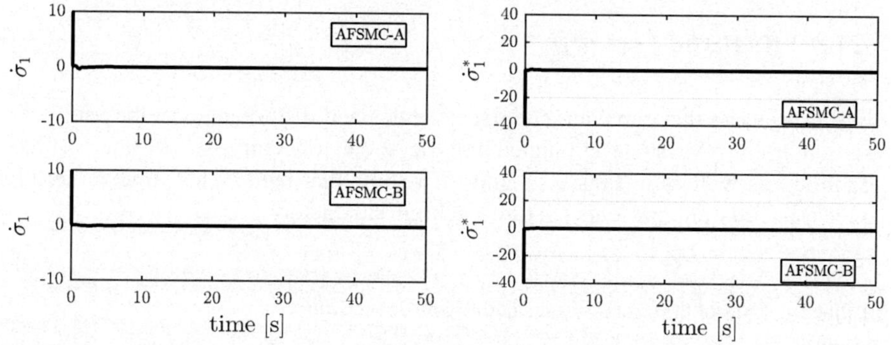

Fig. 7.12 Derivatives of sliding surface σ_1 and derivatives of new sliding surface σ_1^* in the ideal scenario

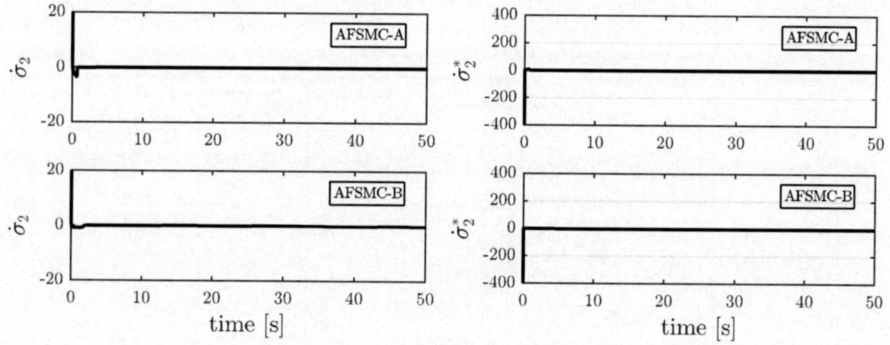

Fig. 7.13 Derivatives of sliding surface σ_2 and derivatives of new sliding surface σ_2^* in the ideal scenario

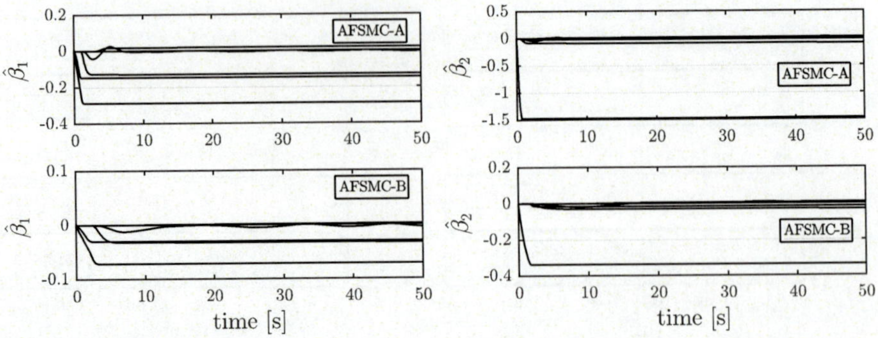

Fig. 7.14 Estimated consequences of the fuzzy rules $\hat{\boldsymbol{\beta}}$ in the ideal scenario

7.4.2 *Realistic Scenario*

In this scenario that considers the uncertainties and disturbances in the MobileSim simulator, AFSMCs are simulated for the same trajectory, whose the results in Table 7.4 as well as in Figs. 7.15 and 7.16 show that the DWMR tracks the reference trajectory and presents posture tracking satisfactory.

Table 7.4 RMS of the errors—results in the realistic scenario

Controller	x_e (m)	y_e (m)	θ_e (rad)
AFSMC-A	0.0238	0.0303	0.0420
AFSMC-B	0.0370	0.0489	0.0482

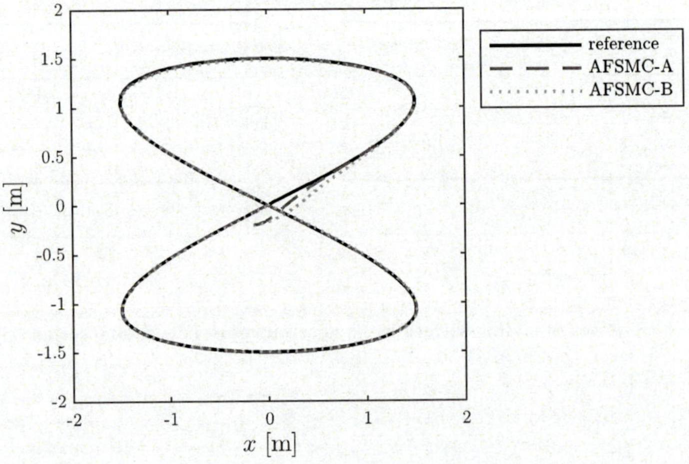

Fig. 7.15 Trajectory tracking in the realistic scenario

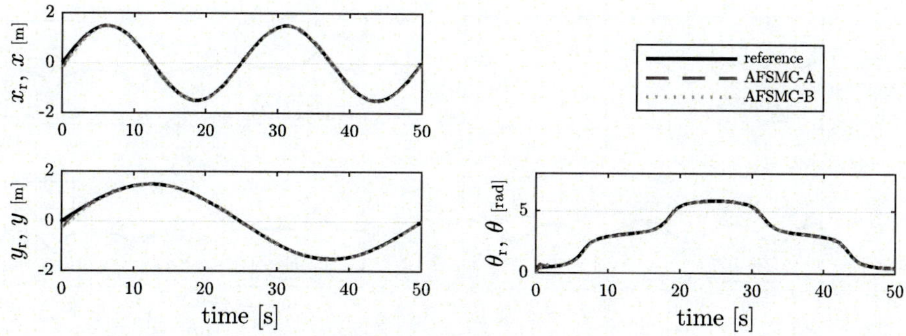

Fig. 7.16 Tracking (x, y, and θ) in the realistic scenario

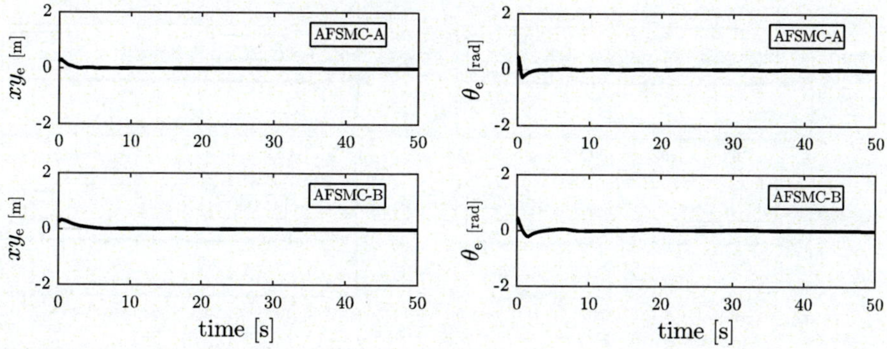

Fig. 7.17 Posture tracking errors, with $xy_e = \sqrt{x_e^2 + y_e^2}$, in the realistic scenario

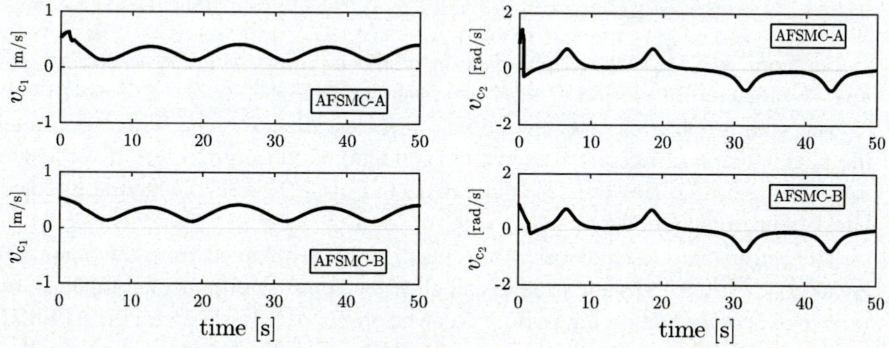

Fig. 7.18 Control velocities in the realistic scenario

Figure 7.17 of the posture tracking and orientation errors verifies that these errors tend to converge to zero.

In Figs. 7.18 and 7.19, all the velocities present smooth behaviors, and that the velocities of the DWMR track the reference velocities.

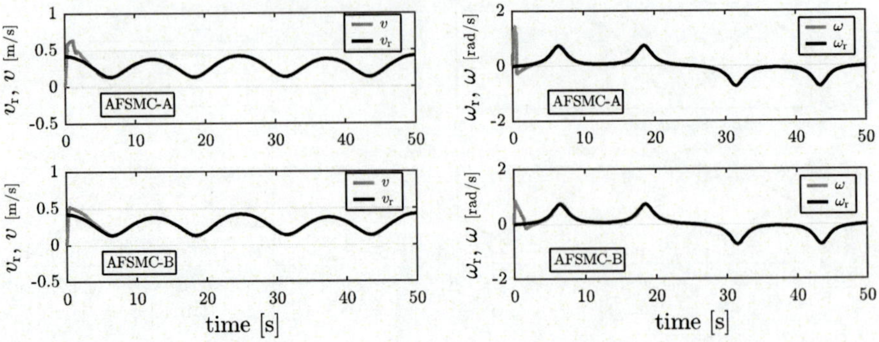

Fig. 7.19 Velocities of the DWMR and reference velocities in the realistic scenario

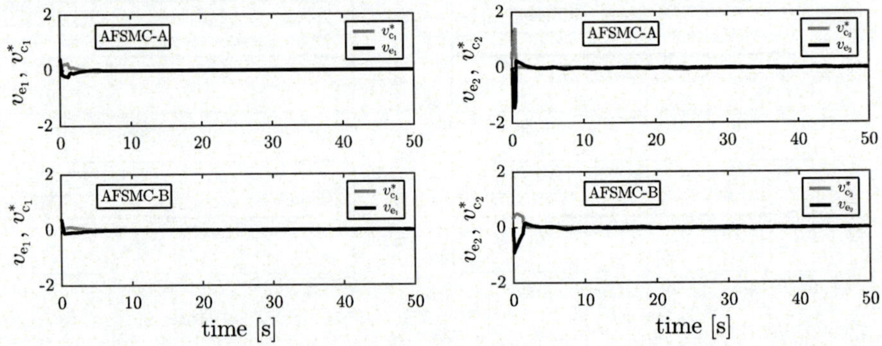

Fig. 7.20 Auxiliary velocity tracking errors and control velocities in the realistic scenario

For both AFSMCs, the compensation of the auxiliary velocity tracking errors given by the control portion $\boldsymbol{v}_{\mathrm{c}}^{*} = \hat{\boldsymbol{F}}(\boldsymbol{\sigma}^{*})$ is shown in Fig. 7.20.

The sliding surfaces and new sliding surfaces are presented with the chattering phenomenon significantly attenuated and tend to converge to zero by observing Figs. 7.21 and 7.22. However, the derivatives in Figs. 7.23 and 7.24 have slight oscillatory behaviors around zero.

The estimated consequences of the fuzzy rules $\hat{\boldsymbol{\beta}}$ obtained from the adaptation laws (Eqs. 7.7 and 7.18) and used to calculate the control portions of compensations $\boldsymbol{v}_{\mathrm{c}}^{*} = \hat{\boldsymbol{F}}(\boldsymbol{\sigma}^{*})$ of the control law in Eq. 7.2 can be observed in Fig. 7.25 for the AFSMCs. As previously related, the difference in the behavior of these consequences is due to the gains γ and ζ in Table 7.2 chosen for the adaptation law of the AFSMC-B.

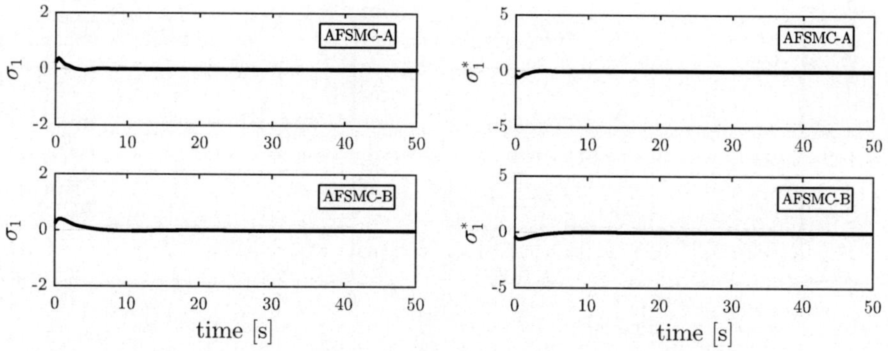

Fig. 7.21 Sliding surface σ_1 and new sliding surface σ_1^* in the realistic scenario

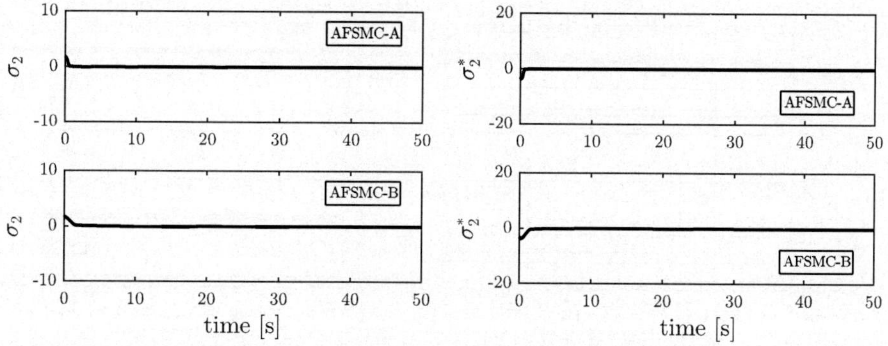

Fig. 7.22 Sliding surface σ_2 and new sliding surface σ_2^* in the realistic scenario

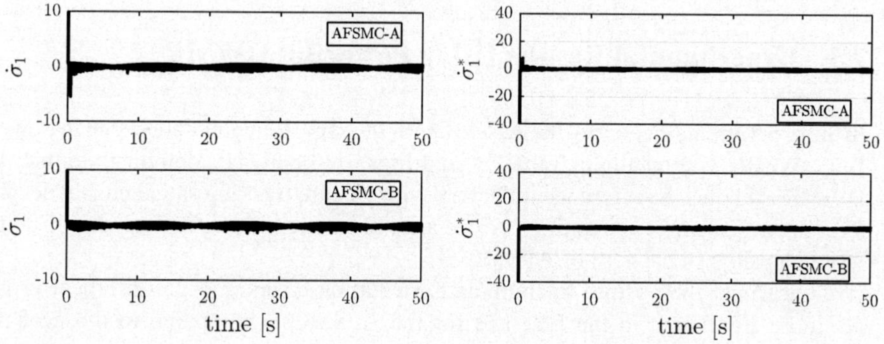

Fig. 7.23 Derivatives of sliding surface σ_1 and derivatives of new sliding surface σ_1^* in the realistic scenario

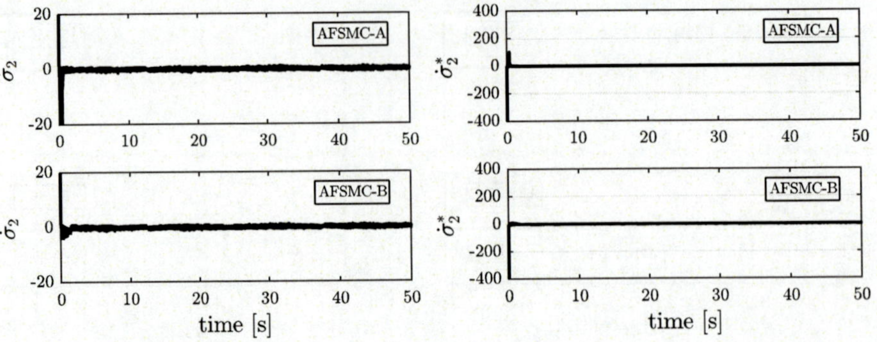

Fig. 7.24 Derivatives of sliding surface σ_2 and derivatives of new sliding surface σ_2^* in the realistic scenario

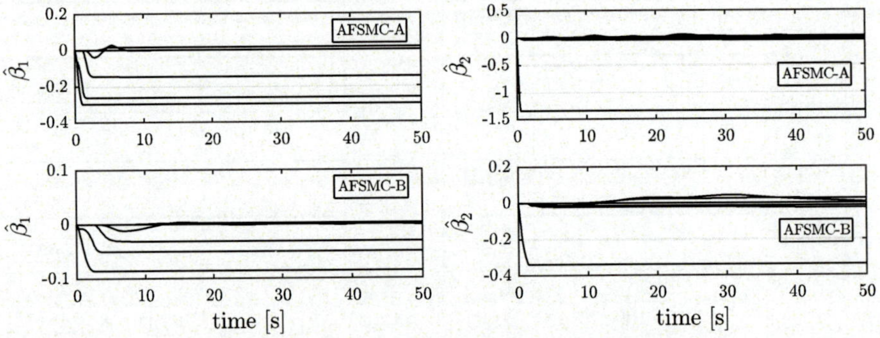

Fig. 7.25 Estimated consequences of the fuzzy rules $\hat{\boldsymbol{\beta}}$ in the realistic scenario

7.5 Experimental Results Using PowerBot DWMR

In the experimental scenario, the AFSMCs for the same trajectory are tested in PowerBot DWMR. The results in Table 7.5 and Figs. 7.26 and 7.27 demonstrate that the DWMR executes the reference trajectory with posture tracking satisfactorily despite a certain degradation of tracking performance in the first 15 s, such degradation being more plausible in the case of AFSMC-B.

Concerning the posture tracking and orientation errors in Fig. 7.28, these errors are more significant in the first 15 s for the AFSMC-B compared to those of the

Table 7.5 RMS of the errors—Results in the experimental scenario

Controller	x_e (m)	y_e (m)	θ_e (rad)
AFSMC-A	0.0644	0.0618	0.1190
AFSMC-B	0.2910	0.1910	0.2540

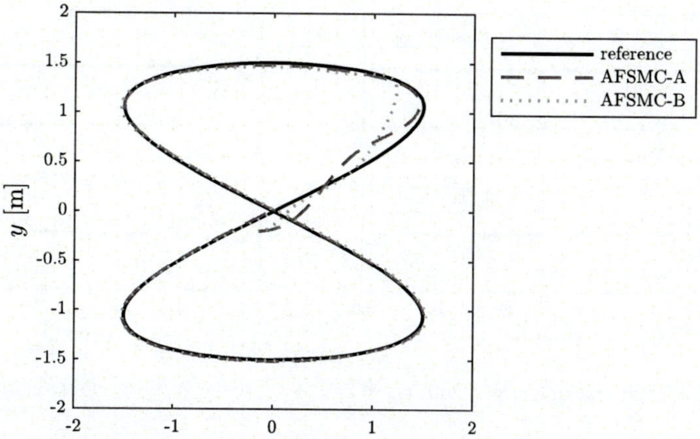

Fig. 7.26 Trajectory tracking in the experimental scenario

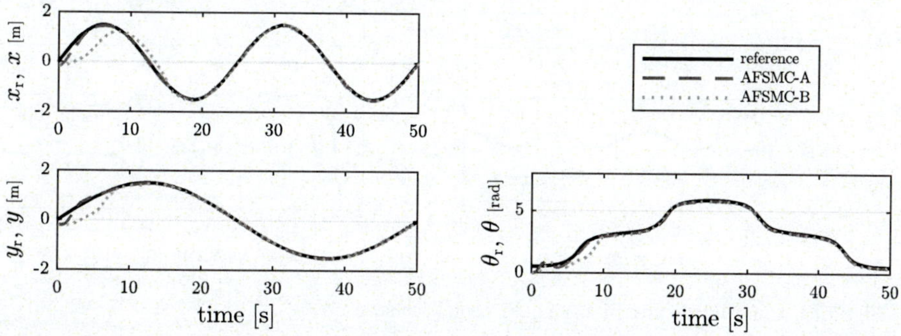

Fig. 7.27 Tracking (x, y, and θ) in the experimental scenario

AFSMC-A. Moreover, after that time, for both AFSMCs these errors tend to zero with a slight variation of the orientation error when the DWMR makes a curve mainly.

Figure 7.29 presents the control velocities. The AFSMC-A presents linear and angular velocities necessary to perform a good tracking performance while the AFSMC-B presents these velocities with the same characteristic only after the first 15 s. These same behaviors can also be observed in Fig. 7.30, in which velocities of the DWMR track the reference velocities. Recalling that the signals from these velocities are all smooth.

The auxiliary velocity tracking errors are duly compensated by the control portion $v_c^* = \hat{F}(\sigma^*)$ for the case of AFSMC-A, however, this compensation of auxiliary velocity tracking errors performed by the control portion of the AFSMC-B occur only after the first 15 s, as can be seen in Fig. 7.31.

For the use of the AFSMC-A, the sliding surfaces and new sliding surfaces tend to converge to zero, as can be seen in Figs. 7.32 and 7.33 with the chattering phenomenon significantly attenuated while these sliding surfaces behave similarly after the first

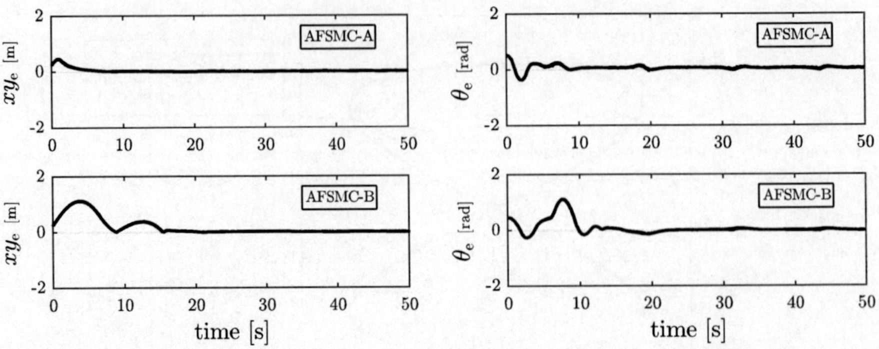

Fig. 7.28 Posture tracking errors, with $xy_e = \sqrt{x_e^2 + y_e^2}$, in the experimental scenario

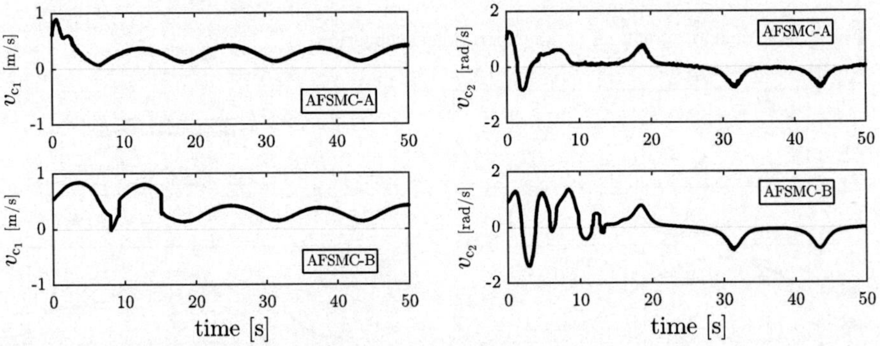

Fig. 7.29 Control velocities in the experimental scenario

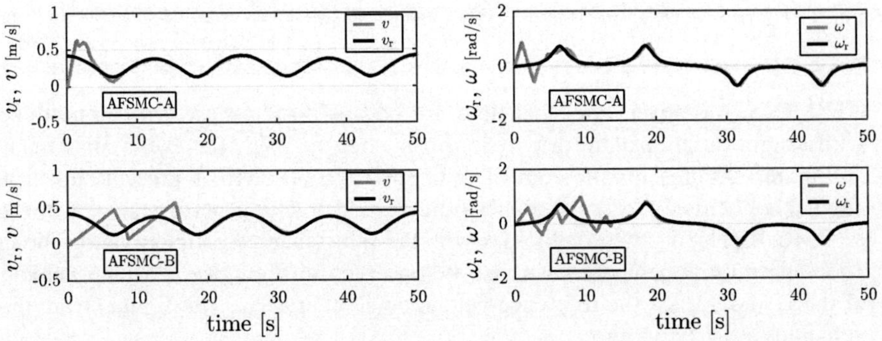

Fig. 7.30 Velocities of the DWMR and reference velocities in the experimental scenario

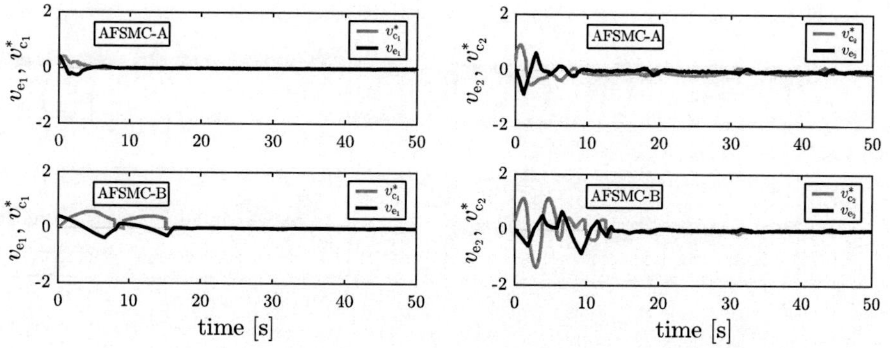

Fig. 7.31 Auxiliary velocity tracking errors and control velocities in the experimental scenario

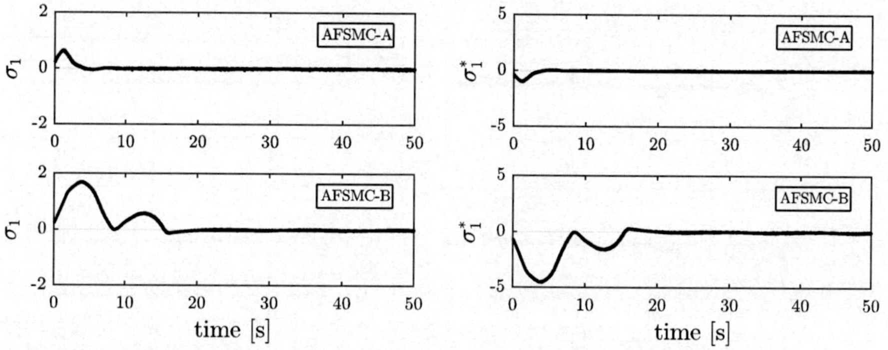

Fig. 7.32 Sliding surface σ_1 and new sliding surface σ_1^* in the experimental scenario

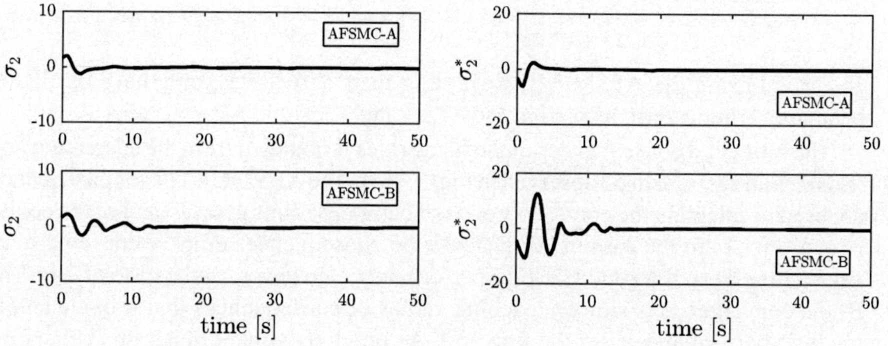

Fig. 7.33 Sliding surface σ_2 and new sliding surface σ_2^* in the experimental scenario

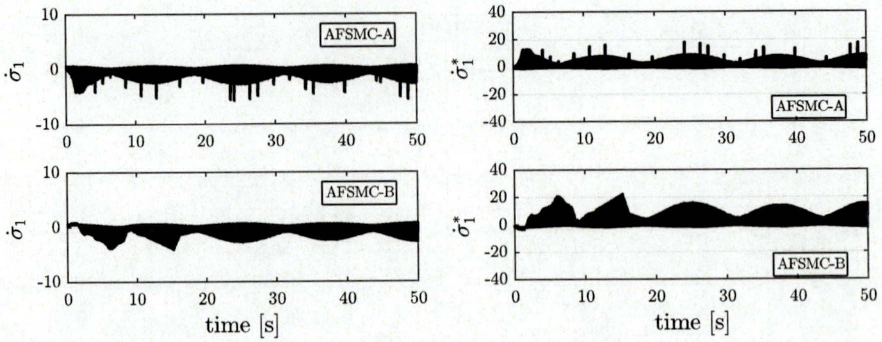

Fig. 7.34 Derivatives of sliding surface σ_1 and derivatives of new sliding surface σ_1^* in the experimental scenario

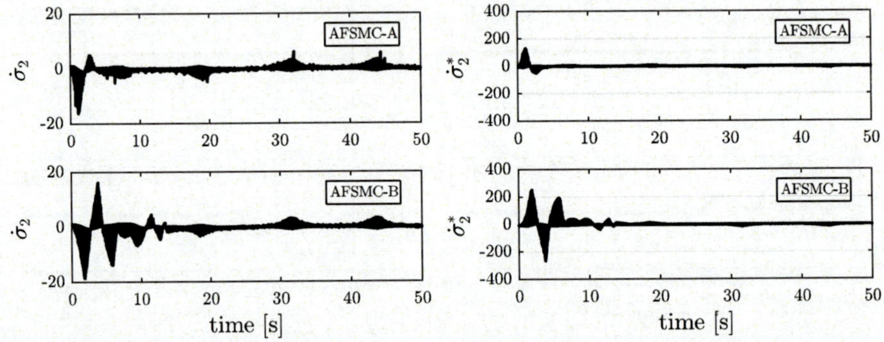

Fig. 7.35 Derivatives of sliding surface σ_2 and derivatives of new sliding surface σ_2^* in the experimental scenario

15 s when using the AFSMC-B. Regarding their derivatives, they have oscillatory behaviors around zero, as seen in Figs. 7.34 and 7.35.

The estimated consequences of the fuzzy rules $\hat{\boldsymbol{\beta}}$ obtained from the adaptation laws (Eqs. 7.7 and 7.18) can be observed in Fig. 7.36 for the AFSMCs. These consequences are used to calculate the control portions of compensations $\boldsymbol{v}_c^* = \hat{\boldsymbol{F}}(\boldsymbol{\sigma}^*)$ of the control law in Eq. 7.2. In the case of AFSMC-A, the consequences quickly converge to the values necessary for good tracking performance. However, in the case of AFSMC-B, the convergence is slower, reaching values of consequences that provide similar tracking performance after the first 15 s. As previously mentioned, this behavior is due to the choice of gains γ and ζ (to see Table 7.2) of the adaptation law of AFSMC-B as well as the incidence of uncertainties and disturbances, as can be verified in Table 7.5 referring to the RMS of the errors.

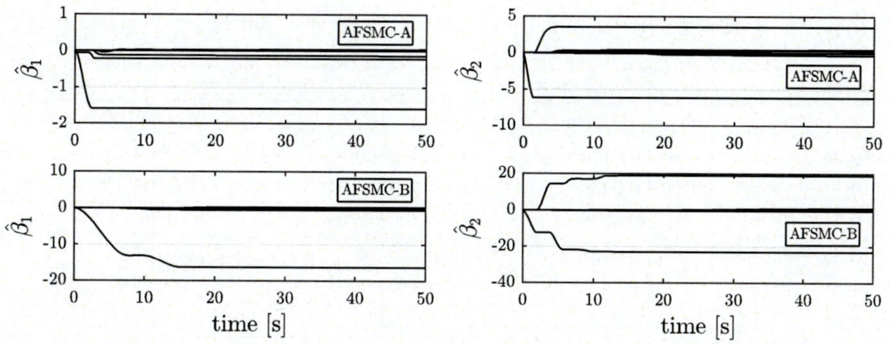

Fig. 7.36 Estimated consequences of the fuzzy rules $\hat{\boldsymbol{\beta}}$ in the experimental scenario

7.6 Analysis and Discussion of Results

Analyzing the results of Sect. 7.5 it is possible to verify that the AFSMC-B presented a performance slightly below the expected concerning the AFSMC-A, but nothing that discredits it in terms of its effectiveness. This AFSMC-B performance is due to the choice of gains γ and ζ in Table 7.2 of the adaptation law in Eq. 7.18 that caused higher RMS of errors and did not provide the control portions of compensation $\boldsymbol{v}_c^* = \hat{\boldsymbol{F}}(\boldsymbol{\sigma}^*)$ needed to overcome incident uncertainties and disturbances of the PowerBot DWMR, mainly in the first 15 s of the tracking of the eight-shape trajectory.

In addition to the choice of gains, other factors can significantly influence the results to demonstrate the effectiveness of the AFSMCs, such as the quantity of the rule base according to the number of inputs (for example, if the inputs are considered $\boldsymbol{\sigma}^*$ and $\dot{\boldsymbol{\sigma}}^*$, these same entries will correspond to 14 if-then rules in the rule base), the type of the membership functions and form of setting the value of the membership functions.

Continuing the analysis different values of the gain γ for the AFSMC-B are used in simulations in the ideal and realistic scenarios to obtain the RMS of the errors slightly smaller and an improvement in the results, as seen in Tables 7.6 and 7.7.

Based on the data obtained from Tables 7.6 and 7.7, the gain $\gamma = 2.0$ is chosen to obtain the results in the realistic scenario only, as shown in sequence.

By observing the results obtained, it can be seen that the DWMR tracks the reference trajectory (Fig. 7.37), the DWMR posture trackings converge to the reference posture trackings (Fig. 7.38), the posture tracking errors tend to converge to zero (Fig. 7.39), the behaviors of the control velocity profiles are smooth (Fig. 7.40), the velocities of the DWMR converge to the reference velocities (Fig. 7.41), the compensation control portions exhibit opposite and similar behaviors to auxiliary velocity tracking errors (Fig. 7.42), the sliding surfaces (Fig. 7.43) and the new sliding surfaces (Fig. 7.44) tend to converge to zero, the behaviors of the derivatives of the

Table 7.6 RMS of the errors—simulation results in the ideal scenario using different values of γ for the AFSMC-B

Gains	RMS errors		
γ	x_e (m)	y_e (m)	θ_e (rad)
0.02	0.0257	0.0586	0.0566
0.1	0.0236	0.0483	0.0468
0.2	0.0219	0.0433	0.0481
0.5	0.0189	0.0365	0.0468
1.0	0.0167	0.0316	0.0445
2.0	0.0148	0.0271	0.0427
5.0	0.0126	0.0220	0.0391
10	0.0113	0.0188	0.0366
50	0.0087	0.0136	0.0323

*All use fixed $\lambda_1 = 1.5$, $\lambda_2 = 6.0$, $\lambda_3 = 1.0$, $k_1 = 0.1$, $k_2 = 0.1$ and $\zeta = 0.1$

Table 7.7 RMS of the errors—simulation results in the realistic scenario using different values of γ for the AFSMC-B

Gains	RMS errors		
γ	x_e (m)	y_e (m)	θ_e (rad)
0.02	0.0392	0.0604	0.0602
0.1	0.0373	0.0490	0.0485
0.2	0.0333	0.0437	0.0463
0.5	0.0293	0.0366	0.0441
1	0.0264	0.0320	0.0415
2	0.0244	0.0292	0.0404
5	0.0227	0.0275	0.0415
10	0.0230	0.0275	0.0409
50	0.0240	0.0278	0.0398

*All use fixed $\lambda_1 = 1.5$, $\lambda_2 = 6.0$, $\lambda_3 = 1.0$, $k_1 = 0.1$, $k_2 = 0.1$ and $\zeta = 0.1$

sliding surfaces (Fig. 7.45) and the new sliding surfaces (Fig. 7.46) oscillate close to zero, the estimated consequences of the fuzzy rules $\hat{\boldsymbol{\beta}}$ (Fig. 7.47) contribute to the compensation control portions to attempt to cancel auxiliary velocity tracking errors.

7.7 General Considerations

This chapter has been approached the AFSMCs, which were derived from FOSMC and lost the invariance principle to mitigate chattering but guaranteeing robustness admissible in the suppression of the uncertainties and disturbances by using a fuzzy

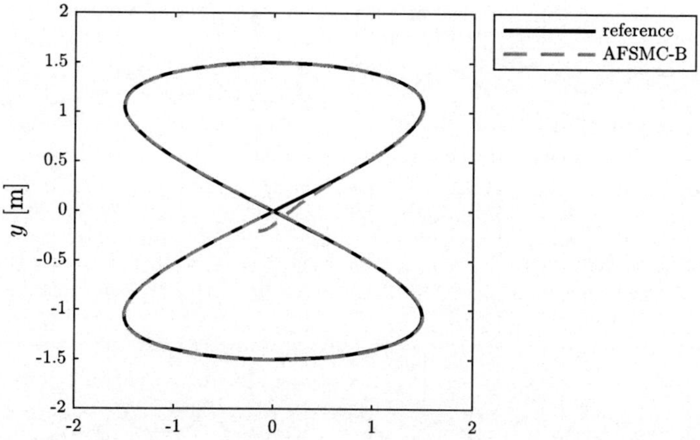

Fig. 7.37 Trajectory tracking using AFSMC-B with $\gamma = 2.0$ in the realistic scenario

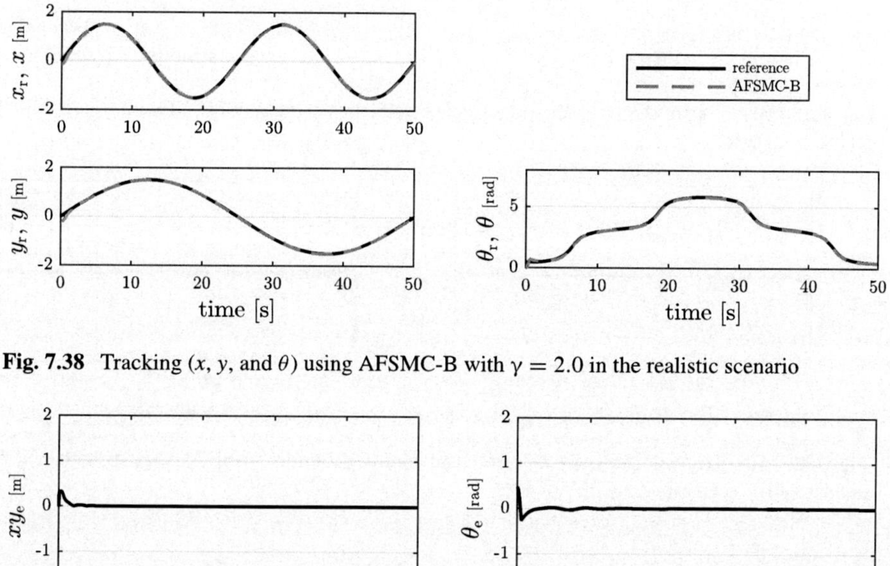

Fig. 7.38 Tracking (x, y, and θ) using AFSMC-B with $\gamma = 2.0$ in the realistic scenario

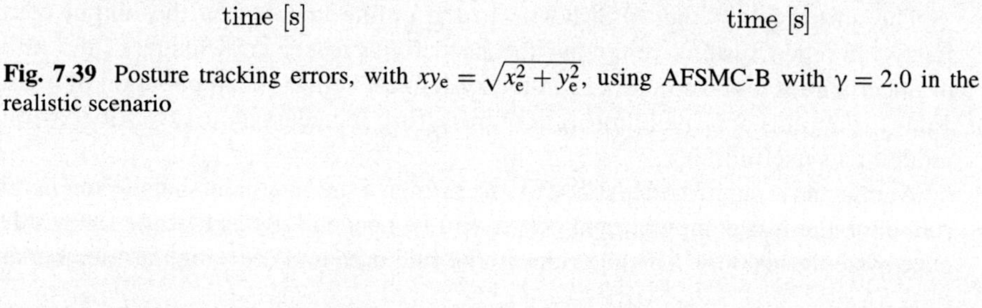

Fig. 7.39 Posture tracking errors, with $xy_{\mathrm{e}} = \sqrt{x_{\mathrm{e}}^2 + y_{\mathrm{e}}^2}$, using AFSMC-B with $\gamma = 2.0$ in the realistic scenario

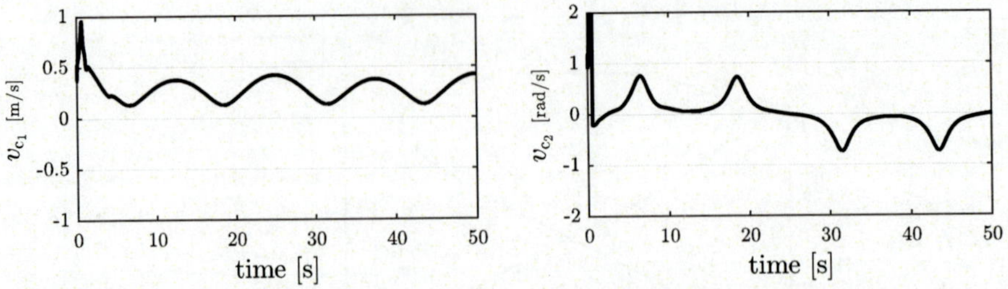

Fig. 7.40 Control velocities using AFSMC-B with $\gamma = 2.0$ in the realistic scenario

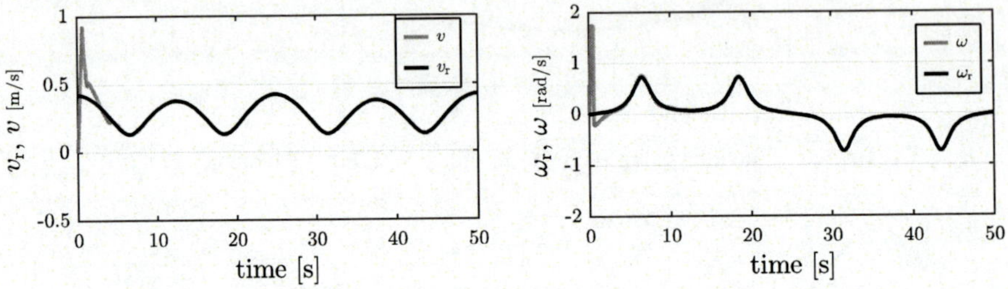

Fig. 7.41 Velocities of the DWMR and reference velocities using AFSMC-B with $\gamma = 2.0$ in the realistic scenario

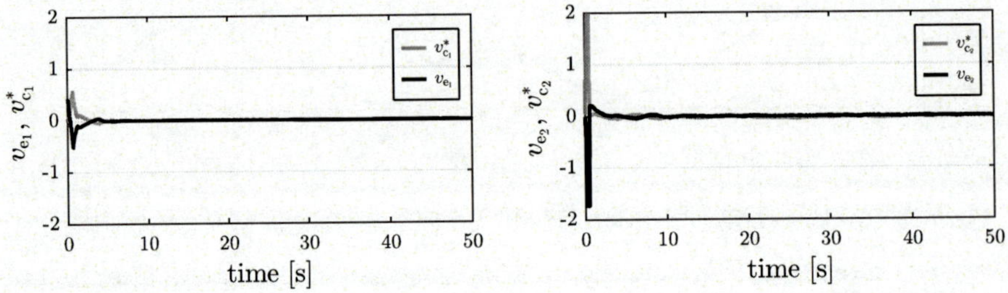

Fig. 7.42 Auxiliary velocity tracking errors and control velocities using AFSMC-B with $\gamma = 2.0$ in the realistic scenario

inference system providing thus a smooth control signal. By the results obtained, it was evidenced that the invariance has little practical meaning [21].

The advantage of the AFSMCs is in the online updates on the output of the fuzzy inference system, removing the need for *a priori* knowledge of the limits of uncertainties and disturbances in the system as well as the application of a large value to the gains of **G** avoiding high control efforts unnecessary to DWMR actuators, reducing its useful life.

Another advantage of the AFSMCs, according to the statement that the smaller the rule base, the less computational power will be needed [16], is that the fuzzy inference system uses only 7 if-then rules in the rule base and of triangular membership

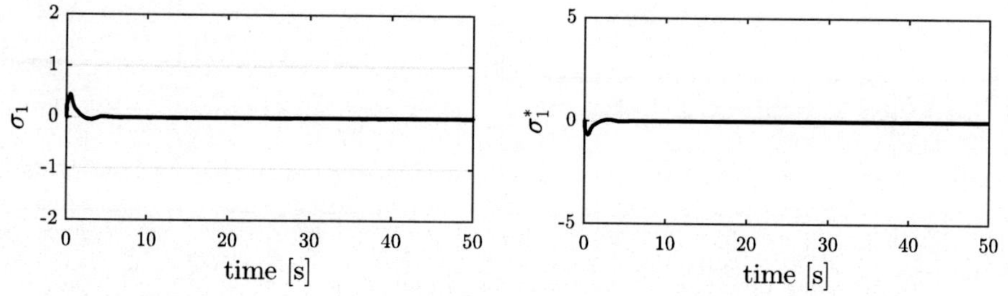

Fig. 7.43 Sliding surface σ_1 and new sliding surface σ_1^* using AFSMC-B with $\gamma = 2.0$ in the realistic scenario

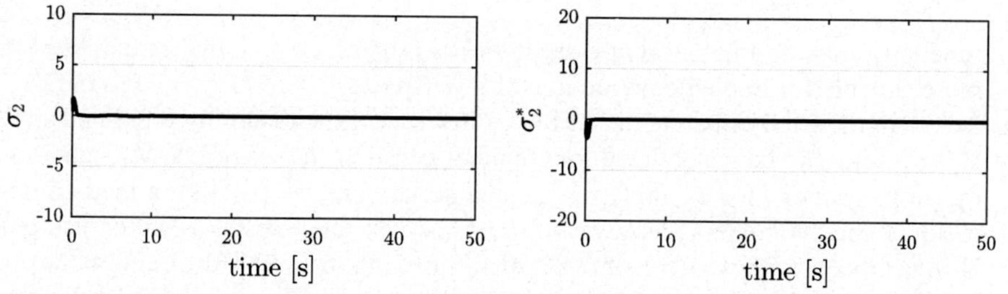

Fig. 7.44 Sliding surface σ_2 and new sliding surface σ_2^* using AFSMC-B with $\gamma = 2.0$ in the realistic scenario

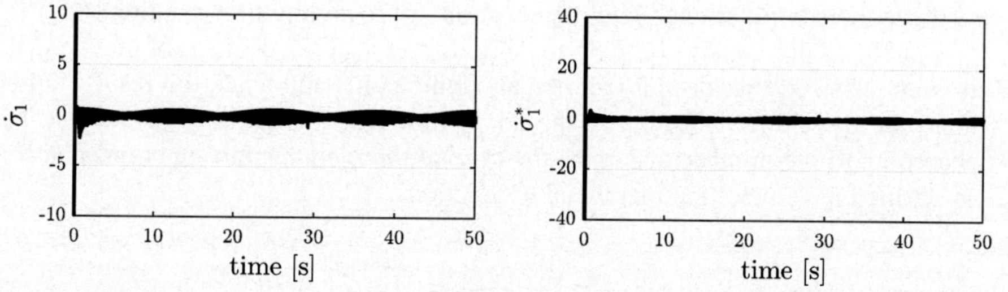

Fig. 7.45 Derivatives of sliding surface σ_1 and derivatives of new sliding surface σ_1^* using AFSMC-B with $\gamma = 2.0$ in the realistic scenario

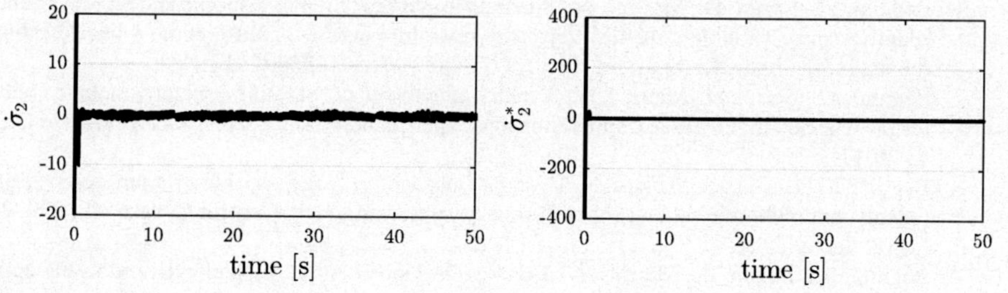

Fig. 7.46 Derivatives of sliding surface σ_2 and derivatives of new sliding surface σ_2^* using AFSMC-B with $\gamma = 2.0$ in the realistic scenario

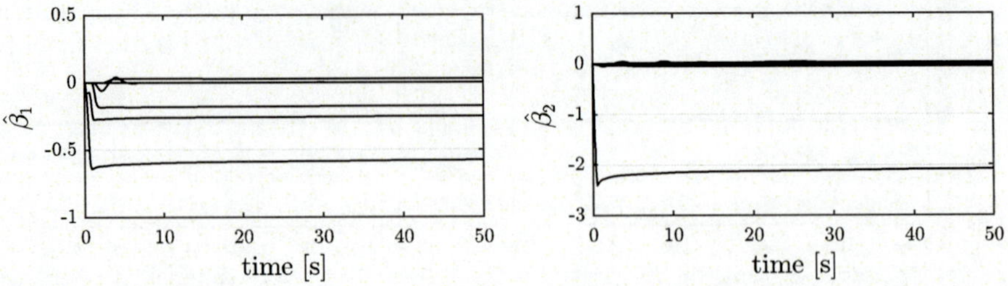

Fig. 7.47 Estimated consequences of the fuzzy rules $\hat{\boldsymbol{\beta}}$ using AFSMC-B with $\gamma = 2.0$ in the realistic scenario

functions, making their structure as simple as possible, which makes the AFSMCs more suitable for implementation in real DWMRs.

To continue the research, the AFSMCs in terms of their stability proof and analysis of results can be generalized for the other possible classes of DWMRs and other dynamic systems, for example, unmanned aerial vehicles (UAVs); a methodology to adjust the parameters and gains of the AFSMCs can be established by using the Design Optimization toolbox of the Matlab/Simulink software [6], aiming at the optimization of performance and the desired trajectory; and a comparative study between the AFSMCs and other kinematic fuzzy controllers considering chattering reduction approaches, such as the works developed by [8–10, 16, 22], can be performed to evaluate the computational load of execution or processing time required.

Finally, as previously reported, to demonstrate the effectiveness of the AFSMCs, it is emphasized that design factors can significantly influence the results, among them are the choice of gains of the adaptation law, the quantity of the rule base according to the number of inputs, the type of the membership functions and form of setting the value of the membership functions.

References

1. Begnini, M., Bertol, D., Martins, N.: Practical implementation of a simple and effective robust adaptive fuzzy variable structure trajectory tracking control for differential wheeled mobile robots. Int. J. Innov. Comput. Inf. Control (IJICIC) **13**, 341–364 (2017)
2. Begnini, M., Bertol, D., Martins, N.: A robust adaptive fuzzy variable structure tracking control for the wheeled mobile robot: simulation and experimental results. Control Eng. Pract. **64**, 27–43 (2017)
3. Begnini, M., Bertol, D., Martins, N.: Design of an adaptive fuzzy variable structure compensator for the nonholonomic mobile robot in trajectory tracking task. Control Cybern. **47**, 239–275 (2018, September)
4. Begnini, M., Bertol, D., Martins, N.: Practical implementation of an effective robust adaptive fuzzy variable structure tracking control for a wheeled mobile robot. J. Intell. Fuzzy Syst. **35**, 1087–1101 (2018, July)
5. Elyoussef, E.S., Martins, N.A., De Pieri, E.R., Moreno, U.F.: PD-super-twisting second order sliding mode tracking control for a nonholonomic wheeled mobile robot. In: Proceedings of the 19th World Congress of the International Federation of Automatic Control—IFAC World Congress, vol. 11, pp. 3827–3832 (2014)

6. Freire, F., Martins, N., Splendor, F.: A simple optimization method for tuning the gains of PID controllers for the autopilot of cessna 182 aircraft using model-in-the-loop platform. J. Control Autom. Electri. Syst. **29**, 441–450 (2018)
7. Guo, Y., Woo, P.-Y.: An adaptive fuzzy sliding mode controller for robotic manipulators. IEEE Trans. Syst. Man Cybern. Part A: Syst. Hum. **33**(2), 149–159 (2003)
8. Keighobadi, J., Menhaj, M.B.: From nonlinear to fuzzy approaches in trajectory tracking control of wheeled mobile robots. Asian J. Control **14**(4), 960–973 (2012)
9. Keighobadi, J., Mohamadi, Y.: Fuzzy sliding mode control of a nonholonomic wheeled mobile robot. In: Proceedings of the 2011 International MultiConference of Engineers and Computer Scientists (IMECS'2011)
10. Keighobadi, J., Mohamadi, Y.: Fuzzy sliding mode control of nonholonomic wheeled mobile robot. In: Proceedings of the 9th IEEE International Symposium on Applied Machine Intelligence and Informatics—SAMI'2011, pp. 273–278. IEEE (2011)
11. Lewis, F.L., Jagannathan, S., Yesildirek, A.: Neural Network Control of Robot Manipulators and Nonlinear Systems. Taylor & Francis, Ltd., 1 Gunpowder Square, London, EC4A 3DE (1999)
12. Lewis, F.L., Dawson, D.M., Abdallah. C.T., Robot Manipulator Control: Theory and Practice, 2 edn . Marcel Dekker, Inc. (2003)
13. Li, Y., Qiang, S., Zhuang, X., Kaynak, O.: Robust and adaptive backstepping control for nonlinear systems using RBF neural networks. IEEE Trans. Neural Netw. **15**(3), 693–701 (2004)
14. Martins, N.A., Alencar, M., Lombardi, W.C., Bertol, D.W., De Pieri, E.R., Ferasoli Filho, H.: Trajectory tracking of a wheeled mobile robot with uncertainties and disturbances, Robust and adaptive backstepping control for nonlinear systems using RBF neural networks. Control Cybern. **44**, 47–98 (2015)
15. Martins, N.A., Elyoussef, E.S., Bertol, D.W., De Pieri, E.R., Moreno, U.F., Castelan, E.B.: Nonholonomic mobile robot with kinematic disturbances in the trajectory tracking: a variable structure controller. Learn. Nonlinear Models **8**(1), 23–40 (2010)
16. Mishra. E.A.: Trajectory tracking of differential drive wheeled mobile robot. Int. J. Mech. Eng. Robot. (IJMER'2014)
17. Slotine, J.-J.E., Li, W.: Applied Nonlinear Control. Prentice Hall, Upper Saddle River, New Jersey, USA (1991)
18. Utkin, V.I.: Sliding Modes in Control Optimization. Springer, Berlin (1992)
19. Utkin, V.I., Guldner, J., Shi, J .: Sliding Mode Control in Electro-Mechanical Systems, 2nd edn. CRC Press (2009)
20. Wang, L.-X.: A Course in Fuzzy Systems and Control. Prentice-Hall Press, Englewood Cliffs, USA (1996)
21. Wang, S., Gao, W.: Robustness and invariance of variable structure systems with multiple inputs. In: Proceedings of the American Control Conference (ACC'1995), vol. 1, pp. 1035–1039. IEEE (1995)
22. Xie, M.-J. Li, L.-T., Wang, Z.-Q..: Trajectory tracking control for mobile robot based on the fuzzy sliding mode. In: Proceedings of the 10th World Congress on Intelligent Control and Automation (WCICA'2012), pp. 2706–2709. IEEE (2012)

Chapter 8
Adaptive Robust Control: Adaptive Neural Sliding Mode Control Technique

8.1 Introduction

Due to its simplicity of design without requiring a precise model and its property of robustness against imprecision in modeling, uncertainties, and disturbances, sliding mode control has become widely popular and is used in many application areas [1, 4, 7, 22]. However, this control technique has important disadvantages that limit its practical applicability, such as high-frequency switching (chattering phenomenon) and the vast control authority, which deteriorate the performance of the system [21]. The first disadvantage mentioned is due to the control actions, which are discontinuous on the sliding surfaces, which causes high-frequency switching at their bound. The chattering phenomenon can cause [22, 27]:

- The excitation of unmodulated dynamics (for example, the fast dynamics of actuators and sensors that are normally neglected during the control design), which in turn can cause low amplitude high-frequency oscillations, imposing unnecessary wear on the actuators and even destabilizing the system, making it impossible to apply the control law in practice;
- Switching imperfections that cause high-frequency oscillations. These oscillations can include small-time delays due to sampling and/or the execution time required to calculate the control, and, more recently, transmission delays in networked control systems.

The second disadvantage mentioned is based on the requirement of prior knowledge of the bound of uncertainties and disturbances in the compensators. If the bound is unknown, a high value must be applied empirically to the gain of the discontinuous part of the control signal, and this high control gain can intensify the high-frequency switching on the sliding surfaces.

Electronic supplementary material The online version of this chapter (https://doi.org/10.1007/978-3-030-77912-2_8) contains supplementary material, which is available to authorized users.

N. A. Martins and D. W. Bertol, *Wheeled Mobile Robot Control*, Studies in Systems, Decision and Control 380, https://doi.org/10.1007/978-3-030-77912-2_8

Thus, the nonlinearities of a real system imply that the ideal condition for the sliding mode is not achieved. In other words, it is not possible to switch control instantly from one value to another. Because of this, controllers in Chap. 5 will not always be able to maintain the trajectory of states slipping on the sliding surfaces, causing the chattering phenomenon. For controllers in Chap. 5, this phenomenon is more evident the higher the value of **G** mainly, as can be seen in simulation and experimental results.

The undesirable chattering phenomenon can be mitigated through some techniques proposed by [8, 22, 27]. A technique applied in the controllers to reduce the chattering phenomenon is the use of the fractional continuous approximation or proper continuous function as described in Chap. 6. However, from a theoretical point of view, the invariance principle is not verified anymore and can cause loss of robustness of the controllers. Therefore, techniques that guarantee a reduction of the chattering phenomenon without loss of robustness should be sought.

In this chapter, to attenuate the chattering phenomenon, the discontinuous term of the controllers in Chap. 5 is replaced by single-input single-output (SISO) radial basis function neural networks (RBFNNs), which are non-linear and continuous functions, thus resulting in adaptive neural sliding mode control (ANSMC). And, to relax the demands of knowing the bounds of the uncertainties and disturbances suffered by the system, the output layer weights of the RBFNNs are updated online to compensate for these uncertainties and disturbances as well as to guarantee the stability of the system without any prior knowledge of them and its bounds [14]. The development of the control design of the ANSMC is made similarly to [12–16].

8.2 Background

This section describes the succinct theory about GL (Ge-Lee) notation, approximation by RBFNNs, and neural networks modeling by RBFNNs necessary for the development of the control design.

8.2.1 GL Notation

This subsection briefly discusses the definition of GL matrix, which is denoted by $\{\cdot\}$, and its product operator "•". Readers are referred to [5], for a detailed discussion on the motivation for using the GL matrix. To avoid any possible confusion, $[\cdot]$ is used to denote the ordinary vector and matrix.

The GL row vector $\{\boldsymbol{\partial}\}$ and its transpose $\{\boldsymbol{\partial}\}^{\mathrm{T}}$ are defined as

$$\left\{\boldsymbol{\partial}\right\} = \left\{\boldsymbol{\partial}_{i1} \ \ldots \ \boldsymbol{\partial}_{\mathrm{n}}\right\}, \quad \text{and} \quad \left\{\boldsymbol{\partial}\right\}^{\mathrm{T}} = \left\{\boldsymbol{\partial}_{1}^{\mathrm{T}} \ \ldots \ \boldsymbol{\partial}_{\mathrm{n}}^{\mathrm{T}}\right\}, \tag{8.1}$$

where $\boldsymbol{\partial}_i \in \mathbb{R}^{\eta_i}$, being $\eta_i \in \mathbb{Z}_{+}^{*}$.

The GL matrix $\{\boldsymbol{\Pi}\}$ and its transpose $\{\boldsymbol{\Pi}\}^{\mathrm{T}}$ are defined accordingly as

$$\{\boldsymbol{\Pi}\} = \begin{Bmatrix} \partial_{11} & \dots & \partial_{1\mathrm{p}} \\ \vdots & \ddots & \vdots \\ \partial_{\mathrm{n}1} & \dots & \partial_{\mathrm{np}} \end{Bmatrix} = \begin{Bmatrix} \{\partial_1\} \\ \vdots \\ \{\partial_{\mathrm{n}}\} \end{Bmatrix}, \tag{8.2}$$

$$\{\boldsymbol{\Pi}\}^{\mathrm{T}} = \begin{Bmatrix} \partial_{11}^{\mathrm{T}} & \dots & \partial_{1\mathrm{p}}^{\mathrm{T}} \\ \vdots & \ddots & \vdots \\ \partial_{\mathrm{n}1}^{\mathrm{T}} & \dots & \partial_{\mathrm{np}}^{\mathrm{T}} \end{Bmatrix}. \tag{8.3}$$

For a given GL matrix $\{\boldsymbol{\Xi}\}$ the GL product ,"$\bullet$", with $\{\boldsymbol{\Pi}\}^{\mathrm{T}}$ is defined as

$$\left[\{\boldsymbol{\Pi}\}^{\mathrm{T}} \bullet \{\boldsymbol{\Xi}\}\right] = \begin{bmatrix} \partial_{11}^{\mathrm{T}}\zeta_{11} & \dots & \partial_{1\mathrm{p}}^{\mathrm{T}}\zeta_{1\mathrm{p}} \\ \vdots & \ddots & \vdots \\ \partial_{\mathrm{n}1}^{\mathrm{T}}\zeta_{\mathrm{n}1} & \dots & \partial_{\mathrm{np}}^{\mathrm{T}}\zeta_{\mathrm{np}} \end{bmatrix}. \tag{8.4}$$

The GL product of a square matrix $\boldsymbol{A} = [\boldsymbol{a}_1 \ \dots \ \boldsymbol{a}_{\mathrm{n}}]$, where $\boldsymbol{a}_i \in \mathbb{R}^{\iota \times \eta_i}$, and a GL row vector $\{\boldsymbol{\zeta}\} = \{\boldsymbol{\zeta}_1 \ \dots \ \boldsymbol{\zeta}_{\mathrm{n}}\}$, where $\boldsymbol{\zeta}_i \in \mathbb{R}^{\eta_i}$ and $\iota = \sum_{i=1}^{\mathrm{n}} \eta_i$, is defined as

$$\boldsymbol{A} \bullet \{\boldsymbol{\zeta}\} = \{\boldsymbol{A}\} \bullet \{\boldsymbol{\zeta}\} = \left[\boldsymbol{a}_1\boldsymbol{\zeta}_1 \ \dots \ \boldsymbol{a}_{\mathrm{n}}\boldsymbol{\zeta}_{\mathrm{n}}\right], \tag{8.5}$$

where $\{\boldsymbol{A}\} \bullet \{\boldsymbol{\zeta}\} \in \mathbb{R}^{\iota \times \mathrm{n}}$.

Note that the GL product should be computed first in a mixed matrix product. For instance, in $\{\boldsymbol{\Pi}\} \bullet \{\boldsymbol{\Xi}\}\boldsymbol{A}$, the matrix $[\{\boldsymbol{\Pi}\} \bullet \{\boldsymbol{\Xi}\}]$ should be computed first, followed by the multiplication of $[\{\boldsymbol{\Pi}\} \bullet \{\boldsymbol{\Xi}\}]$ with matrix $\boldsymbol{A}$.

8.2.2 Approximation by RBFNNs

In the field of control engineering, neural networks are often used to approximate a given nonlinear function $\boldsymbol{f}(\boldsymbol{x})$ up to a small error tolerance. The function approximation problem can be stated formally as follows.

Given that

$$\boldsymbol{f}(\boldsymbol{x}) = \boldsymbol{w}^{\mathrm{T}}\boldsymbol{a}(\boldsymbol{x}), \tag{8.6}$$

and $\boldsymbol{f}(\boldsymbol{x}) : \mathbb{R}^{\mathrm{n}} \to \mathbb{R}^{\mathrm{m}}$ is a continuous function defined on the set $\boldsymbol{x} \in \mathbb{R}^{\mathrm{n}}$, and $\hat{\boldsymbol{f}}(\boldsymbol{w}, \boldsymbol{x}) : \mathbb{R}^{\mathrm{p}\times\mathrm{m}} \times \mathbb{R}^{\mathrm{n}} \to \mathbb{R}^{\mathrm{m}}$ is an approximating function that depends continuously on the parameter matrix $\boldsymbol{w}$ and $\boldsymbol{x}$, the approximation problem is designed to determine the optimal parameter $\boldsymbol{w}^*$ such that, for a certain metric (or distance function) d_{f},

$$d_{\mathrm{f}}\left(\hat{\boldsymbol{f}}(\boldsymbol{w}^*, \boldsymbol{x}),\ \boldsymbol{f}(\boldsymbol{x})\right) \leq \varepsilon_{\mathrm{NN}}, \tag{8.7}$$

for a neural network approximation error ε_{NN} acceptably small [6, 12].

In the development of the ANSMC, the Gaussian RBFNN is considered, which is a particular neural network architecture that uses p numbers of Gaussian radial basis functions (GRBFs) of following the form:

$$\boldsymbol{a}_i(\boldsymbol{x}) = \exp\left(-\frac{(\boldsymbol{x}-\boldsymbol{c}_i)^{\mathrm{T}}(\boldsymbol{x}-\boldsymbol{c}_i)}{2\mu_i^2}\right), \quad i = 1, 2, \cdots, \mathrm{p}, \tag{8.8}$$

where $\boldsymbol{c}_i \in \mathbb{R}^{\mathrm{n}}$ is the vector of centers and $\mu_i^2 \in \mathbb{R}$ is the variance. This particular network architecture is shown in Fig. 8.1, where each Gaussian RBFNN consists of three layers:

- The input layer $\boldsymbol{x}$, which is the connection of the model with the input signals. In this layer, the input space is divided into grids with a basis function at each node defining a receptive field in $\mathbb{R}^{\mathrm{n}}$;
- The hidden layer $\boldsymbol{a}$, which contains the GRBFs. In this layer, a nonlinear transformation of the internal vector space is carried out, which generally has a larger dimension. Moreover, the GBRF produces a non-zero response only when the input pattern is within a small region located in the input space;
- The output layer $\boldsymbol{f}$, which transforms the internal vector space into an output, through a linear process.

It is emphasized that only the connections from the hidden layer to the output are weighted, i.e., contains weights, as can be seen in Fig. 8.1.

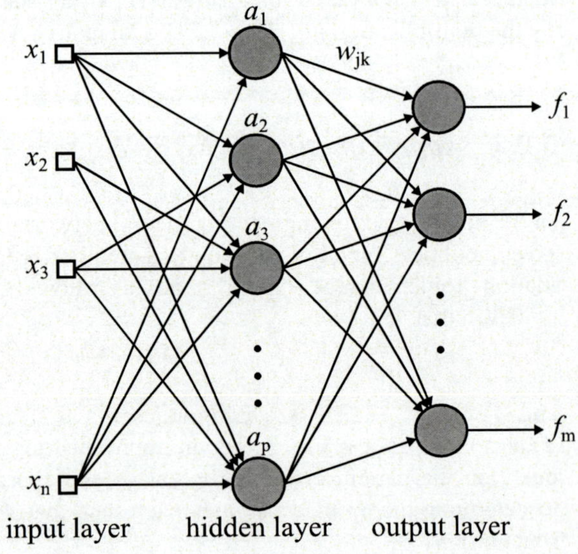

Fig. 8.1 Architecture of an RBFNN

An estimate of the function $\boldsymbol{f}(\boldsymbol{x})$, Eq. (8.6), can be obtained by:

$$\hat{\boldsymbol{f}}(\hat{\boldsymbol{w}}, \boldsymbol{x}) = \hat{\boldsymbol{w}}^{\mathrm{T}} \boldsymbol{a}(\boldsymbol{x}), \tag{8.9}$$

where $\hat{\boldsymbol{f}}(\boldsymbol{x})$ is the estimated output of RBFNN, $\boldsymbol{a}(\boldsymbol{x}) = [a_1(\boldsymbol{x}) \ \ldots \ a_{\mathrm{p}}(\boldsymbol{x})]^{\mathrm{T}}$ is the vector of GRBFs, and $\hat{\boldsymbol{w}}$ corresponds to the estimate of $\boldsymbol{w}$, which is calculated by some online weight update algorithm, i.e., the update will take place during the algorithm execution phase itself, not needing an offline training.

The Gaussian RBFNN has been quite successful in representing the complex nonlinear function. It has been shown that a linear superposition of GRBF gives an optimal mean-square approximation to an unknown function that is infinitely differentiable, the values of which are specified by a finite set of points in $\mathbb{R}^{\mathrm{n}}$. Furthermore, it has been proven that any continuous functions (not necessarily infinitely smooth) can be uniformly approximated by a linear combination of Gaussians [6, 12].

8.2.3 Modeling by RBFNNs

In this subsection, RBFNNs modeling using GL notation (Sect. 8.2.1) is treated, in which the GL matrix or vector is denoted by $\{\cdot\}$ and its term-to-term product operator is referred to by "$\bullet$", while the conventional matrix or vector is denoted by $[\cdot]$. This notation makes it possible to express mathematically a neural network emulator for matrices, as well as for vectors. For convenience and necessity, only the description of the neural network emulator for vectors is considered here, and a description of the GL notation including the neural network emulator for matrices can be found in more detail in [5].

Consider a given vector $\boldsymbol{F}(\boldsymbol{x})$, which depends on any variable expressed by "$\boldsymbol{x}$" only. By using GL notation, their elements $f_i(\boldsymbol{x})$ can be modeled with static neural networks as:

$$f_i(\boldsymbol{x}) = \sum_{j=1}^{\mathrm{p}} w_{ij} \xi_{ij}(\boldsymbol{x}) + \varepsilon_i(\boldsymbol{x}) = \boldsymbol{w}_i^{\mathrm{T}} \boldsymbol{\xi}_i(\boldsymbol{x}) + \varepsilon_i(\boldsymbol{x}), \tag{8.10}$$

with $w_{ij} \in \mathbb{R}$ being a weight of the RBFNN; $\xi_{ij}(\boldsymbol{x})$ being a GRBF; $\varepsilon_i(\boldsymbol{x}) \in \mathbb{R}$ being an approximation or modeling error of $f_i(\boldsymbol{x})$, which is assumed to be bounded. Also, $\boldsymbol{w}_i$ being the weight vector of the RBFNN; $\boldsymbol{\xi}_i(\boldsymbol{x})$ being the vector of GRBFs, and that its ith element expressed similarly to Eq. (8.8) is defined as:

$$\xi_{ij}(\boldsymbol{x}) = \exp\left(\frac{-(\boldsymbol{x} - \boldsymbol{c}_{ij})^{\mathrm{T}} (\boldsymbol{x} - \boldsymbol{c}_{ij})}{2\mu_{ij}^2} \right), \tag{8.11}$$

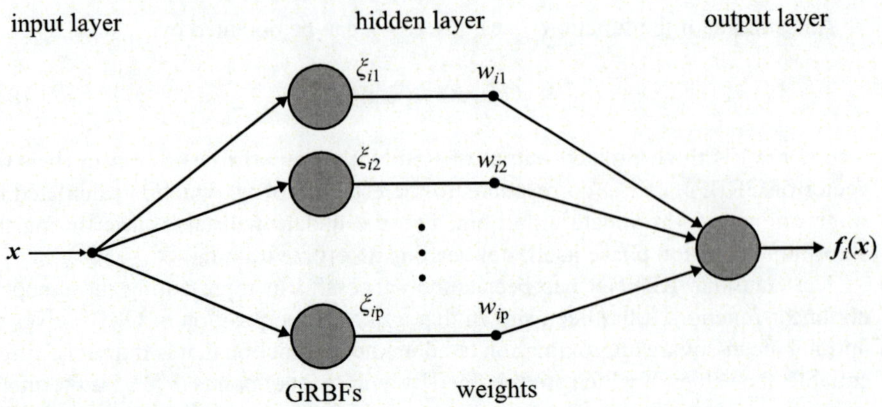

Fig. 8.2 Implementation of $f_i(\boldsymbol{x})$

where $\boldsymbol{c}_{ij} \in \mathbb{R}^{\mathrm{n}}$ is the vector of centers and $\mu_{ij}^2 \in \mathbb{R}$ is the variance. The RBFNN for $f_i(\boldsymbol{x})$, Eq. (8.10), is shown schematically in Fig. 8.2.

Based on the definitions for the GL vector and its operator, one obtains

$$\boldsymbol{F}(\boldsymbol{x}) = \left[\{\boldsymbol{w}\}^{\mathrm{T}} \bullet \{\boldsymbol{\xi}(\boldsymbol{x})\}\right] + \boldsymbol{E}(\boldsymbol{x}), \tag{8.12}$$

where the RBFNN emulator for $\boldsymbol{F}(\boldsymbol{x})$ is conveniently expressed as

$$\left[\{\boldsymbol{w}\}^{\mathrm{T}} \bullet \{\boldsymbol{\xi}(\boldsymbol{x})\}\right] = \begin{bmatrix} \boldsymbol{w}_1^{\mathrm{T}}\boldsymbol{\xi}_1(\boldsymbol{x}) \\ \vdots \\ \boldsymbol{w}_{\mathrm{n}}^{\mathrm{T}}\boldsymbol{\xi}_{\mathrm{n}}(\boldsymbol{x}) \end{bmatrix}, \tag{8.13}$$

with $\{\boldsymbol{w}\}$ and $\{\boldsymbol{\xi}(\boldsymbol{x})\}$ being GL vectors expressed as

$$\{\boldsymbol{w}\} = \{\boldsymbol{w}_1 \ \dots \ \boldsymbol{w}_{\mathrm{n}}\}^{\mathrm{T}}, \quad \text{and} \quad \{\boldsymbol{\xi}(\boldsymbol{x})\} = \left\{\boldsymbol{\xi}_1(\boldsymbol{x}) \ \dots \ \boldsymbol{\xi}_{\mathrm{n}}(\boldsymbol{x})\right\}^{\mathrm{T}}, \tag{8.14}$$

where $\boldsymbol{w}_i$ and $\boldsymbol{\xi}_i$ are regular vectors defined as

$$\boldsymbol{w}_i = \left[w_{i1} \ \dots \ w_{i\mathrm{p}}\right]^{\mathrm{T}}, \quad \boldsymbol{\xi}_i(\boldsymbol{x}) = \left[\xi_{i1}(\boldsymbol{x}) \ \dots \ \xi_{i\mathrm{p}}(\boldsymbol{x})\right]^{\mathrm{T}},$$

and $\boldsymbol{E}(\boldsymbol{x})$ being the corresponding approximation or modeling error vector for $\boldsymbol{F}(\boldsymbol{x})$ defined by:

$$\boldsymbol{E}(\boldsymbol{x}) = [\varepsilon_1(\boldsymbol{x}) \ \dots \ \varepsilon_{\mathrm{n}}(\boldsymbol{x})]^{\mathrm{T}}. \tag{8.15}$$

In GL notation, the estimation of the Eq. (8.12) can be written as:

$$\hat{\boldsymbol{F}}(\boldsymbol{x}) = \left[\{\hat{\boldsymbol{w}}\}^{\mathrm{T}} \bullet \{\boldsymbol{\xi}(\boldsymbol{x})\}\right], \tag{8.16}$$

where $\{\hat{\boldsymbol{w}}\}$ and $\{\boldsymbol{\xi}(\boldsymbol{x})\}$ are GL vectors, with their respective elements $\hat{\boldsymbol{w}}_i$ and $\boldsymbol{\xi}_i(\boldsymbol{x})$.

8.3 Control Design

In this section, the ANSMC design is based on the FOSMC discussed in Sect. 5.2.1 and the stability analysis of the closed-loop control system covered in Sect. 5.2.2. The ANSMC synthesis is similar to the FOSMC synthesis described in Sect. 5.2.3, emphasizing only that RBFNNs are used as replacements for the discontinuous components of the classical SMC [22] to mitigate the chattering phenomenon and to suppress uncertainties and disturbances.

8.3.1 Controller Synthesis

The chattering phenomenon in the control presented in Eqs. (5.36) and (5.49) is caused by the constant value of $\mathbf{G}$ and the discontinuous function $\text{sign}(\boldsymbol{\sigma}^*)$. To mitigate the chattering phenomenon an adaptive neural gain $\hat{\boldsymbol{F}}$ by applying the RBFNNs is provided. The RBFNNs are nonlinear and continuous functions used to approximate $\mathbf{G}\,\text{sign}(\boldsymbol{\sigma}^*)$ in Eqs. (5.36) and (5.49) [13, 16]. With this approximation, the ANSMC results in the new control input $\boldsymbol{v}_c$ as,

$$\begin{aligned}\boldsymbol{v}_c &= -\boldsymbol{B}_{0_\sigma}^{-1}\boldsymbol{A}_{0_\sigma} - \hat{\boldsymbol{F}}(\boldsymbol{\sigma}^*) - \mathbf{K}\boldsymbol{\sigma}^*,\\ \boldsymbol{v}_c &= -\boldsymbol{B}_{0_\sigma}^{-1}\boldsymbol{A}_{0_\sigma} - \left[\{\hat{\boldsymbol{w}}\}^{\mathrm{T}} \bullet \{\boldsymbol{\xi}(\boldsymbol{\sigma}^*)\}\right] - \mathbf{K}\boldsymbol{\sigma}^*,\end{aligned} \tag{8.17}$$

with $\hat{F}(\boldsymbol{\sigma}^*)$ being an n $\times$ 1 output vector of the RBFNNs in which $\hat{f}_i(\boldsymbol{\sigma}^*)$ is the output of the ith RBFNN, where $\{\hat{\boldsymbol{w}}\}$ and $\{\boldsymbol{\xi}(\boldsymbol{\sigma}^*)\}$ are GL vectors [5], and their respective elements are $\hat{\boldsymbol{w}}_i$, and $\boldsymbol{\xi}_i(\boldsymbol{\sigma}^*)$. Also, $\hat{\boldsymbol{F}}(\boldsymbol{\sigma}^*)$ can be rewritten as

$$\hat{\boldsymbol{F}}(\boldsymbol{\sigma}^*) = \left[\{\boldsymbol{w}\}^{\mathrm{T}} \bullet \left\{\boldsymbol{\xi}(\boldsymbol{\sigma}^*)\right\}\right] + \boldsymbol{E}(\boldsymbol{\sigma}^*) - [\{\tilde{\boldsymbol{w}}\}^{\mathrm{T}} \bullet \{\boldsymbol{\xi}(\boldsymbol{\sigma}^*)\}], \tag{8.18}$$

where $\boldsymbol{E}(\boldsymbol{\sigma}^*)$ define the neural network modeling errors vector, and their element is $\varepsilon_i(\boldsymbol{\sigma}^*)$, and the weight estimation error $\{\tilde{\boldsymbol{w}}\}$ is given as:

$$\begin{aligned}\{\tilde{\boldsymbol{w}}\} &= \{\boldsymbol{w}\} - \{\hat{\boldsymbol{w}}\},\\ \tilde{\boldsymbol{w}}_i &= \boldsymbol{w}_i - \hat{\boldsymbol{w}}_i.\end{aligned} \tag{8.19}$$

The learning law of RBFNNs is chosen as:

$$\dot{\hat{\boldsymbol{w}}}_i = \boldsymbol{\Gamma}_i\boldsymbol{\xi}_i(\boldsymbol{\sigma}^*)\sigma_i^* - \alpha_i\boldsymbol{\Gamma}_i|\boldsymbol{\sigma}^*|\hat{\boldsymbol{w}}_i, \tag{8.20}$$

where $\boldsymbol{\Gamma}_i = \boldsymbol{\Gamma}_i^{\mathrm{T}} > \mathbf{0}$ is a dimensional compatible symmetric positive definite matrix and $\alpha_i > 0$ is positive constant. The functions of the second term on the right side of Eq. (8.20) are to eliminate the condition of persistent excitation and to ensure bounded neural networks' weights estimates.

It is important to emphasize that control designs can be implemented based on the partitioning of RBFNNs into smaller sub-RBFNNs to obtain more efficient computation. Also, this implementation simplifies the design, provides added controller structure, and contributes to faster weight tuning algorithms (i.e., individual partitioned neural networks can be separately tuned), as can be seen in [12]. In this case, each unknown element, for example, of the matrices and vectors of the DWMR dynamics, may correspond to a sub-RBFNN. With this approach, it is possible to obtain the adjustment of the parameters of each subnet separately, thus producing a fast adaptation velocity. As related in [2, 6], some of the reasons for this fastness in the adaptation velocity of each sub-RBFNN are the result of its parameter adjustment being carried out linearly, assuming that the centers and variances of the GRBFs are fixed a priori; and of the receptive field concepts of the GRBFs. Still, in these works [2, 6] is reported that, for approximation problems, RBFNNs are very efficient when the dimension of the input vectors is reduced. An advantage of these partitioned RBFNNs is that if, for example, some elements of the matrices and vectors of the DWMR dynamics are well known, then their respective partitioned RBFNNs can be replaced by their deterministic equations. Thus, these partitioned RBFNNs can be used to reconstruct only the unknown elements or those more complicated to calculate.

The following mild bounding assumptions always hold in practical applications, in the development of the ANSMC, in the realization of the stability proof, and are required to proceed.

On any compact subset of $\mathbb{R}^{\mathrm{n}}$, the ideal RBFNN weights are bounded by:

$$|\{\boldsymbol{w}\}| \leq \mathrm{w}_{\max}, \tag{8.21}$$

with $\mathrm{w}_{\max} > 0$ being a positive constant.

Defining $\{\boldsymbol{w}\}$ so that $\boldsymbol{F}(\boldsymbol{\sigma}^*) = [\{\boldsymbol{w}\}^{\mathrm{T}} \bullet \{\boldsymbol{\xi}(\boldsymbol{\sigma}^*)\}]$ is the optimal compensation for $\tilde{\boldsymbol{d}}_0$, then, $\boldsymbol{E}(\boldsymbol{\sigma}^*) = \boldsymbol{0}$ in the Eq. (8.18). According to the property of universal approximation of RBFNNs [11, 25], there exists $\psi > 0$ satisfying

$$\begin{aligned} \left|\tilde{\boldsymbol{d}}_0 - \boldsymbol{F}(\boldsymbol{\sigma}^*)\right| &\leq \psi, \\ \left|\boldsymbol{v}_{\mathrm{e}} - \left[\{\boldsymbol{w}\}^{\mathrm{T}} \bullet \{\boldsymbol{\xi}(\boldsymbol{\sigma}^*)\}\right]\right| &\leq \psi, \end{aligned} \tag{8.22}$$

where ψ is arbitrary and can be chosen as small as possible.

8.3.2 Stability Analysis

For the stability analysis, one can choose the Lyapunov function candidate as

$$V = \frac{1}{2}\left(\boldsymbol{\sigma}^{\mathrm{T}}\boldsymbol{\sigma} + \sum_{i=1}^{\mathrm{n}} \tilde{\boldsymbol{w}}_i^{\mathrm{T}} \boldsymbol{\Gamma}_i^{-1} \tilde{\boldsymbol{w}}_i\right). \tag{8.23}$$

Differentiating Eq. (8.23), considering the Eqs. (5.14) and (5.26), $\dot{V}$ is obtained as

$$\dot{V} = \boldsymbol{\sigma}^{*^{\mathrm{T}}} \boldsymbol{B}_{0_\sigma}^{-1} \boldsymbol{A}_{0_\sigma} + \boldsymbol{\sigma}^{*^{\mathrm{T}}} \boldsymbol{v}_{\mathrm{c}} + \boldsymbol{\sigma}^{*^{\mathrm{T}}} \tilde{\boldsymbol{d}}_0 - \sum_{i=1}^{\mathrm{n}} \tilde{\boldsymbol{w}}_i^{\mathrm{T}} \boldsymbol{\Gamma}_i^{-1} \dot{\hat{\boldsymbol{w}}}_i. \tag{8.24}$$

Using the ANSMC, Eq. (8.17), replacing Eq. (8.18) with $\boldsymbol{E}(\boldsymbol{\sigma}^*) = \mathbf{0}$ due to the compensator $\boldsymbol{F}(\boldsymbol{\sigma}^*)$ being optimal, and considering Eq. (8.19), $\dot{V}$ yields to:

$$\dot{V} = -\boldsymbol{\sigma}^{*^{\mathrm{T}}} \Big[\{\boldsymbol{w}\}^{\mathrm{T}} \bullet \{\boldsymbol{\xi}(\boldsymbol{\sigma}^*)\} \Big] + \boldsymbol{\sigma}^{*^{\mathrm{T}}} \Big[\{\tilde{\boldsymbol{w}}\}^{\mathrm{T}} \bullet \{\boldsymbol{\xi}(\boldsymbol{\sigma}^*)\} \Big] - \boldsymbol{\sigma}^{*^{\mathrm{T}}} \mathbf{K} \boldsymbol{\sigma}^* + \boldsymbol{\sigma}^{*^{\mathrm{T}}} \tilde{\boldsymbol{d}}_0 - \sum_{i=1}^{\mathrm{n}} \tilde{\boldsymbol{w}}_i^{\mathrm{T}} \boldsymbol{\Gamma}_i^{-1} \dot{\hat{\boldsymbol{w}}}_i. \tag{8.25}$$

Recall that

$$\boldsymbol{\sigma}^{*^{\mathrm{T}}} \left[\{\tilde{\boldsymbol{w}}\}^{\mathrm{T}} \bullet \{\boldsymbol{\xi}(\boldsymbol{\sigma}^*)\} \right] = \sum_{i=1}^{\mathrm{n}} \tilde{\boldsymbol{w}}_i^{\mathrm{T}} \boldsymbol{\xi}_i(\boldsymbol{\sigma}^*) \sigma_i^*, \tag{8.26}$$

and replacing the learning law, Eq. (8.20), and considering the Eq. (8.22), $\dot{V}$ results in:

$$\dot{V} \le -\boldsymbol{\sigma}^{*^{\mathrm{T}}} \mathbf{K} \boldsymbol{\sigma}^* + \boldsymbol{\sigma}^{*^{\mathrm{T}}} \left| \boldsymbol{v}_{\mathrm{e}} - \Big[\{\boldsymbol{w}\}^{\mathrm{T}} \bullet \{\boldsymbol{\xi}(\boldsymbol{\sigma}^*)\} \Big] \right| + \sum_{i=1}^{\mathrm{n}} \tilde{\boldsymbol{w}}_i^{\mathrm{T}} \alpha_i \, |\sigma_i| \, \hat{\boldsymbol{w}}_i, \tag{8.27}$$

or

$$\dot{V} \le -\boldsymbol{\sigma}^{*^{\mathrm{T}}} \mathbf{K} \boldsymbol{\sigma}^* + \boldsymbol{\sigma}^{*^{\mathrm{T}}} \psi + \delta \left| \boldsymbol{\sigma}^* \right| \sum_{i=1}^{\mathrm{n}} \tilde{\boldsymbol{w}}_i^{\mathrm{T}} \hat{\boldsymbol{w}}_i, \tag{8.28}$$

where $\delta > 0$ is a positive constant.

Observing that $\mathrm{tr}(\{\tilde{\boldsymbol{w}}\}^{\mathrm{T}}\{\hat{\boldsymbol{w}}\}) = \sum_{i=1}^{\mathrm{n}} \tilde{\boldsymbol{w}}_i^{\mathrm{T}} \hat{\boldsymbol{w}}_i$ stays as follows:

$$\dot{V} \le -\boldsymbol{\sigma}^{*^{\mathrm{T}}} \mathbf{K} \boldsymbol{\sigma}^* + \boldsymbol{\sigma}^{*^{\mathrm{T}}} \psi + \delta \left| \boldsymbol{\sigma}^* \right| \mathrm{tr} \left(\{\tilde{\boldsymbol{w}}\}^{\mathrm{T}} \{\hat{\boldsymbol{w}}\} \right), \tag{8.29}$$

with $\mathrm{tr}(\cdot)$ being trace function.

By using the Eq. (8.21) and the Schwartz inequality [11, 25], the trace function can be written as follows:

$$\begin{aligned} \mathrm{tr} \left(\{\tilde{\boldsymbol{w}}\}^{\mathrm{T}} \{\hat{\boldsymbol{w}}\} \right) &= \mathrm{tr} \left(\{\tilde{\boldsymbol{w}}\}^{\mathrm{T}} \left(\{\boldsymbol{w}\} - \{\tilde{\boldsymbol{w}}\} \right) \right) = \langle \{\tilde{\boldsymbol{w}}\}\{\boldsymbol{w}\} \rangle - \left| \{\tilde{\boldsymbol{w}}\}^{\mathrm{T}} \right|^2 \\ &\le |\{\tilde{\boldsymbol{w}}\}| \, |\{\boldsymbol{w}\}| - |\{\tilde{\boldsymbol{w}}\}|^2 \le |\{\tilde{\boldsymbol{w}}\}| \, \mathrm{w}_{\max} - |\{\tilde{\boldsymbol{w}}\}|^2, \end{aligned} \tag{8.30}$$

and replacing the Eq. (8.30) into Eq. (8.29), $\dot{V}$ leads to the following:

$$\dot{V} \leq -\lambda_{\min}\{\mathbf{K}\}\left|\boldsymbol{\sigma}^*\right|^2 + \overline{\psi}_{\max}\left|\boldsymbol{\sigma}^*\right| + \delta\left|\boldsymbol{\sigma}^*\right|\left(\left|\{\tilde{\boldsymbol{w}}\}\right| \mathrm{w}_{\max} - \left|\{\tilde{\boldsymbol{w}}\}\right|^2\right), \tag{8.31}$$

where $\lambda_{\min}\{\mathbf{K}\}$ is the minimum singular value of $\mathbf{K}$ and $\overline{\psi}_{\max} > |\psi|$ is a known scalar that represents the maximum effect of the uncertainties and disturbances.

Making the necessary mathematical manipulations in the Eq. (8.31), $\dot{V}$ yields to:

$$\dot{V} \leq -\left|\boldsymbol{\sigma}^*\right|\left(\lambda_{\min}\{\mathbf{K}\}\left|\boldsymbol{\sigma}^*\right| - \overline{\psi}_{\max} + \delta\left(\left|\{\tilde{\boldsymbol{w}}\}\right| - \frac{\mathrm{w}_{\max}}{2}\right)^2 - \delta\frac{\mathrm{w}_{\max}^2}{4}\right). \tag{8.32}$$

Thus, $\dot{V}$ is guaranteed negative as long as the term in parenthesis in Eq. (8.32) is positive, either:

$$\left|\boldsymbol{\sigma}^*\right| > \frac{\delta\frac{\mathrm{w}_{\max}^2}{4} + \overline{\psi}_{\max}}{\lambda_{\min}\{\mathbf{K}\}}, \tag{8.33}$$

or

$$\left|\{\tilde{\boldsymbol{w}}\}\right| > \frac{\mathrm{w}_{\max}}{2} + \sqrt{\frac{\mathrm{w}_{\max}^2}{4} + \frac{\overline{\psi}_{\max}}{\delta}}. \tag{8.34}$$

Therefore, $\dot{V}$ is negative definite within a particular compact set and is negative semidefinite outside this set, which is defined by Eqs. (8.33) and (8.34). The stability of the global control system is ensured since $\dot{V}$ is guaranteed negative definite. According to a standard Lyapunov theory and LaSalle's Theorem [9, 10], all signals $|\boldsymbol{\sigma}^*|$ and $|\{\tilde{\boldsymbol{w}}\}|$ are UUB. Thus, the proposed control system is stable. Still, from Eq. (8.33) it is possible to verify that $|\boldsymbol{\sigma}^*|$ can be kept arbitrarily small by the convenient choice of the gain $\mathbf{K}$. This gain plays an important role in the control law (Eq. (8.17)) because in the transient state the weights of the RBFNNs are not properly adjusted, which can move the states of the system away from the desired point. Therefore, this gain will prevent states from moving away from the desired point while the RBFNNs are trained or the weights of the RBFNNs are adjusted. On the other hand, Eq. (8.34) reveals that the weight estimation errors of the RNFBRs $\{\tilde{\boldsymbol{w}}\}$ are fundamentally bounded by $\mathrm{w}_{\max}$. The parameter δ offers a design trade-off between the possible relative magnitudes of $|\boldsymbol{\sigma}^*|$ and $|\{\tilde{\boldsymbol{w}}\}|$. Finally, the right sides of Eqs. (8.33) and (8.34) can be taken as practical bounds of $\boldsymbol{\sigma}^*$ and $\{\tilde{\boldsymbol{w}}\}$, respectively.

8.4 Simulations Using Matlab And/or MobileSim Simulator

To verify the performance of the ANSMC given by the control law of Eq. (8.17) and the learning law of Eq. (8.20), results in the three scenarios are analyzed considering the eight-shape reference trajectory described in Sect. 3.1.2 with the gains of the ANSMC being presented in Table 8.1. For science and knowledge only, the

Table 8.1 ANSMC gains and parameters

Controller	λ_1	λ_2	λ_3	k_1	k_2	n	p	Γ	α	μ^2
ANSMC	1.5	6.0	1.0	0.1	0.1	2	25	1	1×10^{-6}	0.001

parameters n and p in Table 8.1 correspond respectively to the number of neural network inputs (in this case, the inputs are σ_1^* and σ_2^*) and numbers of neural network neurons.

It should be noted that for the ANSMC, the centers of the localized GRBFs are evenly distributed to span the input space of the neural networks [20]. Moreover, the estimated weights of the RBFNNs $\hat{\boldsymbol{w}}$ were initialized to zero without any prior knowledge of the system uncertainties and disturbances, and the training of the RBFNNs is carried out online.

8.4.1 Ideal Scenario

After simulations carried out with ANSMC in Matlab/Simulink the RMS of the errors obtained are shown in Table 8.2.

Looking at Figs. 8.3, 8.4, and 8.5, the DWMR tracks the reference trajectory, the DWMR posture trackings also converge to the reference posture trackings and posture tracking errors tend to converge to zero, with the orientation error θ_e there is a slight oscillation around zero during the execution of maneuvers in curves of the reference trajectory.

The control velocities, as well as the velocities promoted by the DWMR that tend to converge to the reference velocities, present smooth behaviors as can be seen in Figs. 8.6 and 8.7.

Analyzing Fig. 8.8 it is possible to verify that the control velocities $\boldsymbol{v}_c^*$ remained very close to zero, or better, they remained at zero, not providing the control efforts required to compensate for auxiliary velocity tracking errors.

The sliding surfaces (Fig. 8.9), the new sliding surfaces (Fig. 8.10), as well as their respective derivatives (Figs. 8.11 and 8.12), converged to zero with the chattering phenomenon well minimized.

Observing Fig. 8.13, it appears that the estimated weights of the RBFNNs $\hat{\boldsymbol{w}}$ were very close to zero or were at zero and, therefore, not making it possible to provide the control efforts to carry out compensation of the auxiliary velocity tracking errors.

Table 8.2 RMS of the errors—results in the ideal scenario

Controller	x_e [m]	y_e [m]	θ_e [rad]
ANSMC	0.0304	0.0656	0.0636

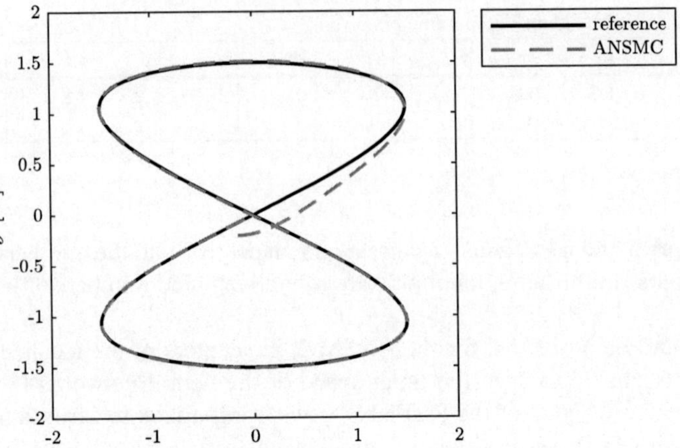

Fig. 8.3 Trajectory tracking in the ideal scenario

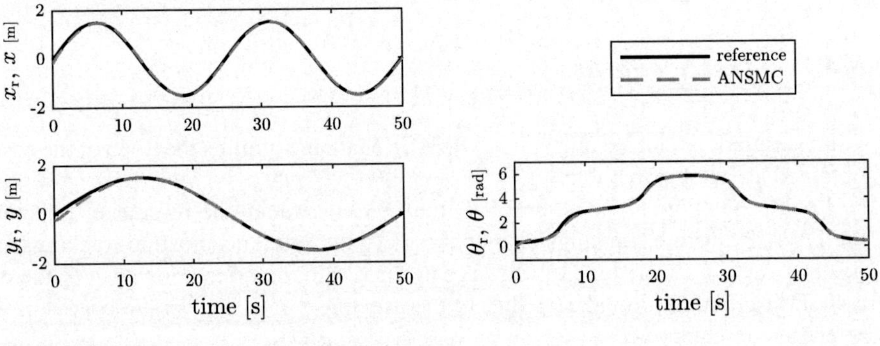

Fig. 8.4 Tracking (x, y, and θ) in the ideal scenario

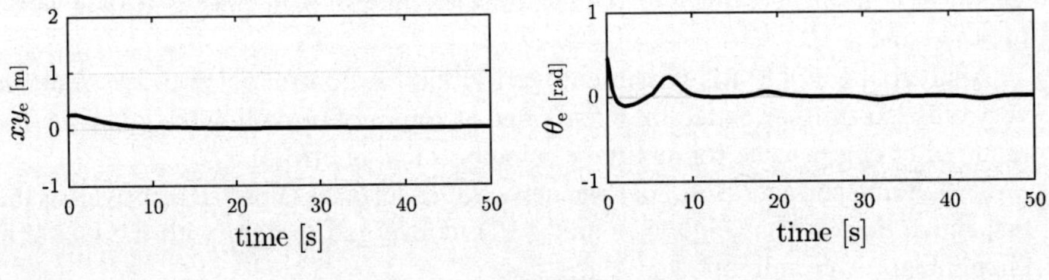

Fig. 8.5 Posture tracking errors, with $xy_e = \sqrt{x_e^2 + y_e^2}$, in the ideal scenario

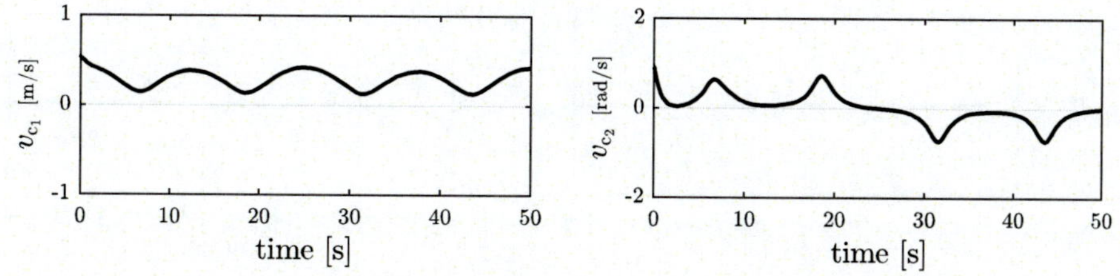

Fig. 8.6 Control velocities in the ideal scenario

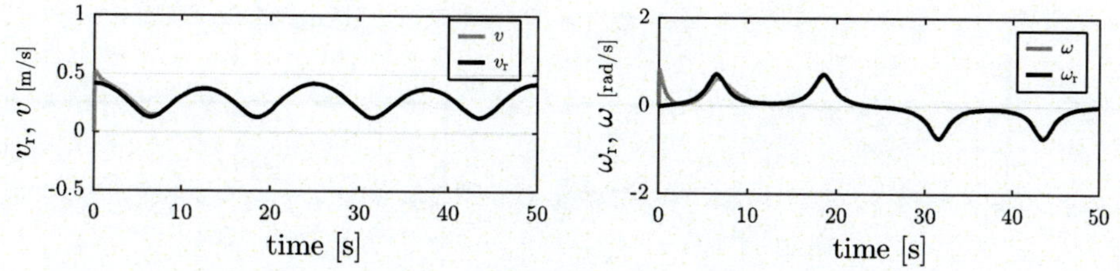

Fig. 8.7 Velocities of the DWMR and reference velocities in the ideal scenario

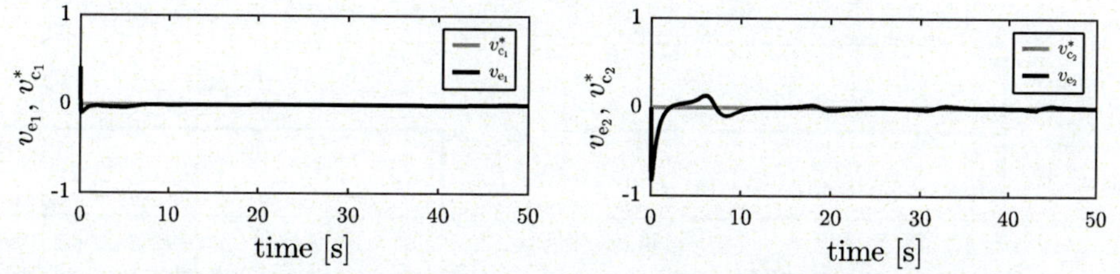

Fig. 8.8 Auxiliary velocity tracking errors and control velocities $\boldsymbol{v}_c^*$ in the ideal scenario

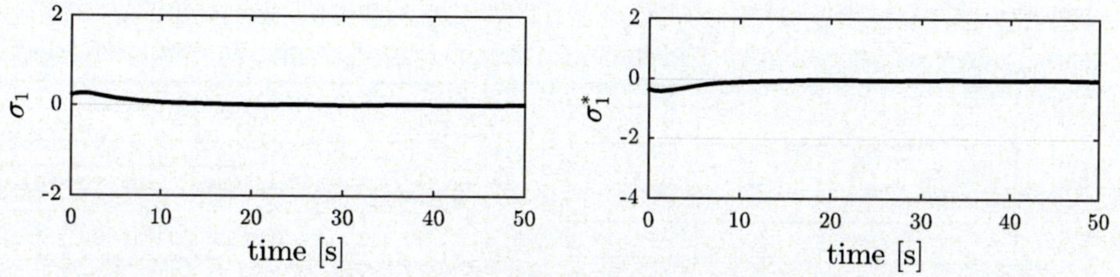

Fig. 8.9 Sliding surface σ_1 and new sliding surface σ_1^* in the ideal scenario

8.4.2 Realistic Scenario

In this scenario, after simulations, the RMS of the errors are presented in Table 8.3, in which there was a gradual increase in these errors when compared to the errors in Table 8.2. This occurrence is reflected: in the trajectory tracking by the DWMR (Fig. 8.14) presenting a certain performance degradation; in the posture trackings (Fig. 8.15) performed by the DWMR with the tendency of convergence to their

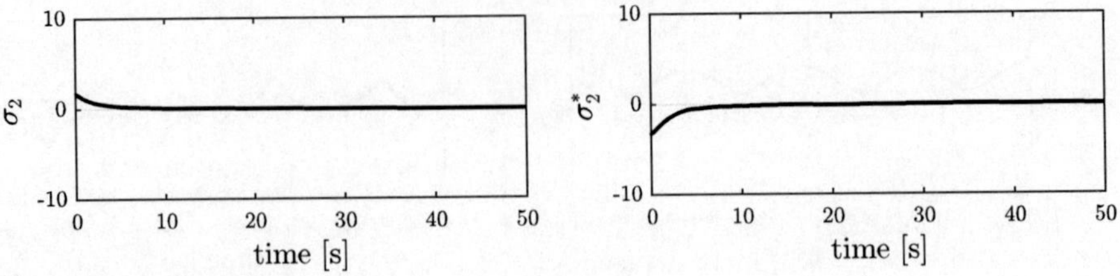

Fig. 8.10 Sliding surface σ_2 and new sliding surface σ_2^* in the ideal scenario

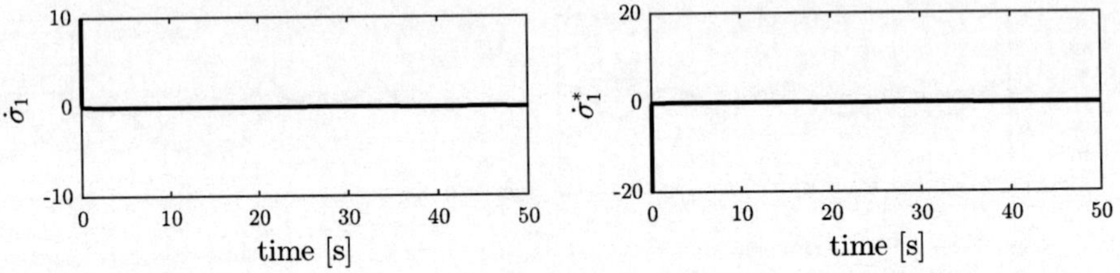

Fig. 8.11 Derivatives of sliding surface σ_1 and derivatives of new sliding surface σ_1^* in the ideal scenario

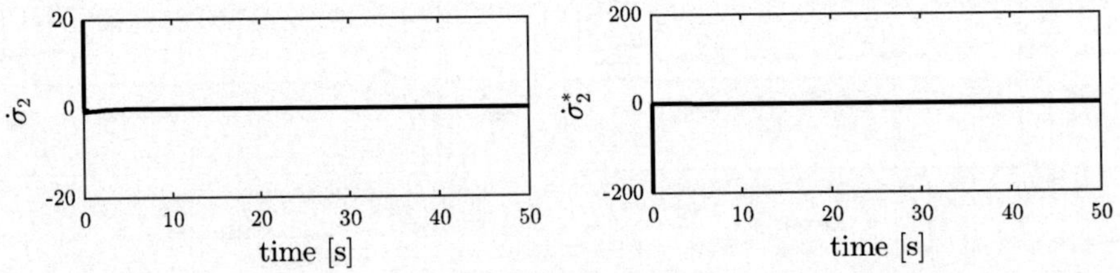

Fig. 8.12 Derivatives of sliding surface σ_2 and derivatives of new sliding surface σ_2^* in the ideal scenario

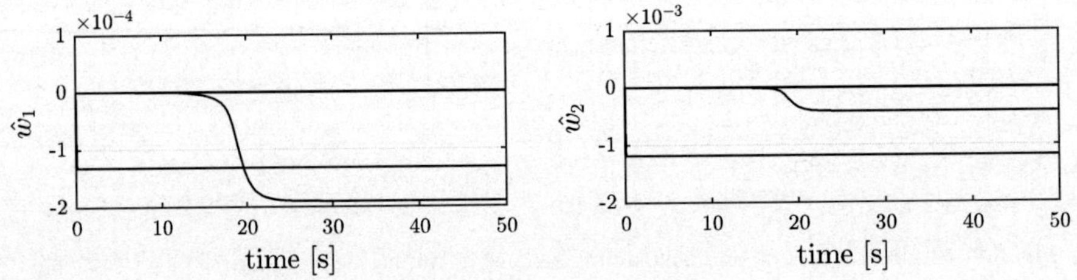

Fig. 8.13 Estimated weights $\hat{\boldsymbol{w}}$ of the RBFNNs in the ideal scenario

Table 8.3 RMS of the errors—results in the realistic scenario

Controller	x_e [m]	y_e [m]	θ_e [rad]
ANSMC	0.0389	0.0687	0.0711

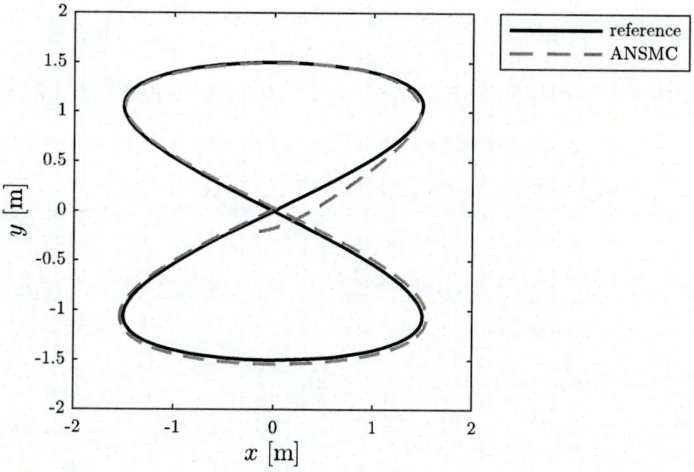

Fig. 8.14 Trajectory tracking in the realistic scenario

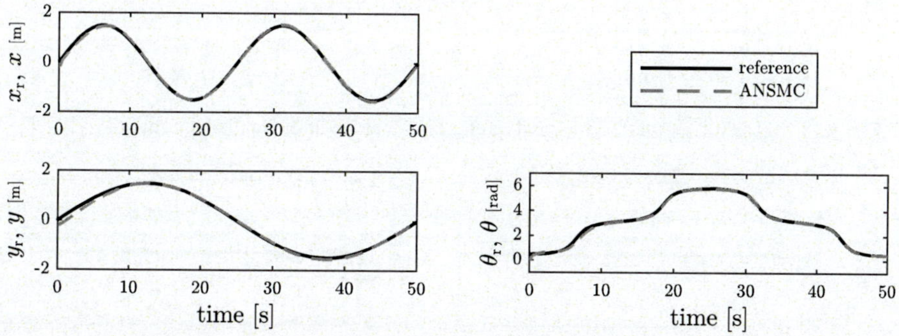

Fig. 8.15 Tracking (x, y, and θ) in the realistic scenario

references; and in the posture tracking errors (Fig. 8.16), being more noticeable in the orientation error θ_e, which presents a slight oscillation around zero during the execution of curves of the reference trajectory.

The smooth behavior with the chattering phenomenon considerably minimized the control velocities, as well as of the velocities performed by the DWMR about the reference velocities can be verified in Figs. 8.17 and 8.18.

As in the ideal scenario, the control velocities $\boldsymbol{v}_c^*$ (Fig. 8.19) remained at zero, that is, no compensation of the auxiliary velocity tracking errors.

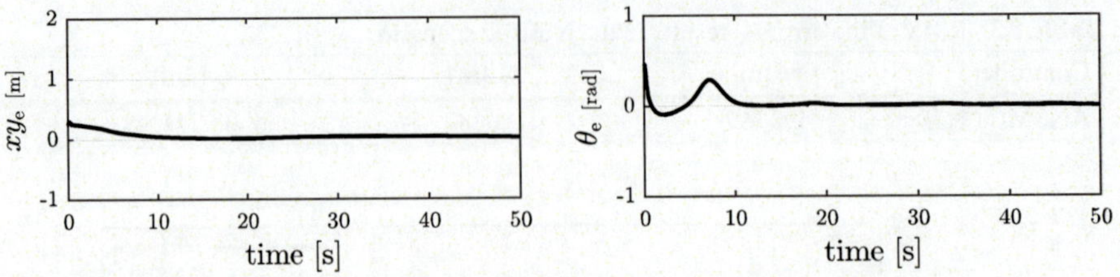

Fig. 8.16 Posture tracking errors, with $xy_e = \sqrt{x_e^2 + y_e^2}$, in the realistic scenario

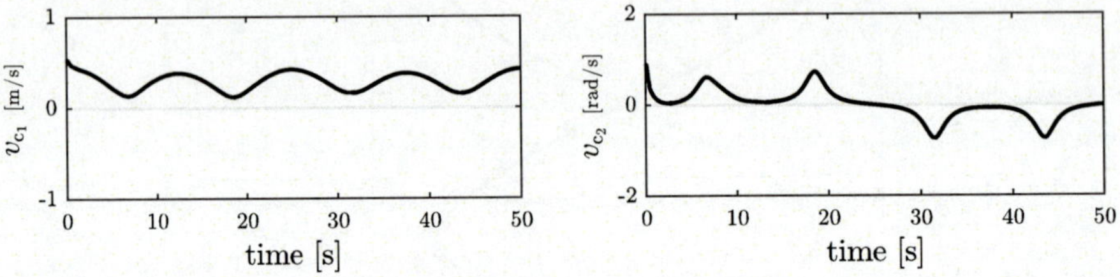

Fig. 8.17 Control velocities in the realistic scenario

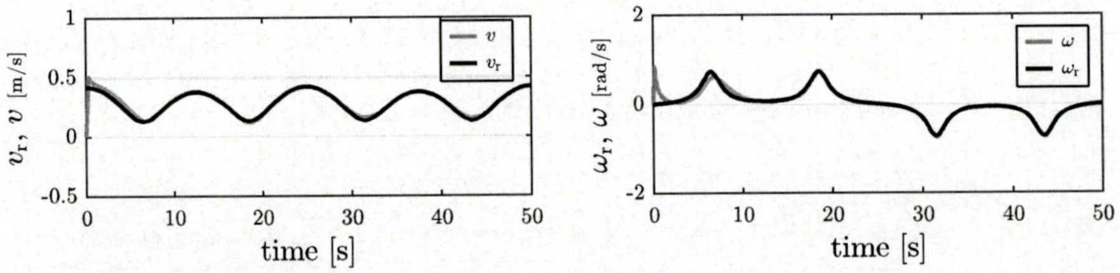

Fig. 8.18 Velocities of the DWMR and reference velocities in the realistic scenario

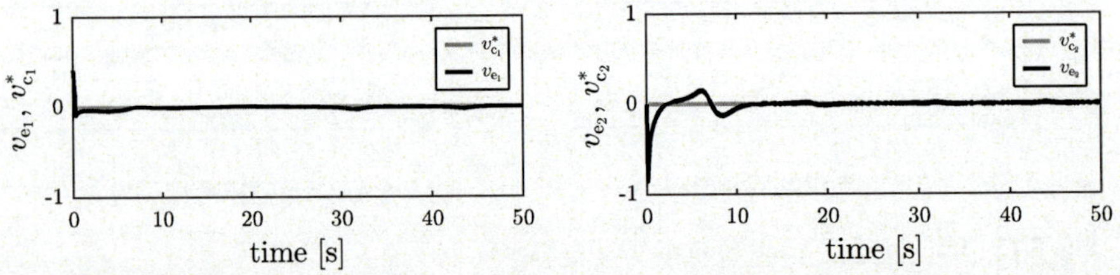

Fig. 8.19 Auxiliary velocity tracking errors and control velocities $\boldsymbol{v}_c^*$ in the realistic scenario

As noted in Figs. 8.20, 8.21, 8.22, and 8.23, the sliding surfaces and the new sliding surfaces tend to converge to zero while their respective derivatives tend to oscillate close to zero due to the chattering phenomenon considerably mitigated.

As already verified in the ideal scenario, the estimated weights of the RBFNNs $\hat{\boldsymbol{w}}$ (Fig. 8.24) also stayed with values very close to zero or stayed with values equal to zero, not providing, therefore, the compensation for the auxiliary velocity tracking errors.

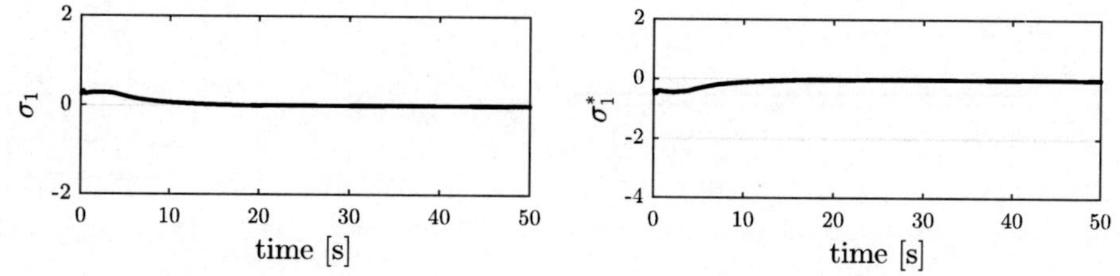

Fig. 8.20 Sliding surface σ_1 and new sliding surface σ_1^* in the realistic scenario

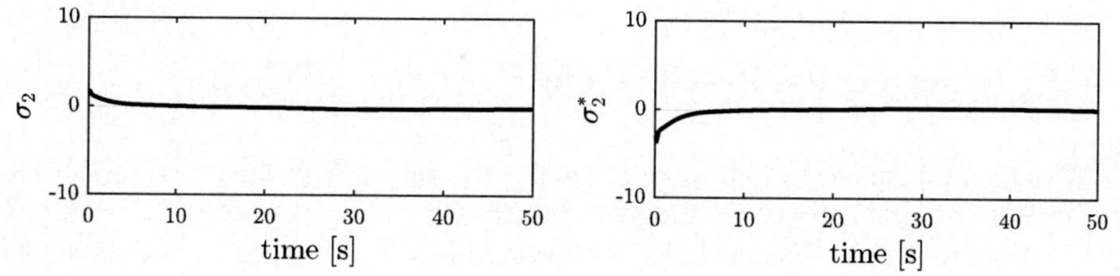

Fig. 8.21 Sliding surface σ_2 and new sliding surface σ_2^* in the realistic scenario

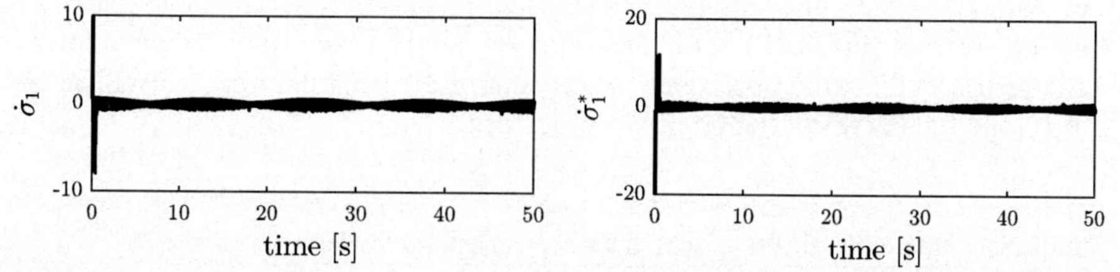

Fig. 8.22 Derivatives of sliding surface σ_1 and derivatives of new sliding surface σ_1^* in the realistic scenario

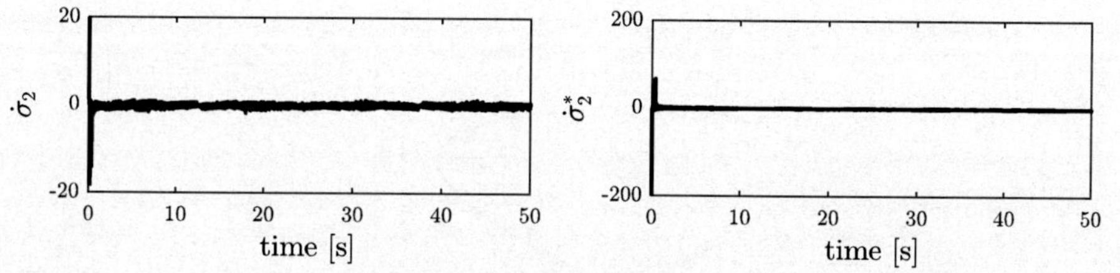

Fig. 8.23 Derivatives of sliding surface σ_2 and derivatives of new sliding surface σ_2^* in the realistic scenario

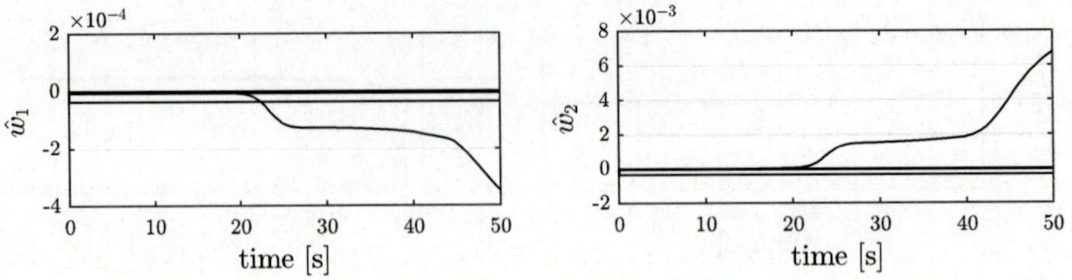

Fig. 8.24 Estimated weights $\hat{\boldsymbol{w}}$ of the RBFNNs in the realistic scenario

8.5 Experimental Results Using PowerBot DWMR

With the performance of the experimental tests, the RMS of the errors in Table 8.4 increased significantly concerning the RMS of the errors obtained in Tables 8.2 and 8.3. With this, it can be concluded, observing Fig. 8.25, that there is a considerable deterioration of the ANSMC's performance in the trajectory tracking performed by the DWMR. This deterioration can also be seen in the first 20 s for both posture trackings (Fig. 8.26) made by DWMR considering the reference trackings and posture tracking errors (Fig. 8.27), which respectively tend to converge to their references and zero after that time. Moreover, the orientation error θ_e continues to oscillate close to zero in the curves of the reference trajectory.

Table 8.4 RMS of the errors—results in the experimental scenario

Controller	x_e [m]	y_e [m]	θ_e [rad]
ANSMC	0.2820	0.2080	0.2650

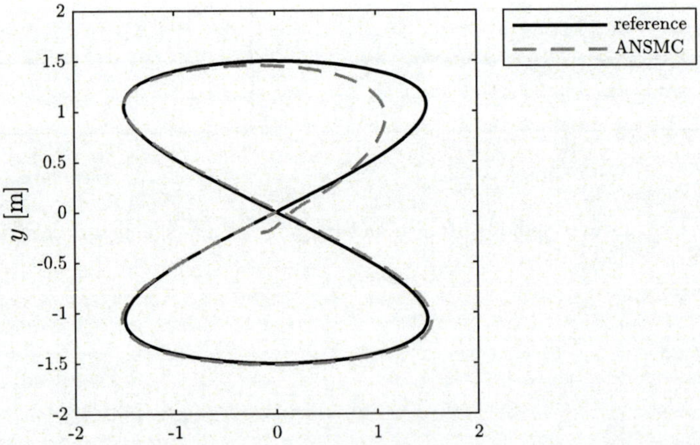

Fig. 8.25 Trajectory tracking in the experimental scenario

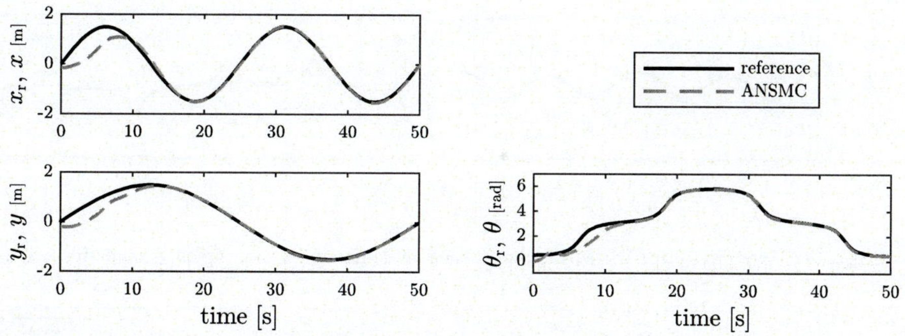

Fig. 8.26 Tracking (x, y, and θ) in the experimental scenario

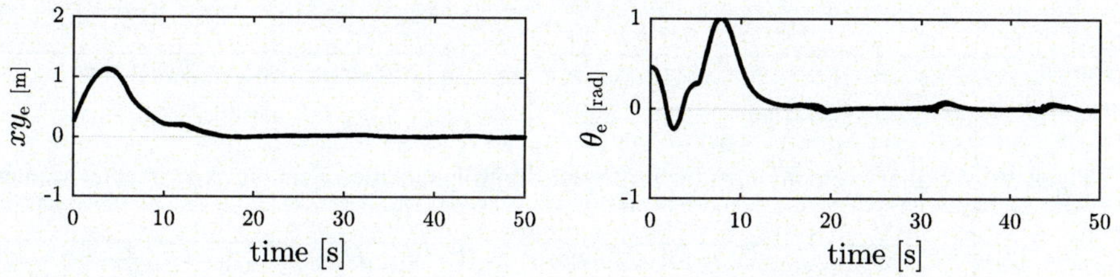

Fig. 8.27 Posture tracking errors, with $xy_e = \sqrt{x_e^2 + y_e^2}$, in the experimental scenario

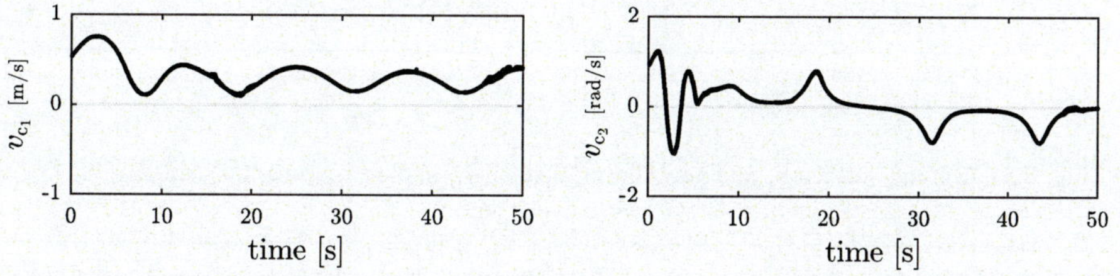

Fig. 8.28 Control velocities in the experimental scenario

By Figs. 8.28 and 8.29, it can be seen that despite the control velocities and the velocities exerted by the DWMR, they have profiles with smooth behaviors. Besides, the control velocities do not provide the necessary control efforts for satisfactory trajectory tracking, especially in the first 20 s. Still, the velocities of the DWMR tend to converge to their references after that time.

The compensations, that are given by the control velocities $\boldsymbol{v}_c^*$, do not provide the necessary control efforts to try to cancel the auxiliary velocity tracking errors as shown in Fig. 8.30 throughout the experiment, but noticeable in the first 20 s.

Looking the Figs. 8.31, 8.32, 8.33, and 8.34, it can be seen that the sliding surfaces and the new sliding surfaces do not converge to zero in the first 20 s and, after that time, they are very slightly oscillating around zero. However, the derivatives of these surfaces have the chattering phenomenon significantly attenuated, oscillating around zero.

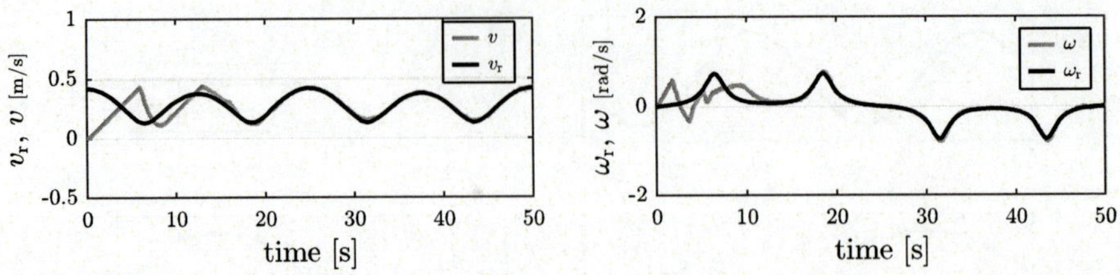

Fig. 8.29 Velocities of the DWMR and reference velocities in the experimental scenario

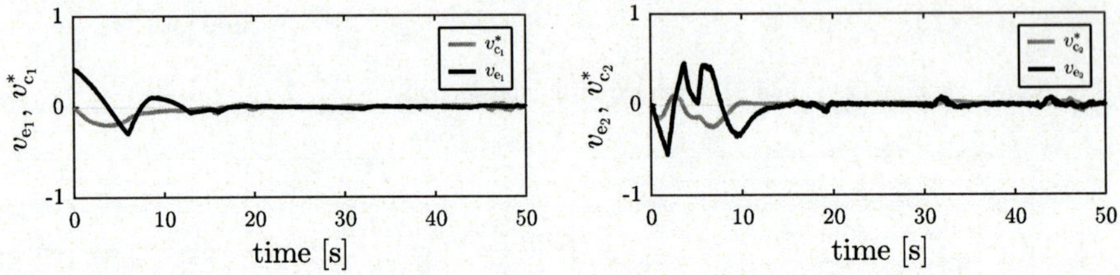

Fig. 8.30 Auxiliary velocity tracking errors and control velocities $\boldsymbol{v}_c^*$ in the experimental scenario

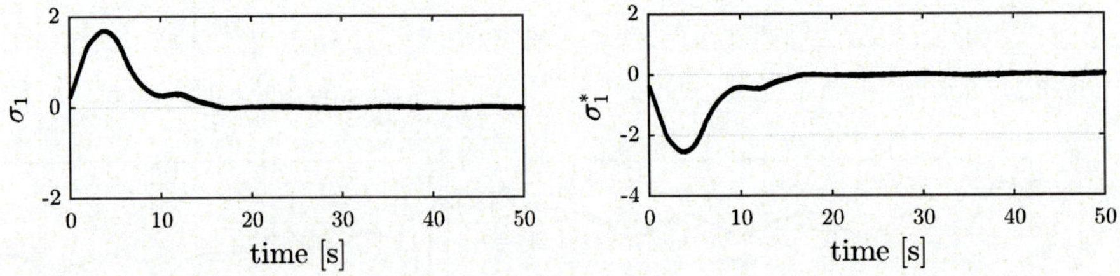

Fig. 8.31 Sliding surface σ_1 and new sliding surface σ_1^* in the experimental scenario

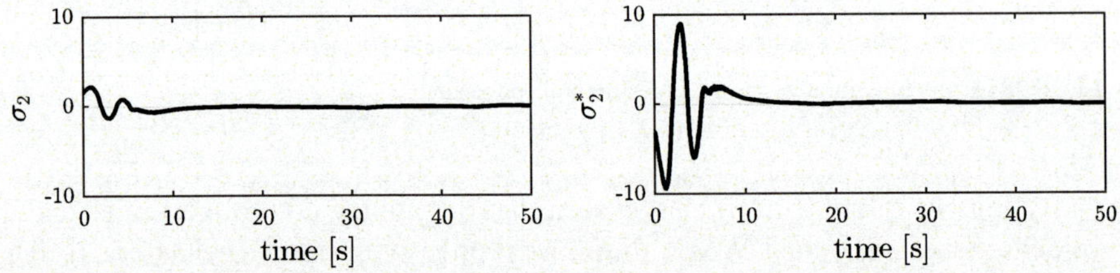

Fig. 8.32 Sliding surface σ_2 and new sliding surface σ_2^* in the experimental scenario

Concerning the estimated weights of the RBFNNs $\hat{\boldsymbol{w}}$ shown in Fig. 8.35, as previously reported in the ideal and realistic scenarios, these remain with values very close to zero or with values equal to zero so that the not to generate the necessary control efforts to compensate the auxiliary velocity tracking errors.

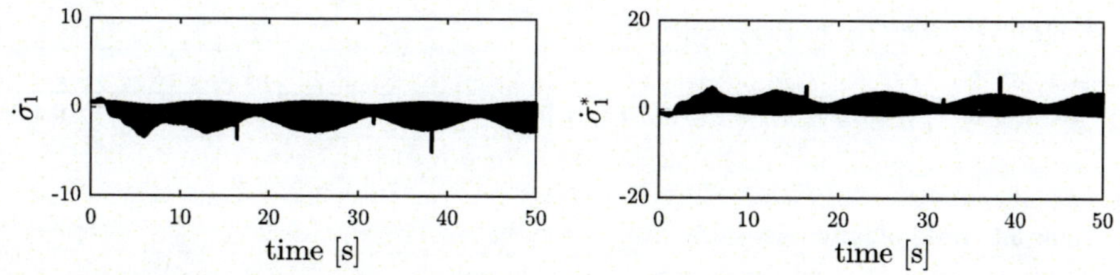

Fig. 8.33 Derivatives of sliding surface σ_1 and derivatives of new sliding surface σ_1^* in the experimental scenario

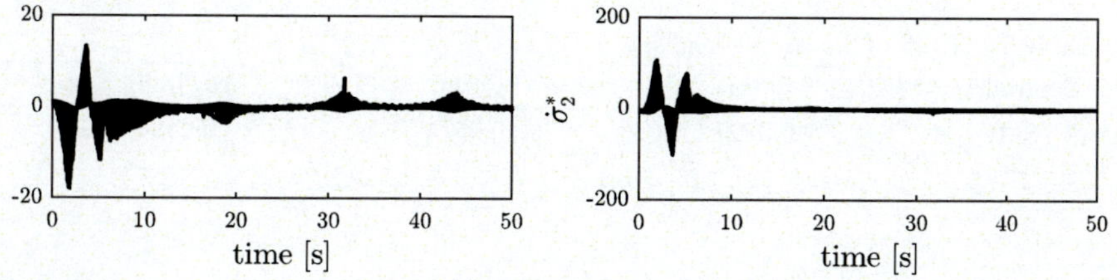

Fig. 8.34 Derivatives of sliding surface σ_2 and derivatives of new sliding surface σ_2^* in the experimental scenario

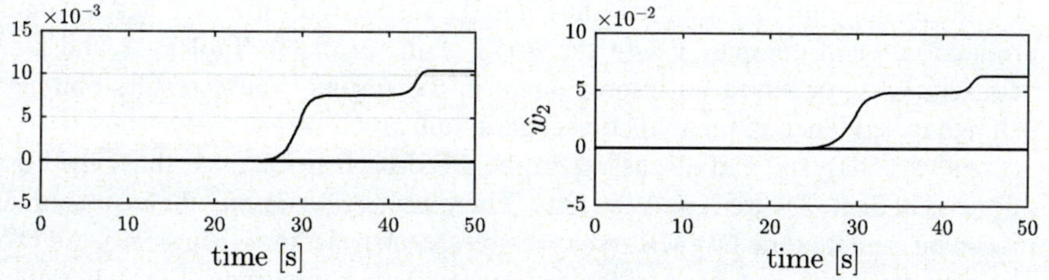

Fig. 8.35 Estimated weights $\hat{\boldsymbol{w}}$ of the RBFNNs in the experimental scenario

8.6 Analysis and Discussion of Results

Given the results obtained for the three scenarios, it is possible to state that the ANSMC's performance suffered considerable and perceptible degradation in tracking the reference trajectory, as can also be seen with the increase in the RMS of the errors in the realistic and experimental scenarios considering the ideal scenario, these errors being more significant when performing the experimental tests.

Several factors can influence the achievement of different tracking performances, such as the adjustment of gains and parameters; the size of the RBFNNs (number of neurons or number of activation functions in the hidden layer); the definition of the centers, and variances of the GRBFs. Thus, one of the possible reasons for the ANSMC performance degradation to occur is the choice of gains $\boldsymbol{\Gamma}$ and α of the learning law in Eq. (8.20), and variance μ^2 of the GRBFs in Eq. (8.11), which did not provide the necessary control efforts to compensate the auxiliary velocity tracking

Table 8.5 ANSMC gains and parameters

Controller	λ_1	λ_2	λ_3	k_1	k_2	n	p	Γ	α	μ^2
ANSMC	1.5	6.0	1.0	0.1	0.1	2	25	0.5	0.1	0.1

Table 8.6 RMS of the errors—ideal

Controller	x_e [m]	y_e [m]	θ_e [rad]
ANSMC	0.0241	0.0427	0.0558

Table 8.7 RMS of the errors—realistic

Controller	x_e [m]	y_e [m]	θ_e [rad]
ANSMC	0.0397	0.0524	0.0584

errors. To show the effectiveness of the ANSMC and improve its performance, these gains and parameters are redefined as can be seen in Table 8.5.

Using these redefined gains and parameters, most of the RMS of the errors in the ideal and realistic scenarios, shown in the respective Tables 8.6 and 8.7, suffered reduction when compared with the RMS of the errors in Tables 8.2 and 8.3 and, consequently, occurred an improvement in the results. These results obtained are shown in sequence in the realistic scenario only.

As was analyzed and discussed for the AFSMC-B in Chap. 7, the same observations as in Sect. 7.6 are replicated here. Therefore, by observing the results obtained, it can be seen that the DWMR tracks the reference trajectory (Fig. 8.36), the DWMR posture trackings converge to the reference posture trackings (Fig. 8.37), the posture tracking errors tend to converge to zero (Fig. 8.38), the behaviors of the control velocity profiles are smooth (Fig. 8.39), the velocities of the DWMR converge to the reference velocities (Fig. 8.40), the compensation control portions exhibit opposite and similar behaviors to auxiliary velocity tracking errors (Fig. 8.41), the sliding surfaces (Fig. 8.42) and the new sliding surfaces (Fig. 8.43) tend to converge to zero, the behaviors of the derivatives of the sliding surfaces (Fig. 8.44) and the new sliding surfaces (Fig. 8.45) oscillate close to zero with the chattering phenomenon well minimized, the estimated weights of the RBFNNs $\hat{\boldsymbol{w}}$ (Fig. 8.46) contribute to the compensation control portions to attempt to cancel auxiliary velocity tracking errors. Again, it must be emphasized that these results can be further improved with the redefinition of gains and parameters, among other factors previously reported.

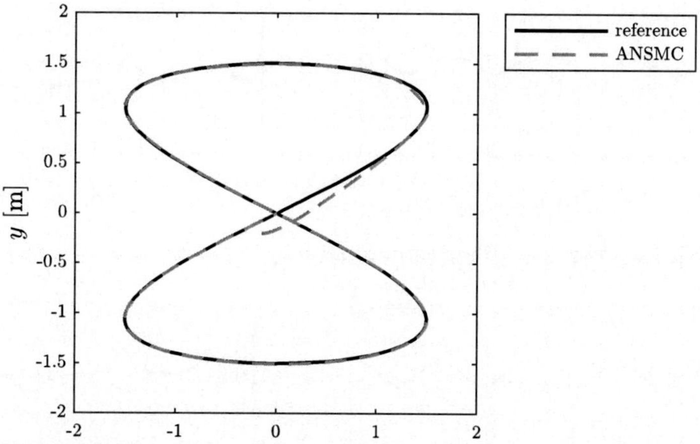

Fig. 8.36 Trajectory tracking in the realistic scenario

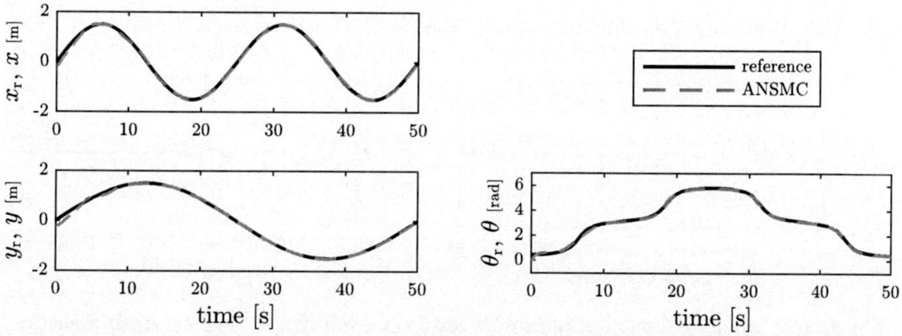

Fig. 8.37 Tracking (x, y, and θ) in the realistic scenario

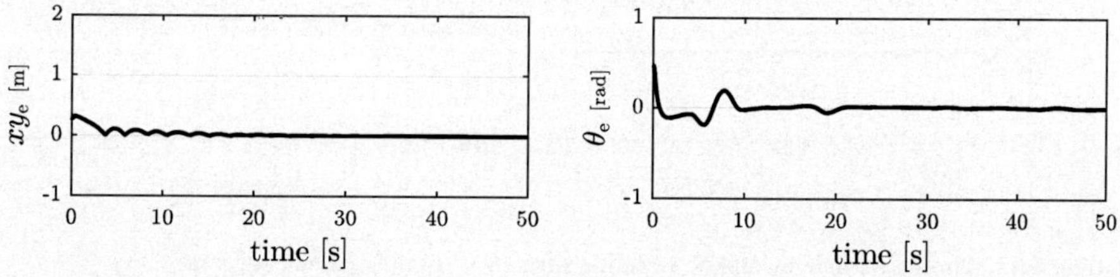

Fig. 8.38 Posture tracking errors, with $xy_e = \sqrt{x_e^2 + y_e^2}$, in the realistic scenario

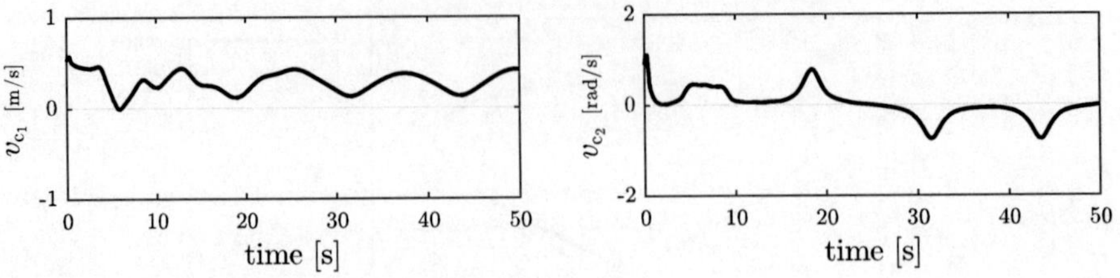

Fig. 8.39 Control velocities in the realistic scenario

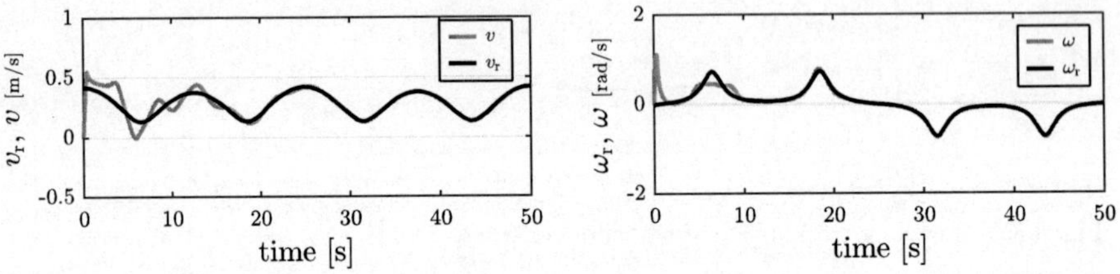

Fig. 8.40 Velocities of the DWMR and reference velocities in the realistic scenario

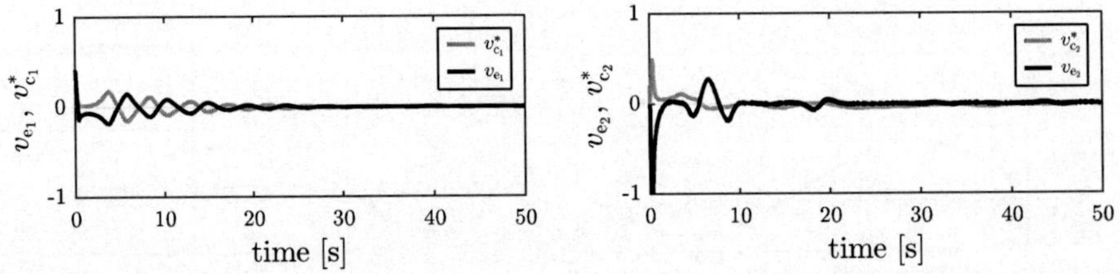

Fig. 8.41 Auxiliary velocity tracking errors and control velocities $\boldsymbol{v}^*_c$ in the realistic scenario

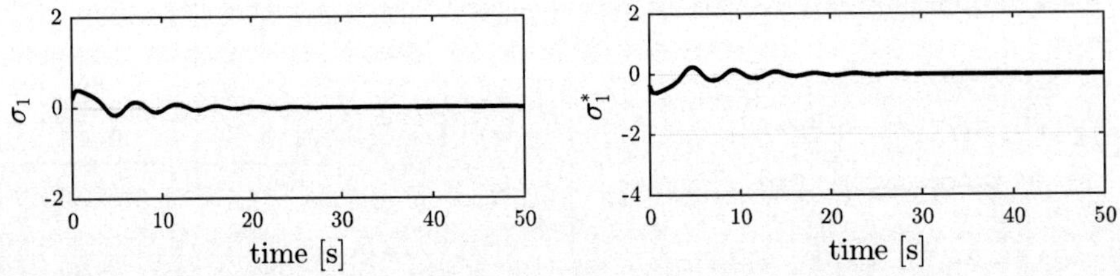

Fig. 8.42 Sliding surface σ_1 and new sliding surface σ^*_1 in the realistic scenario

8.7 General Considerations

In this chapter, to mitigate the chattering phenomenon, not needing to know both the limits of uncertainties and disturbances and the magnitudes of the gains of **G**, the ANSMC was proposed using a neural compensator based on the GL notation [5] to replace the discontinuous control portion so that the control objective is achieved with acceptable precision. To achieve this objective, the ANSMC forces the trajectory

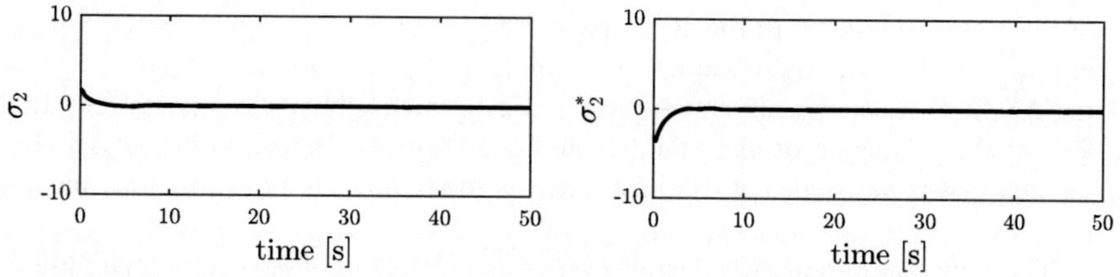

Fig. 8.43 Sliding surface σ_2 and new sliding surface σ_2^* in the realistic scenario

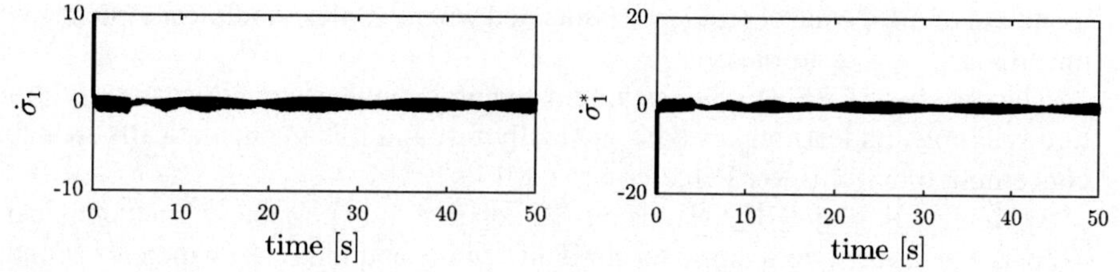

Fig. 8.44 Derivatives of sliding surface σ_1 and derivatives of new sliding surface σ_1^* in the realistic scenario

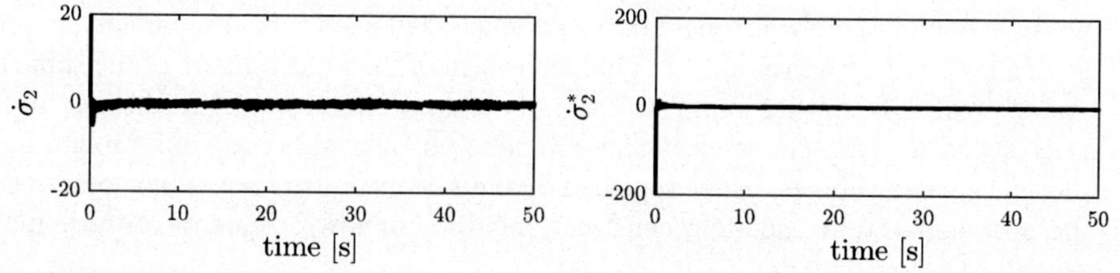

Fig. 8.45 Derivatives of sliding surface σ_2 and derivatives of new sliding surface σ_2^* in the realistic scenario

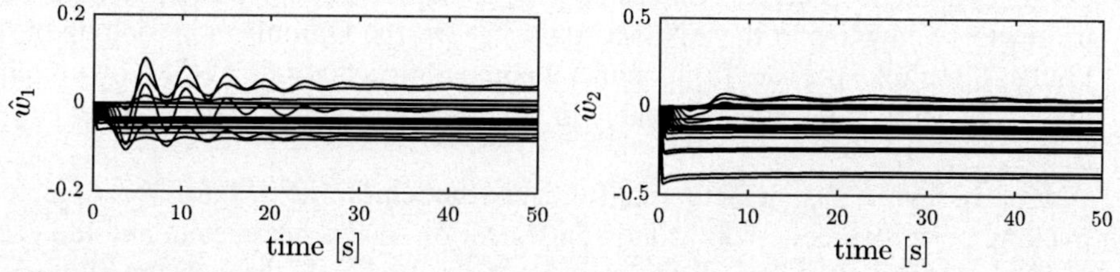

Fig. 8.46 Estimated weights $\hat{\boldsymbol{w}}$ of the RBFNNs in the realistic scenario

of the states to a place in the state space, where the closed-loop system is immune to uncertainties and disturbances, thus significantly reducing its effects. Therefore, the ANSMC was derived from criteria that guarantee robustness and stability in the face of the incidence of uncertainties and disturbances. Based on the experimental results, it was verified that the invariance is ideal and has little practical meaning [26].

The neural compensator design is because neural networks have mapping and approximation properties of nonlinear functions, as well as the availability of online learning algorithms that do not require offline tuning [6], thus an alternative for the treatment of most control problems associated with complex, nonlinear systems with uncertainties and disturbances.

The choice of RBFNN to constitute the neural compensator is because its structure is simple, its learning speed is generally fast and it is mathematically treatable concerning the Multilayer Perceptrons or MLPs [28].

With modeling by RBFNNs, using GL notation, it is possible to mathematically express a neural network emulator for both vectors and matrices, which is of fundamental importance for the development of controller designs, including the ANSMC, given that most dynamic systems involve operations with vectors and matrices. Each unknown element of these vectors and matrices corresponds to a subnet, making it possible to obtain the adjustment of the parameters of each subnet separately, to produce a fast adaptation velocity, to reduce the size of the input vectors of the subnets, and to provide flexibility in the structure of the controllers. However, this flexibility is achieved if some elements of these vectors and matrices are well known, then their respective subnets can be replaced by their deterministic equations, otherwise, these subnets can be used to reconstruct only the unknown elements or those more complicated to calculate.

As already reported for AFSMCs, research can be extended to generalize the ANSMC for the other possible classes of DWMRs and other dynamic systems, for example, unmanned aerial vehicles (UAVs); establish a methodology to adjust the parameters and gains of the ANSMC by using Design Optimization toolbox of the Matlab/Simulink software [3]; conduct a comparative study of ANSMC with other kinematic neural controllers available in the literature, such as the works developed by [19, 23, 24, 29].

Finally, several design factors can influence the achievement of different trajectory tracking performances, such as the adjustment of design parameters and the gains of the learning and control laws; the size of the RBFNNs through the number of neurons or of activation functions in the hidden layer; the definition of the centers and variances of the GRBFs. A comparative study to verify the influence of these factors on the performance through the use of RBFNNs and MLPs can be analyzed in the works of [17, 18].

References

1. DeCarlo, R.A., Zak, S.H., Drakunov, S.V.: Variable structure, sliding mode controller design. Control Handb. **57**, 941–951 (1996)
2. Derks, E.P.P.A., Pastor, M.S.S., Buydens, L.M.C.: Robustness analysis of radial base function and multilayered feedforward neural network models. Chemometr. Intell. Lab. Syst. **28**(1), 49–60 (1995)
3. Freire, F., Martins, N., Splendor, F.: A simple optimization method for tuning the gains of PID controllers for the autopilot of cessna 182 aircraft using model-in-the-loop platform. J. Control Autom. Electr. Syst. **29**, 441–450 (2018)
4. Gao, W., Hung, J.C.: Variable structure control of nonlinear systems: a new approach. IEEE Trans. Ind. Electron. **40**(1), 45–55 (1993)
5. Ge, S.S.: Robust adaptive NN feedback linearization control of nonlinear systems. Int. J. Syst. Sci. **27**(12), 1327–1338 (1996)
6. Haykin, S.O.: Neural Networks and Learning Machines, 3rd edn. Prentice Hall (2008)
7. Hung, J.Y., Gao, W., Hung, J.C.: Variable structure control: a survey. IEEE Trans. Ind. Electron. **40**(1):2–22 (1993, February)
8. Leung, F.H.F., Wong, L.K., Tam, P.K.S.: Algorithm for eliminating chattering in sliding mode control. Electron. Lett. **32**(6), 599–601 (1996)
9. Lewis, F.L., Jagannathan, S., Yesildirek, A.: Neural Network Control of Robot Manipulators and Nonlinear Systems. Taylor & Francis, Ltd., 1 Gunpowder Square, London, EC4A 3DE (1999)
10. Lewis, F.L., Dawson, D.M., Abdallah, C.T.: Robot Manipulator Control: Theory and Practice, 2 edn. Marcel Dekker, Inc. (2003)
11. Li, Y., Qiang, S., Zhuang, X., Kaynak, O.: Robust and adaptive backstepping control for nonlinear systems using RBF neural networks. IEEE Trans. Neural Netw. **15**(3), 693–701 (2004)
12. Martins, N.A., Alencar, M., Lombardi, W.C,, Bertol, D.W., De Pieri, E.R., Ferasoli Filho, H.: Trajectory tracking of a wheeled mobile robot with uncertainties and disturbances: proposed adaptive neural control. Control Cybern. **44**, 47–98 (2015)
13. Martins, N.A., de Alencar, M., Lombardi, W.C., Bertol, D.W., De Pieri, E.R., Ferasoli Filho. H.: A proposed neural control for the trajectory tracking of a nonholonomic mobile robot with disturbances. In: Lecture Notes in Computer Science (LNCS)—Artificial Neural Networks and Machine Learning - Proceedings of the International Conference on Artificial Neural Networks (ICANN), vol. 7552, pp. 330–338, Berlin, Heidelberg, Germany, 11–14 September 2012. Springer, Berlin
14. Martins, N.A., Elyoussef, E.S., Bertol, D.W., De Pieri, E.R., Moreno, U.F., Castelan, E.B.: Nonholonomic mobile robot with kinematic disturbances in the trajectory tracking: a variable structure controller. Learn. Nonlinear Models **8**(1), 23–40 (2010)
15. Martins, N.A., Elyoussef, E.S., Bertol, D.W., De Pieri, E.R., Moreno, U.F., Castelan, E.B.: Trajectory tracking of a nonholonomic mobile robot with kinematic disturbances: a variable structure control design. IEEE Latin Am. Trans. **9**(3), 276–283 (2011)
16. Martins, N.A., Bertol, D., Lombardi, W., Pieri, E.R., Castelan, E.: Trajectory tracking of a nonhonomic mobile robot with parametric an nonparametric uncertainties: a proposed neural control. In: 16h Mediterranean Conference on Control an Automation (2008, June)
17. Oh, C., Kim, M., Lee, J.: Control of a nonholonomic mobile robot using an RBF network. J. Artif. Life Rob. **8**(1), 14–19 (2004)
18. Oh, C., Kim, M., Lee, J., Lee, J.: Control of mobile robots using RBF network. In: Proceedings of the IEEE/RSJ International Conference on Intelligent Robots and Systems, pp. 3528–3533. IEEE (2003)
19. Oubbati, M., Schanz, M., Levi, P.: Kinematic and dynamic adaptive control of a nonholonomic mobile robot using a RNN. In: CIRA 2005, International Symposium on Computational Intelligence in Robotics and Automation, June 27–30, 2005, Espoo, Finland, pp. 27–33. IEEE (2005)

20. Passold, F.: Applying RBF neural nets for position control of an inter/scara robot. Int. J. Comput. Commun. Control **4**(2), 148–157 (2009, June)
21. Shuwen, P., Hongye, S., Xiehe, Jian, C.: Variable structure control theory and application: a survey. In: Proceedings of the 3rd World Congress on Intelligent Control and Automation, vol. 4, pp. 2977–2981. IEEE (2000)
22. Utkin, V.I., Guldner, J., Shi, J.: Sliding Mode Control in Electro-Mechanical Systems, 2nd edn. CRC Press (2009)
23. Velagic, J., Osmic, N., Lacevic, B.: Design of Neural Network Mobile Robot Motion Controller, pp. 187–200. InTech Published (2010)
24. Velagic, J., Osmic, N., Lacevic, B.: Neural network controller for mobile robot motion control. Int. J. Intell. Syst. Technol. **3**(3), 127–132 (2008)
25. Wang, L.-X.: A course in fuzzy systems and control. Prentice-Hall Press, Englewood Cliffs, USA (1996)
26. Wang, S., Gao, W.: Robustness and invariance of variable structure systems with multiple inputs. In: Proceedings of the American Control Conference (ACC'1995) (1995)
27. Young, K.D., Utkin, V.I., Özgüner, U.: A control engineer's guide to sliding mode control. IEEE Trans. Control Syst. Technol. **7**(3), 328–342 (1999)
28. Zhihong, M., Yu, X., Wu, H.: An RBF neural network-based adaptive control for SISO linearisable nonlinear systems. Neural Comput. Appl. **7**, 71–77 (1998)
29. Zou, A., Hou, Z., Tan, M., Chen, X., Zhang, Y.: Tracking control of a mobile robot with kinematic uncertainty using neural networks. In: King, I., Wang, J., Chan, L.W., Wang, D. (eds.) Neural Information Processing, Lecture Notes in Computer Science, vol. 4234, pp. 721–730. Springer, Berlin, Heidelberg (2005)

Chapter 9
Formation Control of DWMRs: Sliding Mode Control Techniques

9.1 Introduction

Over the last four decades, the application of mobile robots in several areas has become popular, including industry, services, health, transportation, food production, military, domestic, social. This popularity was supported by pillars such as the development of electronic devices, new technologies, processing capacity, as well as cost reduction, coupled with theoretical and experimental advances applied to the design and control of these mechanisms [5, 26]. With the increase in this popularity, it became clear that there are many difficult and sometimes impossible tasks to be accomplished by applying only a single mobile robot. The idea that several mobile robots would perform better when performing these tasks is currently accepted and well regarded by the technical-scientific community, in addition to the possibility of applying simpler and cheaper mobile robots performing the same task as a single more expensive and complex mobile robot, thus providing development of solutions for the formation control.

A control problem related to mobile robots is to accomplish tracking them over a feasible trajectory, in a defined time interval. This control system must be robust to minimize uncertainties and disturbances that act over them [29, 40]. A nonholonomic wheeled mobile robot has movement constraints, but it presents advantages as low weight and good energy efficiency [33]. A typical example of this kind of system is DWMR, which due to its simplicity, is widely applied in control experiments [6, 7].

The formation problem consists of coordinating a set of two or more DWMRs so that DWMRs enter a defined formation and, once in formation, remain in formation for all time t [25]. A formation is a geometric shape for which each of its vertices represents the spatial position of a DWMR.

Electronic supplementary material The online version of this chapter (https://doi.org/10.1007/978-3-030-77912-2_9) contains supplementary material, which is available to authorized users.

N. A. Martins and D. W. Bertol, *Wheeled Mobile Robot Control*, Studies in Systems, Decision and Control 380, https://doi.org/10.1007/978-3-030-77912-2_9

To ensure formation, it is necessary to use a control strategy. The most common strategies in the literature are leader-follower, behavioral, virtual structure, and generalized coordinates. Descriptions, advantages, and disadvantages of these control strategies and other control strategies based on predictive control, fuzzy logic, and neural networks can be found in the literature at [8, 25]. In this chapter, only the leader-follower formation control strategy is used, then it is necessary to make a brief description of it.

In the leader-follower formation control strategy, some DWMRs are considered as leaders and others as followers. The main advantage of the approach is that at least one leader follows some reference, while the other DWMRs follow the coordinates of that leader with some compensation [8, 25, 27, 36, 37]. The most common techniques of the leader-follower formation control strategy are Separation-Bearing and Separation-Separation [8, 11, 12, 14, 15, 21, 25, 32]. There is yet another less popular technique, a variation of Separation-Bearing, designed to avoid obstacles called Separation Distance-to-Obstacles [21].

In the Separation-Bearing technique, o follower DWMR follows leader DWMR with a separation distance and bearing angle. The main advantage of this technique is the simplicity of mathematical modeling for each follower. As a disadvantage, it is highlighted the direct propagation of errors [25].

The formation can be classified according to the control structure, architecture, and communication structure. The control structure can be considered centralized or decentralized, i.e., when there is only one controller in the formation responsible for processing the data of all DWMRs present in the formation or when each DWMR has its controller, responsible for decision making in a way independent. There is also a hybrid control structure that brings together the characteristics of both structures, as in [38]. The architecture can be homogeneous, when DWMRs are composed of the same control hardware and software, or heterogeneous if DWMRs have different hardware or software. Systems with homogeneous DWMRs are more robust since any isolated DWMR is not critical to the formation. The communication structure is divided into reach, topology, and communication bandwidth, whose further details can be found in [8, 25].

In this chapter, the Separation-Bearing technique is employed for simplicity and the fact that one of its main disadvantages, the direct error propagation, does not significantly affect formations with few DWMRs. Decentralized structure and homogeneous architecture are also adopted, but details of communication have not been defined, since there was no experimentation with real DWMRs.

Therefore, this chapter aims to describe leader-follower formation controllers based on SMC applied to DWMRs that are robust to the incidence of uncertainties and disturbances in solving the trajectory tracking control problem. To ensure this, the FOSMC-B discussed in Chap. 5 is used, however, this controller causes the chattering phenomenon, and to mitigate it the FOQSMC-B and AFSMC-A treated in Chaps. 6 and 7 respectively are also used. Moreover, stability analyzes of the closed-loop control system are guaranteed based on the Lyapunov theory, and simulation results in the ideal and realistic scenarios are explored to confirm the effectiveness of used leader-follower formation control strategies.

9.2 Problem Formulation

The purpose is accomplished by describing a leader-follower formation control structure based on SMC treated in Chaps. 5–7 for trajectory tracking in the design of robust kinematic controllers that also mitigates chattering phenomenon and makes the posture tracking errors (or posture error) tend to zero quickly, and the PD control described in Chap. 2 is used as the design of dynamic controller that makes the auxiliary velocity tracking errors also tend to zero quickly. Thus, it is deemed necessary to define the tracking control problem for a single DWMR, which is the leader DWMR i, and to extend this problem to the formation control with at least one follower DWMR j. To achieve this aim, the kinematic and dynamic models for the DWMR must firstly be defined.

9.2.1 *Kinematic and Dynamic Models of the DWMR*

A suitable mathematical model, that represents the main kinematics and dynamics of the DWMR, the posture model, is extensively treated in literature [1–4, 9, 18–20, 28, 30]. Thus, to design the formation control, it is important to emphasize that the posture model of the DWMR is used as the basis for the synthesis of the controllers only. Therefore, the motion equations representative of the PowerBot DWMR, Eqs. (1.21), (1.25), (1.35), (1.66), and (1.67), as well as their terms defined and described in Chap. 1, are repeated here for convenience:

- Kinematics

$$\dot{\boldsymbol{q}} = \boldsymbol{S}(\boldsymbol{q})\boldsymbol{v}, \tag{9.1}$$

with

$$\boldsymbol{S}(\boldsymbol{q}) = \begin{bmatrix} \cos(\theta) & -\mathrm{d}\sin(\theta) \\ \sin(\theta) & \mathrm{d}\cos(\theta) \\ 0 & 1 \end{bmatrix}.$$

- Dynamics

$$\boldsymbol{H}(\boldsymbol{q})\ddot{\boldsymbol{q}} + \boldsymbol{C}(\boldsymbol{q},\dot{\boldsymbol{q}})\dot{\boldsymbol{q}} + \boldsymbol{\tau}_{\mathrm{d}} = -\boldsymbol{A}^{\mathrm{T}}(\boldsymbol{q})\boldsymbol{\rho} + \boldsymbol{E}(\boldsymbol{q})\boldsymbol{\tau}. \tag{9.2}$$

- Formulation of dynamics in the state-space

$$\bar{\boldsymbol{H}}(\boldsymbol{q})\dot{\boldsymbol{v}} + \bar{\boldsymbol{C}}(\boldsymbol{q},\dot{\boldsymbol{q}})\boldsymbol{v} + \bar{\boldsymbol{\tau}}_{\mathrm{d}} = \bar{\boldsymbol{E}}(\boldsymbol{q})\boldsymbol{\tau} = \bar{\boldsymbol{\tau}}, \tag{9.3}$$

$$\dot{\boldsymbol{\chi}} = f(\boldsymbol{\chi}) + g(\boldsymbol{\chi})\boldsymbol{\tau},$$

$$\dot{\chi} = \begin{bmatrix} S(q)v \\ -\bar{H}^{-1}(q)\left(\bar{C}(q,\dot{q})v + \bar{\tau}_{\mathrm{d}}\right) \end{bmatrix} + \begin{bmatrix} 0 \\ \bar{H}^{-1}(q)\bar{E}(q) \end{bmatrix}\tau,$$

with $\chi = [q \;\; v]^{\mathrm{T}}$.

Equation (9.3) is transformed into an appropriate representation for control purposes with the elimination of the term containing the constraint matrix, *Properties 1.3* (matrix $\bar{H}(q)$, norms of the matrix $\bar{C}(q,\dot{q})$ and the vector $\bar{\tau}_{\mathrm{d}}$ are bounded) and *1.4* (the matrix $\dot{\bar{H}}(q) - 2\bar{C}(q,\dot{q})$ is skew-symmetric) are maintained and considered in the stability analysis of the control system.

9.2.2 Trajectory Tracking

The trajectory tracking control problem for a single DWMR is posed as follows. Let there be a reference DWMR that is prescribed, in which the structure of DWMR is given and the reference posture and velocities are known, with $v_{\mathrm{r}} > 0$ for all t, find a smooth velocity control v_{c} such that $\lim_{t\to\infty}(q_{\mathrm{e}}) = 0$, where q_{e} and v_{r} are the posture error and the reference velocity vector, respectively. Then compute the control torque τ for Eq. (9.3), such that $v \to v_{\mathrm{c}}$ as $t \to \infty$ [22].

9.2.3 Leader-Follower Formation Control

To extend the trajectory tracking control problem to formation control, the virtual reference DWMR must be replaced by as leader DWMR i.

Among the strategies in leader-follower formation control, only separation-bearing [16] is considered. Separation-bearing formation control aims to find a smooth velocity control input $v_{\mathrm{c}j}$ for follower DWMR j such that

$$\lim_{t\to\infty}(\delta_{\mathrm{d}ij} - \delta_{ij}) = 0, \quad \lim_{t\to\infty}(\psi_{\mathrm{d}ij} - \psi_{ij}) = 0 \tag{9.4}$$

where δ_{ij} and ψ_{ij} are the measured separation distance and bearing angle of the follower DWMR j to leader DWMR i, with $\delta_{\mathrm{d}ij}$ and $\psi_{\mathrm{d}ij}$ representing desired separation distance and bearing angle respectively.

In addition to meeting these conditions, Eq. (9.4), this same smooth velocity input $v_{\mathrm{c}j}$ for follower DWMR j must also satisfy $\lim_{t\to\infty}(q_{\mathrm{r}j} - q_j) = 0$. Then compute the torque τ_j for the dynamic system, Eq. (9.3), so that $\lim_{t\to\infty}(v_{\mathrm{c}j} - v_j) = 0$. Achieving this for every leader DWMR i and follower DWMR $j = 1, \ldots,$ h ensures that the entire formation tracks the formation trajectory.

9.3 Control Design

This section presents the control structures of the leader and the follower are developed and individual stability analyzes are verified as well as formation stability analysis is described.

9.3.1 Leader Control Structure

In every formation, it is assumed there is leader DWMR i such that the following assumptions hold [16]:

Assumption 1: The formation leader DWMR i follows no physical DWMRs but follows the virtual reference leader DWMR.

Assumption 2: The formation leader DWMR i is capable of measuring its absolute position via instrumentation like GPS so that tracking the virtual reference leader DWMR is possible.

Without loss of generality, the subscript i that represents the leader DWMR is omitted in mathematical formulations.

The kinematics of the formation leader DWMR i is defined by Eq. (9.1) while the kinematics of the virtual reference leader DWMR is modeled by Eqs. (2.16) and (2.17) through which posture tracking errors in the inertial frame to the DWMR frame are obtained and, consequently, the posture error of the leader DWMR i as in Eq. (2.18). The error dynamics of the closed-loop system for trajectory tracking results from the time derivative of Eq. (2.18) and that, after mathematical manipulations, is given by Eq. (2.22), which is rewritten below:

$$\dot{\boldsymbol{q}}_{\mathrm{e}} = \begin{bmatrix} \dot{x}_{\mathrm{e}} \\ \dot{y}_{\mathrm{e}} \\ \dot{\theta}_{\mathrm{e}} \end{bmatrix} = \begin{bmatrix} \omega y_{\mathrm{e}} - \upsilon + \upsilon_{\mathrm{r}} \cos(\theta_{\mathrm{e}}) \\ -\omega(\mathrm{d} + x_{\mathrm{e}}) + \upsilon_{\mathrm{r}} \sin(\theta_{\mathrm{e}}) \\ \omega_{\mathrm{r}} - \omega \end{bmatrix}. \tag{9.5}$$

Emphasizing that, under robustness considerations, in practical situations, the perfect velocity tracking does not hold. Thus, the closed-loop kinematic model, based on Eq. (9.1), leads in Eq. (2.25), i.e.,

$$\begin{cases} \dot{x} = (\upsilon_{\mathrm{c}} + \upsilon_{\mathrm{e}}) \cos(\theta) - (\omega_{\mathrm{c}} + \omega_{\mathrm{e}})\mathrm{d} \sin(\theta) \\ \dot{y} = (\upsilon_{\mathrm{c}} + \upsilon_{\mathrm{e}}) \sin(\theta) + (\omega_{\mathrm{c}} + \omega_{\mathrm{e}})\mathrm{d} \cos(\theta) \\ \dot{\theta} = (\omega_{\mathrm{c}} + \omega_{\mathrm{e}}) \end{cases}. \tag{9.6}$$

It is also known that the effect of the uncertainties and disturbances affect the system can be considered and, after the appropriate mathematical manipulations, the error dynamics, Eq. (9.5), yields to:

$$\dot{\boldsymbol{q}}_e = \begin{bmatrix} \dot{x}_e \\ \dot{y}_e \\ \dot{\theta}_e \end{bmatrix} = \begin{bmatrix} (\omega_c + \omega_e)y_e - (\upsilon_c + \upsilon_e) + \upsilon_r \cos(\theta_e) \\ -(\omega_c + \omega_e)(d + x_e) + \upsilon_r \sin(\theta_e) \\ \omega_r - (\omega_c + \omega_e) \end{bmatrix}, \tag{9.7}$$

that corresponds to Eq. (2.27). Due to the nonholonomic constraint [17] and $d \neq 0$, the orientation of the leader DWMR i in the trajectory tracking will not be equal to the orientation of virtual reference leader DWMR while the execution is turning maneuvers, and thus, the reference orientation of the leader DWMR i cannot be chosen as $\theta_r = \theta$. However, choosing the reference orientation $\dot{\theta}_r^*$ relative to the virtual reference leader DWMR satisfying the differential equation

$$\dot{\theta}_r^* = \omega_r^* = \frac{1}{d}\Big(\upsilon_r \sin(\theta_e) + y_e\Big), \tag{9.8}$$

the asymptotic stability of all three error states can be shown provided the control velocity is tracked with zero error.

Replacing Eq. (9.8) in place of ω_r into the Eq. (9.7), the error dynamics can be expressed as:

$$\dot{\boldsymbol{q}}_e = \begin{bmatrix} \dot{x}_e \\ \dot{y}_e \\ \dot{\theta}_e^* \end{bmatrix} = \begin{bmatrix} (\omega_c + \omega_e)y_e - (\upsilon_c + \upsilon_e) + \upsilon_r \cos(\theta_e) \\ -(\omega_c + \omega_e)(d + x_e) + \upsilon_r \sin(\theta_e) \\ \omega_r^* - (\omega_c + \omega_e) \end{bmatrix}, \tag{9.9}$$

$$\dot{\boldsymbol{q}}_e = \begin{bmatrix} \dot{x}_e \\ \dot{y}_e \\ \dot{\theta}_e^* \end{bmatrix} = \begin{bmatrix} (\omega_c + \omega_e)y_e - (\upsilon_c + \upsilon_e) + \upsilon_r \cos(\theta_e) \\ -(\omega_c + \omega_e)(d + x_e) + \upsilon_r \sin(\theta_e) \\ \frac{1}{d}\Big(\upsilon_r \sin(\theta_e) + y_e\Big) - (\omega_c + \omega_e) \end{bmatrix}.$$

emphasizing that $\theta_e = \theta_r - \theta$ and $\theta_e^* = \theta_r^* - \theta$. The transformed error dynamics, Eq. (9.9), now acts as a trajectory tracking controller that will achieve a relative orientation concerning the virtual reference leader DWMR. Furthermore, the orientation of the leader DWMR i will become the orientation of the virtual reference leader DWMR when $\omega_r = 0$.

With basis on the generic modeling of nonlinear systems described in Sect. 2.5, the error dynamics in Eq. (9.9) can be rewritten as in Eq. (2.30), i.e.,

$$\dot{\tilde{z}} = \boldsymbol{A}_0(\tilde{z}, t) + \boldsymbol{B}_0(\tilde{z}, t)\boldsymbol{\upsilon}_c(\tilde{z}, t) + d_b(t), \tag{9.10}$$

since there are not parametric uncertainties and that $\tilde{z}$, $\boldsymbol{A}_0$, $\boldsymbol{B}_0$, d_b are similarly defined as in Eq. (5.22), whose representation in matrix form is given as:

$$\begin{bmatrix} \dot{x}_e \\ \dot{y}_e \\ \dot{\theta}_e^* \end{bmatrix} = \begin{bmatrix} \upsilon_r \cos(\theta_e) \\ \upsilon_r \sin(\theta_e) \\ \omega_r^* \end{bmatrix} + \begin{bmatrix} -1 & y_e \\ 0 & -(d + x_e) \\ 0 & -1 \end{bmatrix} \begin{bmatrix} \upsilon_c \\ \omega_c \end{bmatrix} + \begin{bmatrix} -1 & y_e \\ 0 & -(d + x_e) \\ 0 & -1 \end{bmatrix} \begin{bmatrix} \upsilon_e \\ \omega_e \end{bmatrix}. \tag{9.11}$$

To perform the controller synthesis, which is similar on described in Sect. 5.2.3, from the error dynamics, Eq. (9.11), are selected the following sliding surfaces:

$$\boldsymbol{\sigma} = \begin{bmatrix} \lambda_1 x_e \\ \lambda_2 y_e + \lambda_3 \theta_e^* \end{bmatrix}, \tag{9.12}$$

and, consequently:

$$\boldsymbol{\sigma}^* = \boldsymbol{B}_{0_\sigma}^{\mathrm{T}} \boldsymbol{\sigma} = \begin{bmatrix} -\lambda_1^2 x_e \\ \lambda_1^2 x_e y_e - [\lambda_2(\mathrm{d} + x_e) + \lambda_3](\lambda_2 y_e + \lambda_3 \theta_e^*) \end{bmatrix}. \tag{9.13}$$

with λ_1, λ_2, and λ_3 being positive constants and $\boldsymbol{B}_{0_\sigma}$ similar as in Eq. (5.38).

For the stability analysis, it is chosen the Lyapunov function candidate in the form:

$$V = \frac{1}{2} \boldsymbol{\sigma}^{\mathrm{T}} \boldsymbol{\sigma}, \tag{9.14}$$

which is positive definite. The sliding surfaces will be attractive since the control law $\boldsymbol{v}_c$ referring to FOSMC-B, Eq. (5.49),

$$\boldsymbol{v}_c = -\boldsymbol{B}_{0_\sigma}^{-1} \boldsymbol{A}_{0_\sigma} - (\mathbf{G}\,\mathrm{sign}(\boldsymbol{\sigma}^*) + \mathbf{K}\boldsymbol{\sigma}^*), \tag{9.15}$$

with

$$\boldsymbol{B}_{0_\sigma}^{-1} = \begin{bmatrix} -\frac{1}{\lambda_1} & -\frac{y_e}{\lambda_2(\mathrm{d}+x_e)+\lambda_3} \\ 0 & -\frac{1}{\lambda_2(\mathrm{d}+x_e)+\lambda_3} \end{bmatrix}, \quad \text{and} \quad \boldsymbol{A}_{0_\sigma} = \begin{bmatrix} \lambda_1 v_r \cos(\theta_e) \\ \lambda_2 v_r \sin(\theta_e) + \lambda_3 \omega_r^* \end{bmatrix}, \tag{9.16}$$

ensures that

$$\dot{V} = \boldsymbol{\sigma}^{\mathrm{T}} \dot{\boldsymbol{\sigma}} = -\boldsymbol{\sigma}^{*\mathrm{T}} \mathbf{G}\,\mathrm{sign}(\boldsymbol{\sigma}^*) - \boldsymbol{\sigma}^{*\mathrm{T}} \mathbf{K} \boldsymbol{\sigma}^* + \boldsymbol{\sigma}^{*\mathrm{T}} \tilde{\mathrm{d}}_0, \tag{9.17}$$

is negative definite since Eq. (9.17) is similar to Eqs. (5.9) and (5.21). Therefore, the same conclusions on the stability analysis are valid as described in Sect. 5.2.2, where further details on control law and stability analysis can be found.

As already discussed previously, for chattering phenomenon attenuation the replacement of the discontinuous function $\mathrm{sign}(\boldsymbol{\sigma}^*)$ over a continuous approximation is required.

A commonly used solution is to use the fractional continuous approximation or proper continuous function given by $\mathrm{sign}(\boldsymbol{\sigma}^*) \approx \boldsymbol{\sigma}^* \oslash |\boldsymbol{\sigma}^*|_\circ + \boldsymbol{\epsilon}$, Eq. (6.4), where $\boldsymbol{\epsilon}$ is a positive constant. Replacing the discontinuous function by the fractional continuous approximation or proper continuous function on the control law, Eq. (9.15), one obtains:

$$\boldsymbol{v}_c = -\boldsymbol{B}_{0\sigma}^{-1} \boldsymbol{A}_{0\sigma} - [\mathbf{G}\left(\boldsymbol{\sigma}^* \oslash \left(|\boldsymbol{\sigma}^*|_\circ + \boldsymbol{\epsilon}\right)\right) + \mathbf{K}\boldsymbol{\sigma}^*], \tag{9.18}$$

which is similar to Eq. (9.15) from FOQSMC-B, whose stability analysis is identical to the described in Sect. 6.2.2.

Another solution for the chattering phenomenon attenuation in the control caused by the discontinuous control gain $\mathbf{G}\,\mathrm{sign}(\boldsymbol{\sigma}^*)$ is the replacement of the same by a fuzzy gain $\hat{\boldsymbol{F}}(\boldsymbol{\sigma}^*)$, so that the control law, Eq. (9.15), becomes:

$$\begin{aligned}\boldsymbol{v}_{\mathrm{c}} &= -\boldsymbol{B}_{0_\sigma}^{-1}\boldsymbol{A}_{0_\sigma} - \hat{\boldsymbol{F}}(\boldsymbol{\sigma}^*) - \mathbf{K}\boldsymbol{\sigma}^*,\\ \boldsymbol{v}_{\mathrm{c}} &= -\boldsymbol{B}_{0_\sigma}^{-1}\boldsymbol{A}_{0_\sigma} - \hat{\boldsymbol{\beta}}^{\mathrm{T}}\boldsymbol{\phi}(\boldsymbol{\sigma}^*) - \mathbf{K}\boldsymbol{\sigma}^*,\end{aligned} \tag{9.19}$$

where $\hat{\boldsymbol{\beta}}$ is the vector of online updated consequences and $\boldsymbol{\phi}(\boldsymbol{\sigma}^*)$ is the vector of rule weights, as can be verified in Eqs. (7.2) and (7.3). This control law is similar to the AFSMC-A and their adaptation law is given by Eq. (7.7), i.e.,

$$\dot{\hat{\boldsymbol{\beta}}} = \boldsymbol{\sigma}^*\boldsymbol{\phi}(\boldsymbol{\sigma}^*), \tag{9.20}$$

while the rule base, triangular-shaped membership functions and their parameters are the same considered in Sect. 7.3.4. Also, the stability analysis to be addressed here can be the same as found in Sect. 7.3.2.

9.3.2 *Leader-Follower Trajectory Tracking Control*

To distinguish the variables involved in mathematical formulations are used the subscripts i and j for the leader and follower DWMRs, respectively.

In the separation-bearing strategy, the follower DWMR j aims to move in a measured separation distance and bearing angle from the leader DWMR i as reference follower DWMR, thus avoiding collisions as can be verified in Fig. 9.1. Moreover, it is necessary to find auxiliary velocities that meet the conditions described in Eq. (9.4).

To achieve these objectives, firstly, the kinematic modeling of the reference follower DWMR must be provided as:

$$\dot{\boldsymbol{q}}_{\mathrm{r}j} = \boldsymbol{S}_{\mathrm{r}j}(\boldsymbol{q}_{\mathrm{r}j})\boldsymbol{v}_{\mathrm{r}j}, \tag{9.21}$$

with

$$\boldsymbol{S}_{\mathrm{r}j}(\boldsymbol{q}_{\mathrm{r}j}) = \begin{bmatrix}\cos(\theta_{\mathrm{r}j}) & 0\\ \sin(\theta_{\mathrm{r}j}) & 0\\ 0 & 1\end{bmatrix}, \quad \boldsymbol{q}_{\mathrm{r}j} = \begin{bmatrix}x_{\mathrm{r}j}\\ y_{\mathrm{r}j}\\ \theta_{\mathrm{r}j}\end{bmatrix}, \quad \boldsymbol{v}_{\mathrm{r}j} = \begin{bmatrix}\upsilon_{\mathrm{r}j}\\ \omega_{\mathrm{r}j}\end{bmatrix},$$

which are identical to the Eqs. (2.16) and (2.17). Secondly, the kinematic model for the front of the follower DWMR j must be defined similarly as in Eqs. (1.21) and (1.25), i.e.,

$$\dot{\boldsymbol{q}}_j = \boldsymbol{S}_j(\boldsymbol{q}_j)\boldsymbol{v}_j, \tag{9.22}$$

with

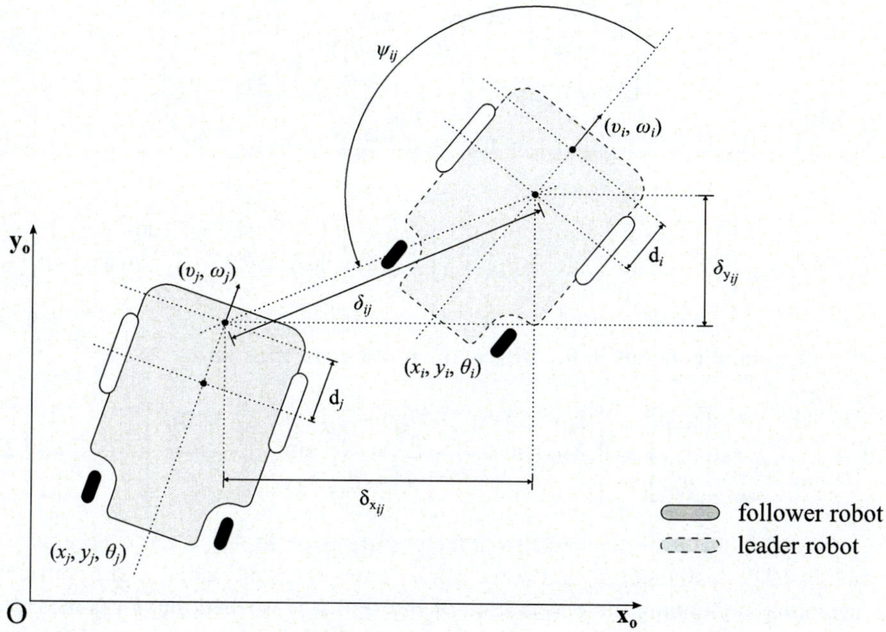

Fig. 9.1 Modeling of the leader-follower formation control using separation-bearing strategy

$$\boldsymbol{S}_j(\boldsymbol{q}_j) = \begin{bmatrix} \cos(\theta_j) & -\mathrm{d}_j \sin(\theta_j) \\ \sin(\theta_j) & \mathrm{d}_j \cos(\theta_j) \\ 0 & 1 \end{bmatrix},$$

where d_j is the distance from the rear axle to the front of the follower DWMR j.

Observing at Fig. 9.1, in which the leader DWMR i is tracked by the follower DWMR j, it is possible to obtain the posture vector of the reference follower DWMR based on the leader DWMR i as follows:

$$\begin{cases} x_{\mathrm{r}j} = x_i - \mathrm{d}_i \cos(\theta_i) + \delta_{\mathrm{d}ij} \cos(\psi_{\mathrm{d}ij} + \theta_i) \\ y_{\mathrm{r}j} = y_i - \mathrm{d}_i \sin(\theta_i) + \delta_{\mathrm{d}ij} \sin(\psi_{\mathrm{d}ij} + \theta_i) \\ \theta_{\mathrm{r}j} = \theta_i \end{cases}, \tag{9.23}$$

and, consequently, the posture vector of the follower DWMR j is then defined as;

$$\begin{cases} x_j = x_i - \mathrm{d}_i \cos(\theta_i) + \delta_{\mathrm{d}ij} \cos(\psi_{\mathrm{d}ij} + \theta_i) \\ y_j = y_i - \mathrm{d}_i \sin(\theta_i) + \delta_{\mathrm{d}ij} \sin(\psi_{\mathrm{d}ij} + \theta_i) \\ \theta_j = \theta_j \end{cases}. \tag{9.24}$$

With the Eqs. (9.23) and (9.24), the posture tracking errors $(\boldsymbol{q}_\mathrm{r} - \boldsymbol{q})$ are obtained as well as the following posture error of the follower DWMR j:

$$\boldsymbol{q}_{\mathrm{e}j} = \begin{bmatrix} x_{\mathrm{e}j} \\ y_{\mathrm{e}j} \\ \theta_{\mathrm{e}j} \end{bmatrix} = \begin{bmatrix} \cos(\theta_j) & \sin(\theta_j) & 0 \\ -\sin(\theta_j) & \cos(\theta_j) & 0 \\ 0 & 0 & 1 \end{bmatrix} \begin{bmatrix} x_{\mathrm{r}j} - x_j \\ y_{\mathrm{r}j} - y_j \\ \theta_{\mathrm{r}j} - \theta_j \end{bmatrix}, \tag{9.25}$$

such as in Eq. (2.18). Using some trigonometrical identities, the Eq. (9.25) can be rewritten as:

$$\boldsymbol{q}_{\mathrm{e}j} = \begin{bmatrix} x_{\mathrm{e}j} \\ y_{\mathrm{e}j} \\ \theta_{\mathrm{e}j} \end{bmatrix} = \begin{bmatrix} \cos(\theta_j) & \sin(\theta_j) & 0 \\ -\sin(\theta_j) & \cos(\theta_j) & 0 \\ 0 & 0 & 1 \end{bmatrix} \begin{bmatrix} \delta_{\mathrm{d}ij}\cos(\psi_{\mathrm{d}ij} + \theta_i) - \delta_{ij}\cos(\psi_{ij} + \theta_i) \\ \delta_{\mathrm{d}ij}\sin(\psi_{\mathrm{d}ij} + \theta_i) - \delta_{ij}\sin(\psi_{ij} + \theta_i) \\ \theta_i - \theta_j \end{bmatrix}, \tag{9.26}$$

and after simplifications in Eq. (9.26), the posture error results in:

$$\boldsymbol{q}_{\mathrm{e}j} = \begin{bmatrix} x_{\mathrm{e}j} \\ y_{\mathrm{e}j} \\ \theta_{\mathrm{e}j} \end{bmatrix} = \begin{bmatrix} \delta_{\mathrm{d}ij}\cos(\psi_{\mathrm{d}ij} + \theta_{\mathrm{e}j}) - \delta_{ij}\cos(\psi_{ij} + \theta_{\mathrm{e}j}) \\ \delta_{\mathrm{d}ij}\sin(\psi_{\mathrm{d}ij} + \theta_{\mathrm{e}j}) - \delta_{ij}\sin(\psi_{ij} + \theta_{\mathrm{e}j}) \\ \theta_i - \theta_j \end{bmatrix}. \tag{9.27}$$

To obtain the error dynamics from the posture error, Eq. (9.27), it is necessary to obtain the derivatives of δ_{ij} and ψ_{ij}, and it is assumed that $\delta_{\mathrm{d}ij}$ and $\psi_{\mathrm{d}ij}$ are constants. Therefore, continuing the observation of Fig. 9.1, it is verified that δ_{ij} is measured from the center of the wheel axis of the leader DWMR i to the center of the front part of the follower DWMR j, and their x and y components can be defined as:

$$\begin{aligned} \delta_{\mathrm{x}ij} &= x_{\mathrm{axis}\,i} - x_{\mathrm{front}\,j} = x_i - \mathrm{d}_i\cos(\theta_i) - x_j, \\ \delta_{\mathrm{y}ij} &= y_{\mathrm{axis}\,i} - y_{\mathrm{front}\,j} = y_i - \mathrm{d}_i\sin(\theta_i) - y_j, \end{aligned} \tag{9.28}$$

and the derivatives of $\delta_{\mathrm{x}ij}$ e $\delta_{\mathrm{y}ij}$ are given by:

$$\begin{aligned} \dot{\delta}_{\mathrm{x}ij} &= \upsilon_i\cos(\theta_i) - \upsilon_j\cos(\theta_j) - \mathrm{d}_j\omega_j\sin(\theta_j), \\ \dot{\delta}_{\mathrm{y}ij} &= \upsilon_i\sin(\theta_i) - \upsilon_j\sin(\theta_j) - \mathrm{d}_j\omega_j\cos(\theta_j). \end{aligned} \tag{9.29}$$

Still analyzing the Fig. 9.1, definitions $\delta_{ij}^2 = \delta_{\mathrm{x}ij}^2 + \delta_{\mathrm{y}ij}^2$ and $\psi_{ij} = \arctan\left(\frac{\delta_{\mathrm{y}ij}}{\delta_{\mathrm{x}ij}}\right) - \theta_i + \pi$ are obtained, it can be shown that derivatives of the separation and bearing are consistent with [16] even when using the kinematics described in Eq. (9.22) such that

$$\begin{aligned} \dot{\delta}_{ij} &= \upsilon_j\cos(\zeta_j) - \upsilon_i\cos(\psi_{ij}) + \mathrm{d}_j\omega_j\sin(\zeta_j), \\ \dot{\psi}_{ij} &= \frac{1}{\delta_{ij}}\Big[\upsilon_i\sin(\psi_{ij}) - \upsilon_j\sin(\zeta_j) + \mathrm{d}_j\omega_j\cos(\zeta_j) - \delta_{ij}\omega_i\Big], \end{aligned} \tag{9.30}$$

where $\zeta_j = \psi_{ij} + \theta_{\mathrm{e}j}$. Thus, the closed-loop error dynamics $\dot{\boldsymbol{q}}_{\mathrm{e}j}$ of the follower DWMR j can be expressed by using the derivative of the posture error of Eq. (9.27) and (9.30) after the application of some trigonometric identities, i.e.,

$$\dot{\boldsymbol{q}}_{\mathrm{e}j} = \begin{bmatrix} \dot{x}_{\mathrm{e}j} \\ \dot{y}_{\mathrm{e}j} \\ \dot{\theta}_{\mathrm{e}j} \end{bmatrix} = \begin{bmatrix} -\upsilon_j + \upsilon_i \cos(\theta_{\mathrm{e}j}) + \omega_j y_{\mathrm{e}j} - \omega_i \delta_{\mathrm{d}ij} \sin(\psi_{\mathrm{d}ij} + \theta_{\mathrm{e}j}) \\ -\omega_j x_{\mathrm{e}j} + \upsilon_i \sin(\theta_{\mathrm{e}j}) - \mathrm{d}_j \omega_j + \omega_i \delta_{\mathrm{d}ij} \cos(\psi_{\mathrm{d}ij} + \theta_{\mathrm{e}j}) \\ \omega_i - \omega_j \end{bmatrix}. \tag{9.31}$$

As already discussed for the leader DWMR i, perfect velocity tracking is also not maintained in practice for the follower DWMR j, therefore, under robustness considerations, the closed-loop kinematic model leads to:

$$\begin{cases} \dot{x}_j = (\upsilon_{\mathrm{c}j} + \upsilon_{\mathrm{e}j}) \cos(\theta_j) - (\omega_{\mathrm{c}j} + \omega_{\mathrm{e}j})\, \mathrm{d_j} \sin(\theta_j) \\ \dot{y}_j = (\upsilon_{\mathrm{c}j} + \upsilon_{\mathrm{e}j}) \sin(\theta_j) + (\omega_{\mathrm{c}j} + \omega_{\mathrm{e}j})\, \mathrm{d_j} \cos(\theta_j) \\ \dot{\theta}_j = (\omega_{\mathrm{c}j} + \omega_{\mathrm{e}j}) \end{cases}. \tag{9.32}$$

Thus, the occurrence of auxiliary velocity tracking errors should be taken into account as one of the effects of uncertainties and disturbances in the system so that the closed-loop error dynamics, Eq. (9.31), is as follows:

$$\begin{aligned} \dot{\boldsymbol{q}}_{\mathrm{e}j} &= \begin{bmatrix} \dot{x}_{\mathrm{e}j} \\ \dot{y}_{\mathrm{e}j} \\ \dot{\theta}_{\mathrm{e}j} \end{bmatrix} \\ &= \begin{bmatrix} -(\upsilon_{\mathrm{c}j} + \upsilon_{\mathrm{e}j}) + \upsilon_i \cos(\theta_{\mathrm{e}j}) + (\omega_{\mathrm{c}j} + \omega_{\mathrm{e}j}) y_{\mathrm{e}j} - \omega_i \delta_{\mathrm{d}ij} \sin(\psi_{\mathrm{d}ij} + \theta_{\mathrm{e}j}) \\ -(\omega_{\mathrm{c}j} + \omega_{\mathrm{e}j}) x_{\mathrm{e}j} + \upsilon_i \sin(\theta_{\mathrm{e}j}) - \mathrm{d}_j (\omega_{\mathrm{c}j} + \omega_{\mathrm{e}j}) + \omega_i \delta_{\mathrm{d}ij} \cos(\psi_{\mathrm{d}ij} + \theta_{\mathrm{e}j}) \\ \omega_i - (\omega_{\mathrm{c}j} + \omega_{\mathrm{e}j}) \end{bmatrix}. \end{aligned} \tag{9.33}$$

The reference orientation for follower DWMR j is defined relative to the leader DWMR i that satisfies the differential equation as:

$$\dot{\theta}_{\mathrm{r}j} = \omega_{\mathrm{r}j} = \frac{1}{\mathrm{d}_j} \left(\upsilon_i \sin(\theta_{\mathrm{e}j}) + \omega_i \delta_{\mathrm{d}ij} \cos(\psi_{\mathrm{d}ij} + \theta_{\mathrm{e}j}) + y_{\mathrm{e}j} \right), \tag{9.34}$$

whose such definition of the Eq. (9.34) is due to the nonholonomic constraint [17], $\mathrm{d} \neq 0$ and the separation-bearing formation control objective, since the orientations of follower DWMR j in the formation will not be equal orientations of leader DWMR i while the formation is turning maneuvers, and thus, the reference orientation of follower DWMR j cannot be chosen such that $\theta_{\mathrm{r}j} = \theta_i$ as in Eq. (9.27). However, defining $\theta_{\mathrm{r}j}$ by Eq. (9.34) allows the asymptotic stability of all three error states to be shown. It can be shown that the reference orientation $\theta_{\mathrm{r}j}$ converges to the orientation of leader DWMR i when $\omega_i = 0$ (when the leader DWMR i travels in a straight path) and formation errors have converged to zero.

Replacing θ_i (Eq. (9.27)) and ω_i [Eq. (9.33)] by θ_{rj} and ω_{rj} [Eq. (9.34)] respectively, the posture error and the error dynamics are transformed as follows:

$$\boldsymbol{q}_{ej} = \begin{bmatrix} x_{ej} \\ y_{ej} \\ \theta^*_{ej} \end{bmatrix} = \begin{bmatrix} \delta_{dij}\cos(\psi_{dij}+\theta_{ej}) - \delta_{ij}\cos(\psi_{ij}+\theta_{ej}) \\ \delta_{dij}\sin(\psi_{dij}+\theta_{ej}) - \delta_{ij}\sin(\psi_{ij}+\theta_{ej}) \\ \theta_{rj}-\theta_j \end{bmatrix}, \tag{9.35}$$

$$\begin{aligned}\dot{\boldsymbol{q}}_{ej} &= \begin{bmatrix} \dot{x}_{ej} \\ \dot{y}_{ej} \\ \dot{\theta}^*_{ej} \end{bmatrix} \\ &= \begin{bmatrix} -(\upsilon_{cj}+\upsilon_{ej}) + \upsilon_i\cos(\theta_{ej}) + (\omega_{cj}+\omega_{ej})y_{ej} - \omega_i\delta_{dij}\sin(\psi_{dij}+\theta_{ej}) \\ -(\omega_{cj}+\omega_{ej})x_{ej} + \upsilon_i\sin(\theta_{ej}) - \mathrm{d}_j(\omega_{cj}+\omega_{ej}) + \omega_i\delta_{dij}\cos(\psi_{dij}+\theta_{ej}) \\ \omega_{rj} - (\omega_{cj}+\omega_{ej}) \end{bmatrix},\end{aligned}$$

$$\begin{aligned}\dot{\boldsymbol{q}}_{ej} &= \begin{bmatrix} \dot{x}_{ej} \\ \dot{y}_{ej} \\ \dot{\theta}^*_{ej} \end{bmatrix} \\ &= \begin{bmatrix} -(\upsilon_{cj}+\upsilon_{ej}) + \upsilon_i\cos(\theta_{ej}) + (\omega_{cj}+\omega_{ej})y_{ej} - \omega_i\delta_{dij}\sin(\psi_{dij}+\theta_{ej}) \\ -(\omega_{cj}+\omega_{ej})x_{ej} + \upsilon_i\sin(\theta_{ej}) - \mathrm{d}_j(\omega_{cj}+\omega_{ej}) + \omega_i\delta_{dij}\cos(\psi_{dij}+\theta_{ej}) \\ \frac{1}{\mathrm{d}_j}\left(\upsilon_i\sin(\theta_{ej}) + \omega_i\delta_{dij}\cos(\psi_{dij}+\theta_{ej}) + y_{ej}\right) - (\omega_{cj}+\omega_{ej}) \end{bmatrix},\end{aligned} \tag{9.36}$$

remembering that $\theta_{ej} = \theta_i - \theta_j$ and $\theta^*_{ej} = \theta_{rj} - \theta_j$. Thus, the transformed error dynamics, Eq. (9.36), can act as a formation tracking controller, which not only seeks to remain at a fixed desired distance δ_{dij} with a desired angle ψ_{dij} relative to the leader DWMR i but also will achieve a relative orientation concerning the leader DWMR i [17].

The transformed error dynamics, Eq. (9.36), can be rewritten by using generic modeling of nonlinear systems such as:

$$\dot{\tilde{z}}_j = \boldsymbol{A}_{0_j}(\tilde{z}_j, t) + \boldsymbol{B}_{0_j}(\tilde{z}_j, t)\boldsymbol{\upsilon}_{cj}(\tilde{z}_j, t) + \mathrm{d}_{\mathrm{b}_j}(t), \tag{9.37}$$

$$\begin{aligned}\begin{bmatrix} \dot{x}_{ej} \\ \dot{y}_{ej} \\ \dot{\theta}^*_{ej} \end{bmatrix} &= \begin{bmatrix} \upsilon_i\cos(\theta_{ej}) - \omega_i\delta_{dij}\sin(\psi_{dij}+\theta_{ej}) \\ \upsilon_i\sin(\theta_{ej}) + \omega_i\delta_{dij}\cos(\psi_{dij}+\theta_{ej}) \\ \omega_{rj} \end{bmatrix} + \begin{bmatrix} -1 & y_{ej} \\ 0 & -(\mathrm{d}_j + x_{ej}) \\ 0 & -1 \end{bmatrix} \begin{bmatrix} \upsilon_{cj} \\ \omega_{cj} \end{bmatrix} \\ &\quad + \begin{bmatrix} -1 & y_{ej} \\ 0 & -(\mathrm{d}_j + x_{ej}) \\ 0 & -1 \end{bmatrix} \begin{bmatrix} \upsilon_{ej} \\ \omega_{ej} \end{bmatrix},\end{aligned} \tag{9.38}$$

$$\begin{bmatrix} \dot{x}_{ej} \\ \dot{y}_{ej} \\ \dot{\theta}^*_{ej} \end{bmatrix} = \begin{bmatrix} \upsilon_i \cos(\theta_{ej}) - \omega_i \delta_{dij} \sin(\psi_{dij} + \theta_{ej}) \\ \upsilon_i \sin(\theta_{ej}) + \omega_i \delta_{dij} \cos(\psi_{dij} + \theta_{ej}) \\ \frac{1}{d_j}\left(\upsilon_i \sin(\theta_{ej}) + \omega_i \delta_{dij} \cos(\psi_{dij} + \theta_{ej}) + y_{ej}\right) \end{bmatrix}$$
$$+ \begin{bmatrix} -1 & y_{ej} \\ 0 & -(d_j + x_{ej}) \\ 0 & -1 \end{bmatrix} \begin{bmatrix} \upsilon_{cj} \\ \omega_{cj} \end{bmatrix}$$
$$+ \begin{bmatrix} -1 & y_{ej} \\ 0 & -(d_j + x_{ej}) \\ 0 & -1 \end{bmatrix} \begin{bmatrix} \upsilon_{ej} \\ \omega_{ej} \end{bmatrix},$$

from Eqs. (9.37) and (9.38), the choice of sliding surfaces of the FOSMC-B for the follower DWMR j is determined as:

$$\boldsymbol{\sigma}_j = \begin{bmatrix} \lambda_{1j} x_{ej} \\ \lambda_{2j} y_{ej} + \lambda_{3j} \theta^*_{ej} \end{bmatrix}, \tag{9.39}$$

$$\boldsymbol{\sigma}^*_j = \boldsymbol{B}^{\mathrm{T}}_{0_{\sigma_j}} \boldsymbol{\sigma}_j = \begin{bmatrix} -\lambda^2_{1j} x_{ej} \\ \lambda^2_{1j} x_{ej} y_{ej} - [\lambda_{2j}(d_j + x_{ej}) + \lambda_{3j}](\lambda_{2j} y_{ej} + \lambda_{3j} \theta^*_{ej}) \end{bmatrix}. \tag{9.40}$$

thus making it possible to obtain the following control law:

- FOSMC-B

$$\boldsymbol{v}_{cj} = -\boldsymbol{B}^{-1}_{0_{\sigma_j}} \boldsymbol{A}_{0_{\sigma_j}} - (\mathbf{G}_j \operatorname{sign}(\boldsymbol{\sigma}^*_j) + \mathbf{K}_j \boldsymbol{\sigma}^*_j), \tag{9.41}$$

with

$$\boldsymbol{B}^{-1}_{0_{\sigma_j}} = \begin{bmatrix} -\frac{1}{\lambda_{1j}} & -\frac{y_{ej}}{\lambda_{2j}(d_j + x_{ej}) + \lambda_{3j}} 5pt \\ 0 & -\frac{1}{\lambda_{2j}(d_j + x_{ej}) + \lambda_{3j}} \end{bmatrix},$$

$$\boldsymbol{A}_{0_{\sigma_j}} = \begin{bmatrix} \lambda_{1j} \upsilon_i \cos(\theta_{ej}) - \lambda_{1j} \omega_i \delta_{dij} \sin(\psi_{dij} + \theta_{ej}) \\ \lambda_{2j} \upsilon_i \sin(\theta_{ej}) + \lambda_{2j} \omega_i \delta_{dij} \cos(\psi_{dij} + \theta_{ej}) + \lambda_{3j} \omega^*_{rj} \end{bmatrix},$$

however, the control law of the FOSMC-B, Eq. (9.41), causes the chattering phenomenon so that to mitigate it, the control laws for the follower DWMR j are given as follows:

- FOQSMC-B

$$\boldsymbol{v}_{cj} = -\boldsymbol{B}^{-1}_{0_{\sigma_j}} \boldsymbol{A}_{0\sigma_j} - [\mathbf{G}_j \left(\boldsymbol{\sigma}^*_j \oslash \left(|\boldsymbol{\sigma}^*_j|_\circ + \boldsymbol{\epsilon}_j\right)\right) + \mathbf{K}_j \boldsymbol{\sigma}^*_j], \tag{9.42}$$

- AFSMC-A

$$\begin{aligned} \boldsymbol{v}_{cj} &= -\boldsymbol{B}^{-1}_{0_{\sigma_j}} \boldsymbol{A}_{0_{\sigma_j}} - \hat{\boldsymbol{F}}_j(\boldsymbol{\sigma}^*_j) - \mathbf{K}_j \boldsymbol{\sigma}^*_j, \\ \boldsymbol{v}_{cj} &= -\boldsymbol{B}^{-1}_{0_{\sigma_j}} \boldsymbol{A}_{0_{\sigma_j}} - \hat{\boldsymbol{\beta}}_j^{\mathrm{T}} \boldsymbol{\phi}_j(\boldsymbol{\sigma}^*_j), -\mathbf{K}_j \boldsymbol{\sigma}^*_j, \end{aligned} \tag{9.43}$$

with the adaptation law being given as

$$\dot{\hat{\boldsymbol{\beta}}}_j = \boldsymbol{\sigma}_j^* \boldsymbol{\phi}_j(\boldsymbol{\sigma}_j^*). \tag{9.44}$$

Finally, it is important to emphasize that the stability analyzes of closed-loop control systems for the follower DWMR j are similar to the stability analyzes performed for the leader DWMR i. Further details on control laws and stability analyzes can be found for the FOSMC-B, FOQSMC-B, and AFSMC-A in Chaps. 5–7 respectively.

9.3.3 Formation Stability Analysis

To carry out the formation stability analysis [16], the following assumptions are necessary.

1. The dynamics of the follower DWMRs j and the leader DWMR i are fully known.
2. Each follower DWMR j has full knowledge of the dynamics of their leader DWMR i.
3. The follower DWMR j is equipped with sensors capable of measuring the separation distance δ_{ij} and bearing angle ψ_{ij} related to the leader DWMR i, and both are equipped with instrumentation to measure their linear and angular velocities as well as their orientations θ_i and θ_j.
4. There is wireless communication between the follower DWMR j and the leader DWMR i without communication delays.
5. The leader DWMR i communicates their linear and angular velocities υ_i, ω_i as well as their orientation θ_i and control torque $\boldsymbol{\tau}_i$, Eq. 2.12, to their follower DWMRs j at each sampling time.
6. For the nonholonomic system, Eqs. 9.1–9.3, with n generalized coordinates q, m independent constraints, and r actuators, the number of actuators is equal to the number of degrees of freedom $r = n - m$.
7. The linear and angular reference velocities measured from the leader DWMR i are bounded and $\upsilon_{rj}(t) \geq 0$ for every t.
8. The gains $\lambda_1, \lambda_2, \lambda_3, g_1, g_2, \kappa_1, \kappa_2$, and the parameter ϵ are positive constants used in the kinematic controls, Eqs. 9.15 and 9.41 (FOSMC-B), Eqs. 9.18 and 9.42 (FOQSMC-B), Eqs. 9.19, 9.20, 9.43 and 9.44 (AFSMC-A), and the gains $\mathrm{k_{p_\upsilon}}$, $\mathrm{k_{p_\omega}}$, $\mathrm{k_{d_\upsilon}}$, and $\mathrm{k_{d_\omega}}$ are positive constants used in the dynamic control, Eq. 2.12 (PD Control), both for the leader DWMR i and the follower DWMRs j.
9. The perfect velocity tracking hold such that $\boldsymbol{v}_i = \boldsymbol{v}_{ci}$, $\boldsymbol{v}_j = \boldsymbol{v}_{cj}$, $\dot{\boldsymbol{v}}_i = \dot{\boldsymbol{v}}_{ci}$ and $\dot{\boldsymbol{v}}_j = \dot{\boldsymbol{v}}_{cj}$ for the kinematic controls only, but it is relaxed with the integration of the kinematic controls to dynamic control.
10. The formation leader DWMR i does not follow any physical DWMR but follows the virtual reference leader DWMR described in Eqs. 2.16 and 2.17.

11. The formation leader DWMR i is capable to measure their absolute position via instrumentation as a global positioning system so that it is possible to track the virtual reference DWMR.
12. The linear reference velocity υ_{ri} is greater than zero and bounded and the angular reference velocity $\omega_r i$ is bounded for all time t.

The formation stability can be analyzed by using the individual Lyapunov functions as well as their derivatives with their sufficient conditions being required and met, considering the closed-loop control systems, for both the leader DWMR i and the follower DWMR j, i.e.,

- FOSMC-B

$$V = \frac{1}{2}\boldsymbol{\sigma}^{\mathrm{T}}\boldsymbol{\sigma}, \tag{9.45}$$

$$\dot{V} \leq -\boldsymbol{\sigma}^{*^{\mathrm{T}}}\mathbf{K}\,\boldsymbol{\sigma}^* - (\lambda_{\min}\{\mathbf{G}\} - \overline{\psi}_{\max})|\boldsymbol{\sigma}^*|_1 \leq 0, \tag{9.46}$$

- FOQSMC-B

$$V = \frac{1}{2}\boldsymbol{\sigma}^{\mathrm{T}}\boldsymbol{\sigma}, \tag{9.47}$$

$$\dot{V} \leq -\boldsymbol{\sigma}^{*^{\mathrm{T}}}\mathbf{K}\boldsymbol{\sigma}^* - (\mathbf{G}_{\sigma^*} - \overline{\psi}_{\max})|\boldsymbol{\sigma}^*|_1 \leq 0, \tag{9.48}$$

- AFSMC-A

$$V = \frac{1}{2}\left(\boldsymbol{\sigma}^{\mathrm{T}}\boldsymbol{\sigma} + \sum_{m=1}^{\mathrm{n}}\tilde{\boldsymbol{\beta}}_m^{\mathrm{T}}\tilde{\boldsymbol{\beta}}_m\right) = \frac{1}{2}\left(\boldsymbol{\sigma}^{\mathrm{T}}\boldsymbol{\sigma} + \tilde{\boldsymbol{\beta}}^{\mathrm{T}}\tilde{\boldsymbol{\beta}}\right), \tag{9.49}$$

$$\dot{V} \leq -\boldsymbol{\sigma}^{*^{\mathrm{T}}}(\mathbf{K} - \Psi)\boldsymbol{\sigma}^* \leq \left|\boldsymbol{\sigma}^*\right|^2(-\lambda_{\min}\{\mathbf{K}\} + \overline{\psi}_{\max}) \leq 0, \tag{9.50}$$

- PD control

$$V = \frac{1}{2}\boldsymbol{v}_{\mathrm{e}}^{\mathrm{T}}\bar{\boldsymbol{H}}(\boldsymbol{q})\boldsymbol{v}_{\mathrm{e}} + \frac{1}{2}\boldsymbol{v}_{\mathrm{e}}^{\mathrm{T}}\mathbf{K}_{\mathrm{d}}\boldsymbol{v}_{\mathrm{e}}. \tag{9.51}$$

$$\dot{V} \leq -||\boldsymbol{v}_{\mathrm{e}}||\Big(\lambda_{\min}\left\{\mathbf{K}_{\mathrm{p}}\right\}||\boldsymbol{v}_{\mathrm{e}}|| - ||\bar{\boldsymbol{\varphi}}^{\mathrm{T}}\bar{\boldsymbol{\gamma}}^{\mathrm{T}}||\Big). \tag{9.52}$$

For a formation of h + 1 DWMRs composed of a leader DWMR i and h followers DWMRs j, the following considerations must be taken into account:

- The assumptions 1–12 be maintained;
- Gains $\mathrm{k_{p_{\upsilon_i}}}$, $\mathrm{k_{p_{\omega_i}}}$, $\mathrm{k_{d_{\upsilon_i}}}$, $\mathrm{k_{d_{\omega_i}}}$ of the control torque $\boldsymbol{\tau}_i$, Eq. 2.12, be sufficiently large positive constants for the leader DWMR i;
- Gains $\mathrm{k_{p_{\upsilon_j}}}$, $\mathrm{k_{p_{\omega_j}}}$, $\mathrm{k_{d_{\upsilon_j}}}$, $\mathrm{k_{d_{\omega_j}}}$ of the control torque $\boldsymbol{\tau}_j$, Eq. 2.12, be sufficiently large positive constants for the follower DWMR j;
- A smooth control velocity $\boldsymbol{v}_{ci}$, Eqs. 9.15 (FOSMC-B), 9.18 (FOQSMC-B), 9.19 and 9.20 (AFSMC-A), and a control torque $\boldsymbol{\tau}_i$, Eq. 2.12, be applied to the leader DWMR i under the incidence of uncertainties and disturbances;
- A smooth control velocity $\boldsymbol{v}_{cj}$, Eqs. 9.41 (FOSMC-B), 9.42 (FOQSMC-B), 9.43 and 9.44 (AFSMC-A), and a control torque $\boldsymbol{\tau}_j$, Eq. 2.12, be applied to the follower DWMR j under incidence uncertainties and disturbances.

With basis in these considerations the origin $\boldsymbol{q}_{eij} = [\boldsymbol{q}_{ei}^{\mathrm{T}}\ \boldsymbol{v}_{ei}^{\mathrm{T}}\ \boldsymbol{q}_{ej}^{\mathrm{T}}\ \boldsymbol{v}_{ej}^{\mathrm{T}}] = 0$ where $\boldsymbol{q}_{eij} \in \Re^{(n+r)(1+h)\times 1}$ represents augmented posture and auxiliary velocity tracking errors for the leader DWMR i and h followers DWMRs j, respectively is asymptotically stable. Emphasizing that $\boldsymbol{v}_{ei}$ and $\boldsymbol{v}_{ej}$ are the auxiliary velocity tracking errors for the leader DWMR i and follower DWMR j respectively, which is similar to Eq. 2.3.

Consider the Lyapunov function candidate for the formation control as:

$$V_{ij} = (V_i + V_i^{'}) + \sum_{j=1}^{\mathrm{h}} (V_j + V_j^{'}) \tag{9.53}$$

where V_i and V_j are defined by Eqs. 9.45 (FOSMC-B), 9.47 (FOQSMC-B) and 9.49 (AFSMC-A) while $V_i^{'}$ and $V_j^{'}$ are defined by Eq. 9.51 (PD control).

The proofs that $V_i \geq 0$ and $V_j \geq 0$ for all j in h for the kinematic controls are demonstrated in Sects. 5.2.2, 6.2.2 and 7.3.2, and $V_i^{'} \geq 0$ and $V_j^{'} \geq 0$ for all j in h for the dynamic control are verified by Sect. 2.2. Therefore, it can concluded that V_{ij}, Eq. 9.53, is positive definite, i.e., $V_{ij} \geq 0$. Moreover, $V_{ij} < 0$ for $\boldsymbol{\sigma} \neq 0, \boldsymbol{\sigma}^* \neq 0$, $\boldsymbol{q}_{eij} \neq 0$ and $\tilde{\boldsymbol{\beta}}_m \neq 0$ or $\tilde{\boldsymbol{\beta}} \neq 0$ in the AFSMC-A (Eq. 9.49). Also, $V_{ij} = 0$ only when $\boldsymbol{\sigma} = 0, \boldsymbol{\sigma}^* = 0, \boldsymbol{q}_{eij} = 0$ and $\tilde{\boldsymbol{\beta}}_m = 0$ or $\tilde{\boldsymbol{\beta}} \neq 0$ in the AFSMC-A (Eq. 9.49).

Differentiating V_{ij}, Eq. 9.53, yields:

$$\dot{V}_{ij} = (\dot{V}_i + \dot{V}_i^{'}) + \sum_{j=1}^{\mathrm{h}} (\dot{V}_j + \dot{V}_j^{'}), \tag{9.54}$$

with $\dot{V}_i$ and $\dot{V}_j$ being determined by Eqs. 9.46 (FOSMC-B), 9.48 (FOQSMC-B) and 9.50 (AFSMC-A) while $\dot{V}_i^{'}$ and $\dot{V}_j^{'}$ being determined by Eq. 9.52 (PD control). It is also emphasized that all the derivatives are negative definite, $\dot{V}_i \geq 0$ and $\dot{V}_j \geq 0$ for all j in h for the kinematic controls, $\dot{V}_i^{'} \geq 0$ and $\dot{V}_j^{'} \geq 0$ for all j in h for the dynamic control, as can be seen in Sects. 5.2.2, 6.2.2, 7.3.2 and 2.2. Therefore, satisfying the required conditions, then $\dot{V}_{ij}$ is negative definite, i.e., $\dot{V}_{ij} \leq 0$. In addition, $\dot{V}_{ij} < 0$ for $\boldsymbol{\sigma} \neq 0$, $\boldsymbol{\sigma}^* \neq 0$ and $\boldsymbol{q}_{eij} \neq 0$ while that $\dot{V}_{ij} = 0$ only when $\boldsymbol{\sigma} = 0$, $\boldsymbol{\sigma}^* = 0$ and, consequently, $\boldsymbol{q}_{eij} = 0$, which is an asymptotically stable equilibrium point.

Regarding the estimates of the consequences, $\hat{\boldsymbol{\beta}}$, or equivalently $\tilde{\boldsymbol{\beta}}$ in AFSMC-A, these remain bounded.

As in [17], it is important to emphasize that the asymptotic stability of a formation for the case when the follower DWMR j becomes a leader DWMR to follower DWMR *j+1* can be directly demonstrated similarly considering the following positive definite Lyapunov function candidate:

$$V_{jj} = \sum_{j}^{j+1} (V_j + V_j^{'}), \tag{9.55}$$

with V_j being determined by Eqs. 9.45 (FOSMC-B), 9.47 (FOQSMC-B) and 9.49 (AFSMC-A) while $V_j^{'}$ is determined by Eq. 9.51 (PD control). In this case, follower DWMR j becomes the reference for follower DWMR *j+1*, and thus, the dynamics of follower DWMR j must be considered by follower DWMR *j+1*. The dynamics of follower DWMR j incorporates the dynamics of leader DWMR i; thus, follower DWMR *j+1* inherently brings in the dynamics of leader DWMR i by considering the dynamics of follower DWMR j.

9.4 Simulations Using Matlab and/or MobileSim Simulator

In this section, the FOSMC-B, FOQSMC-B, and AFSMC-A are named FCFOSMC, FCFOQSMC, and FCAFSMC, which are simulated, and the results obtained are shown for the ideal and realistic scenarios using the gains defined in Table 9.1.

The eight-shape trajectory as reference trajectory is modified with the linear velocity varying between 0.3 and 1.03 m/s and the angular velocity varying between −0.19 and 0.19 rad/s along the trajectory. With this modification, the formulation of this trajectory is then described by Eq. 9.56 and Fig. 9.2 shows the reference trajectory and velocities, whose execution time is now 200 s. Besides, the initial posture condition of the reference trajectory is $\boldsymbol{q}_{\mathrm{r}_0} = [0.0 \text{ m}, \ 0.0 \text{ m}, \ 0.46 \text{ rad}]^{\mathrm{T}}$ and the initial posture conditions of the leader DWMR and their two followers DWMRs are provided in Table 9.2.

$$\dot{\boldsymbol{q}}_{\mathrm{r}} = \begin{bmatrix} \dot{x}_{\mathrm{r}} \\ \dot{y}_{\mathrm{r}} \\ \dot{\theta}_{\mathrm{r}} \end{bmatrix} = \begin{bmatrix} \dfrac{30 * \sin\left(\left(t + \frac{200}{4}\right)\frac{2\pi}{200}\right)\frac{2\pi}{200}}{2} \\ -30 * \cos\left(2\left(t + \frac{200}{4}\right)\frac{2\pi}{200}\right)\frac{2\pi}{200} \\ \dfrac{\ddot{y}_{\mathrm{r}}\dot{x}_{\mathrm{r}} - \ddot{x}_{\mathrm{r}}\dot{y}_{\mathrm{r}}}{\dot{x}_{\mathrm{r}}^2 + \dot{y}_{\mathrm{r}}^2} \end{bmatrix}, \quad \boldsymbol{v}_{\mathrm{r}} = \begin{bmatrix} \upsilon_{\mathrm{r}} \\ \omega_{\mathrm{r}} \end{bmatrix} = \begin{bmatrix} \sqrt{\dot{x}_{\mathrm{r}}^2 + \dot{y}_{\mathrm{r}}^2} \\ \dot{\theta}_{\mathrm{r}} \end{bmatrix} \tag{9.56}$$

Table 9.1 Gains of the FCFOSMC, FCFOQSMC, and FCAFSMC in the leader-follower formation control strategy

Gain	DWMR		
	Leader	Follower 1	Follower 2
g_1	0.1	0.1	0.1
g_2	0.1	0.1	0.1
κ_1	0.1	0.1	0.1
κ_2	0.1	0.1	0.1
λ_1	2.5	2.5	2.5
λ_2	2.0	2.0	2.0
λ_3	1.4	1.4	1.4
k_{p_υ}	40.0	40.0	40.0
k_{p_ω}	40.0	40.0	40.0
k_{d_υ}	20.0	20.0	20.0
k_{d_ω}	20.0	20.0	20.0
N_υ	1.0	1.0	1.0
N_ω	1.0	1.0	1.0

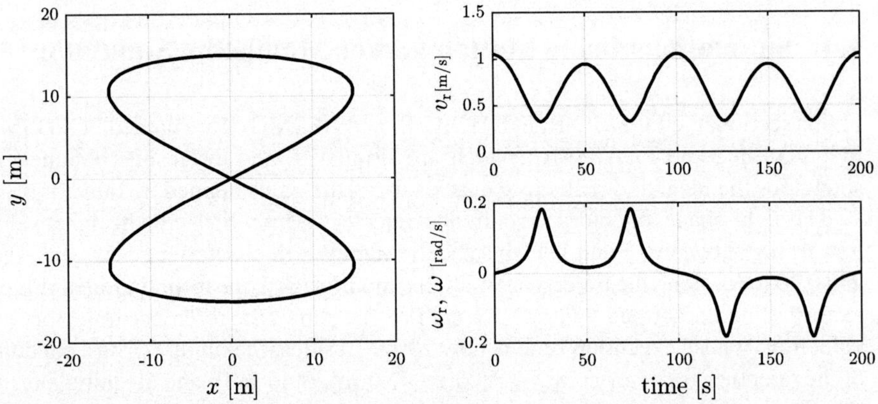

Fig. 9.2 Reference eight-shape trajectory and velocities

Table 9.2 Initial posture conditions of leader DWMR, and follower DWMRs

DWMR	$x(0)$ (m)	$y(0)$ (m)	$\theta(0)$ (rad)
Leader	−1.0	1.0	0.0
Follower 1	−3.5	3.0	0.0
Follower 2	−3.0	−1.0	0.0

Table 9.3 RMS of the errors—Simulation results in the ideal scenario for the formation control

DWMR	Error	FCFOSMC	FCFOQSMC	FCAFSMC
Leader	x_e	0.0656	0.0657	0.0352
	y_e	0.0519	0.0519	0.0461
	θ_e	0.0205	0.0214	0.0211
Follower 1	x_e	0.0874	0.0875	0.0548
	y_e	0.0381	0.0381	0.0372
	θ_e	0.0263	0.0262	0.0148
Follower 2	x_e	0.0580	0.0581	0.0361
	y_e	0.0403	0.0404	0.0360
	θ_e	0.0147	0.0141	0.0251

9.4.1 Ideal Scenario

In the leader-follower formation control strategy, Table 9.3 shows the RMS of the errors for the leader DWMR and follower DWMRs using the FCFOSMC, FCFO-QSMC, and FCAFSMC, in which it is possible to check the satisfactory performance, both in trajectory tracking with minimum errors and in the desired formation.

For the FCFOSMC, FCFOQSMC, and FCAFSMC, Fig. 9.3 shows how the leader DWMR reaches and tracks the reference trajectory and remains on it as well as the follower DWMRs reach and remain in the desired formation, for all subsequent time.

In Figs. 9.4 and 9.5 respectively, it is observed how the posture tracking errors of all DWMRs as well as separation and barrier errors of the follower DWMRs tend to zero and stay there for the remainder of the experiment.

The control velocities generated by the FCFOQSMC and FCAFSMC have smooth control signals while the control velocities provided by FCFOSMC have a significant occurrence of the chattering phenomenon, as illustrated in Fig. 9.6. Concerning the velocities of the DWMRs, it can be observed in Fig. 9.7 that these velocities track the reference velocities, again highlighting the significant occurrence of the chattering phenomenon promoted by the FCFOSMC.

In Figs. 9.8 and 9.9, it can be verified that the sliding surfaces, the new sliding surfaces, and the respective derivatives tend to converge to zero with the chattering phenomenon was considerably mitigated for both FCFOQSMC and FCAFSMC. However, for FCFOSMC, although all of its surfaces are with the chattering phenomenon was considerably attenuated and tend to converge to zero, their derivatives have a significant occurrence of the chattering phenomenon even though they tend to converge to zero, as can be seen in Figs. 9.10, and 9.11.

Figures 9.12, 9.13, and 9.14 show how the FCFOSMC, FCFOQSMC, and FCAF-SMC offer robustness to uncertainties and disturbances, whose opposite behaviors of the auxiliary velocity tracking errors and the compensations provided by the control velocities $\boldsymbol{v}_c^*$ try to cancel themselves. As expected, the control velocities $\boldsymbol{v}_c^*$ promoted by FCFOSMC present the occurrence of the chattering phenomenon.

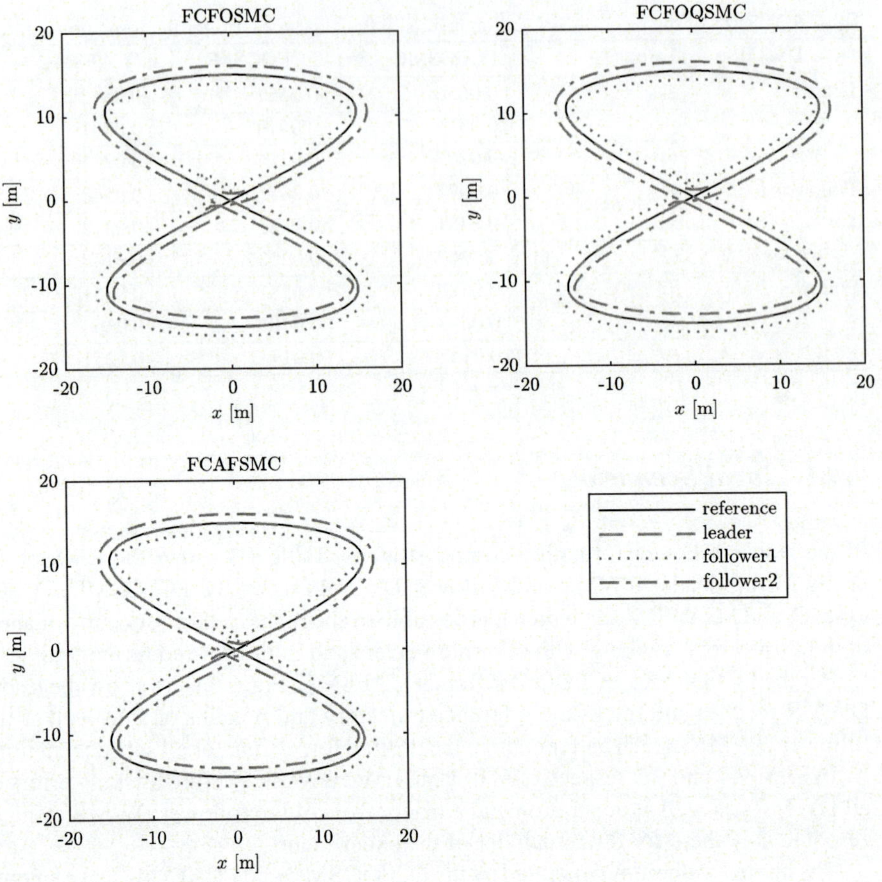

Fig. 9.3 Trajectory tracking in the ideal scenario

The estimated consequences of the fuzzy rules $\hat{\boldsymbol{\beta}}$ obtained from the adaptation laws (Eqs. 9.20 and 9.44) of the FCAFSMC for leader DWMR and the follower DWMRs, shown in Fig. 9.15, contribute to the compensation portions $\boldsymbol{v}_{\mathrm{c}}^{*} = \hat{\boldsymbol{F}}(\boldsymbol{\sigma}^{*})$ of the control laws in Eqs. 9.19 and 9.43 to attempt to cancel auxiliary velocity tracking errors.

9.4.2 Realistic Scenario

In this scenario, FCFOSMC, FCFOQSMC, and FCAFSMC are simulated in the MobileSim simulator considering the uncertainties and disturbances (e.g., modeling imprecisions; surface friction; external disturbances; bounded unknown disturbances; unmodeled and unstructured dynamics of actuators, sensors, and other

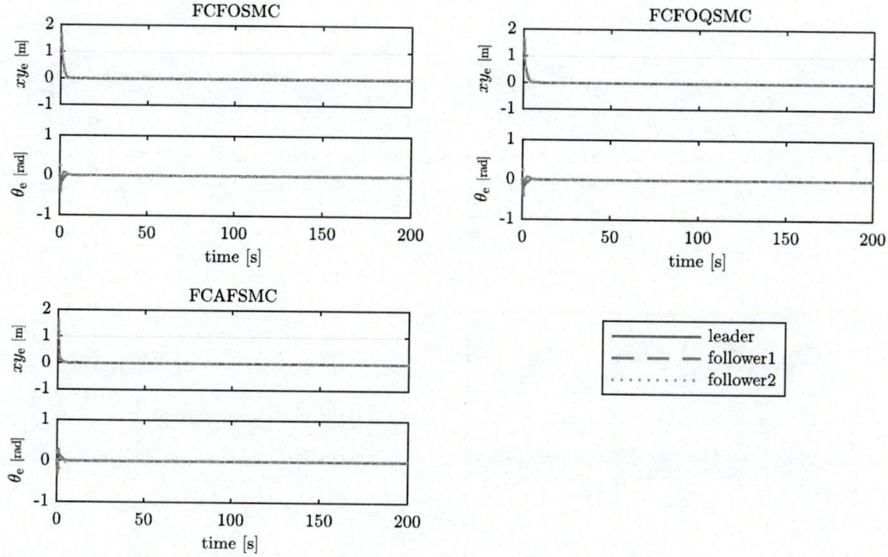

Fig. 9.4 Posture tracking errors, with $xy_e = \sqrt{x_e^2 + y_e^2}$, in the ideal scenario

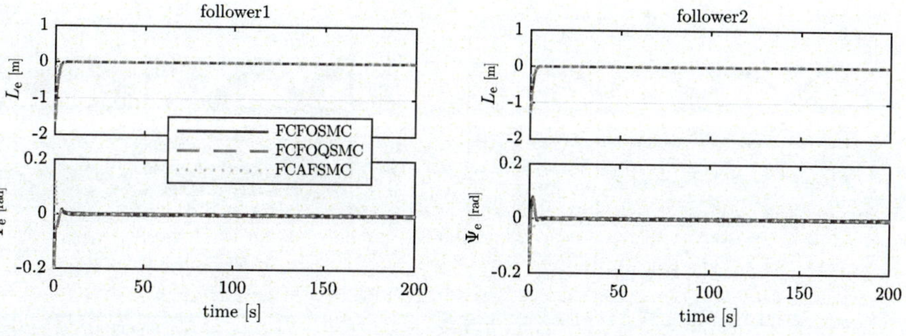

Fig. 9.5 Separation and barrier errors for the Follower DWMRs in the ideal scenario

electronic devices; integration errors; noises; operational, sampling and physical limitations) of the PowerBot DWMR.

Even under the influence of those uncertainties and disturbances in the tracking trajectory, it is possible to verify that the values of RMS of the errors in Table 9.4 remained close to the values of the RMS of the errors in Table 9.3 of the ideal scenario.

Observing and analyzing the results obtained, the same statements as in Sect. 9.4.1 can be considered here concerning trajectory tracking and the desired formation (Fig. 9.16), the posture tracking errors (Fig. 9.17), the separation and barrier errors (Fig. 9.18), the control velocity profiles (Fig. 9.19), the velocities of the DWMRs (Fig. 9.20), the sliding surfaces (Fig. 9.21) and new sliding surfaces (Fig. 9.22), the

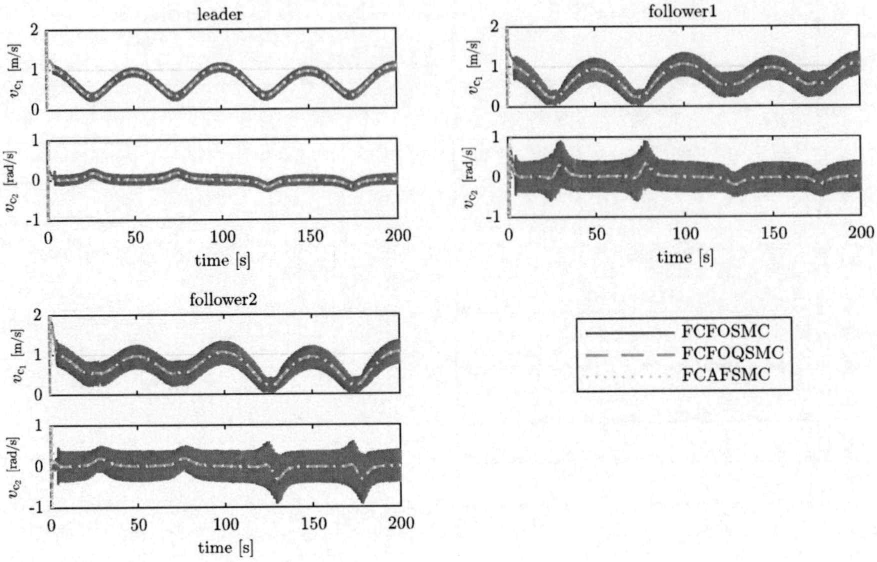

Fig. 9.6 Control velocities in the ideal scenario

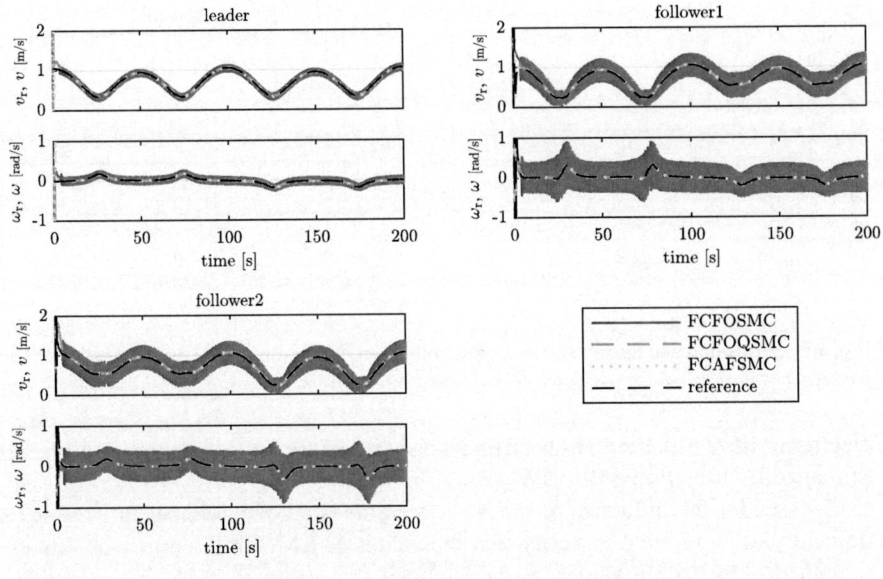

Fig. 9.7 Velocities of the DWMRs and reference velocities in the ideal scenario

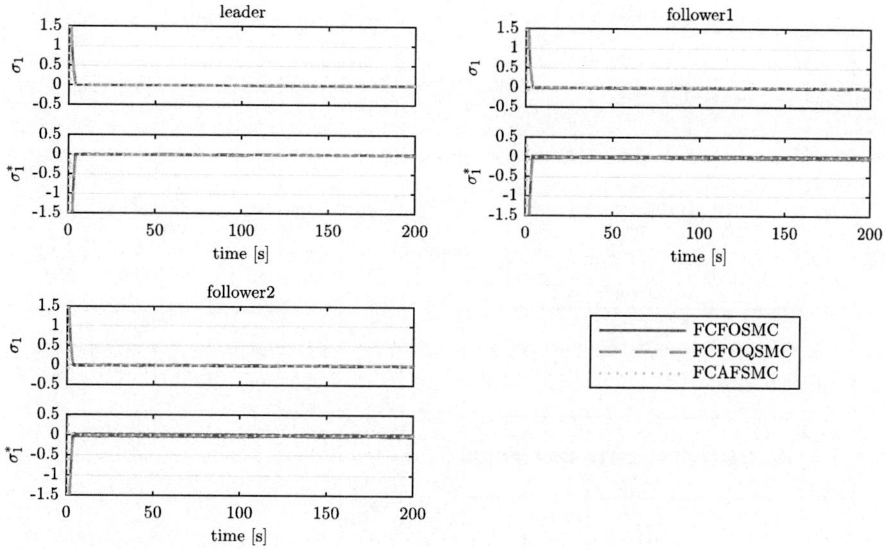

Fig. 9.8 Sliding surfaces σ_1 and new sliding surfaces σ_1^* in the ideal scenario

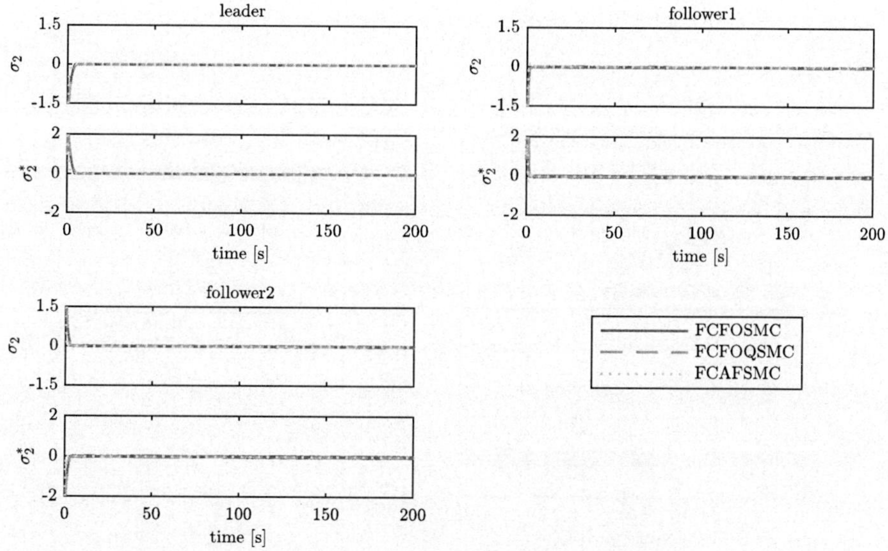

Fig. 9.9 Sliding surfaces σ_2 and new sliding surfaces σ_2^* in the ideal scenario

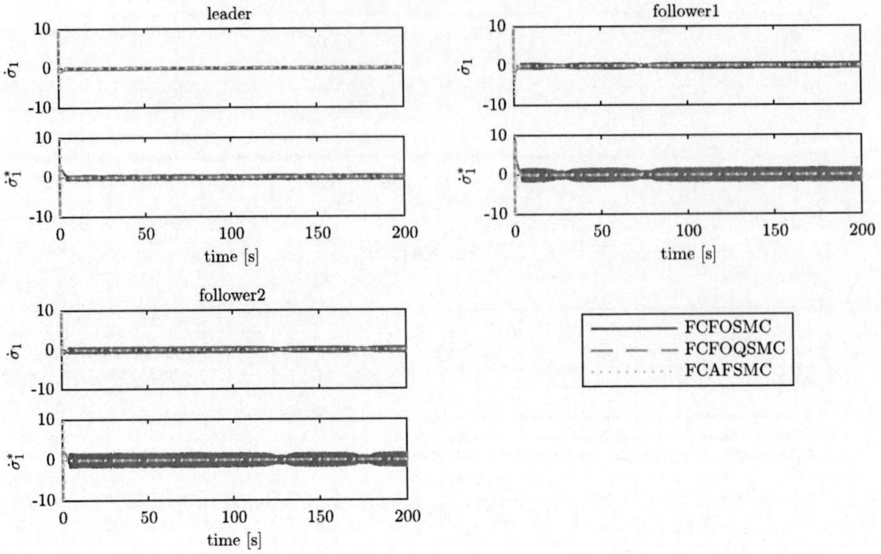

Fig. 9.10 Derivative of sliding surfaces $\dot{\sigma}_1$ and new sliding surfaces $\dot{\sigma}_1^*$ in the ideal scenario

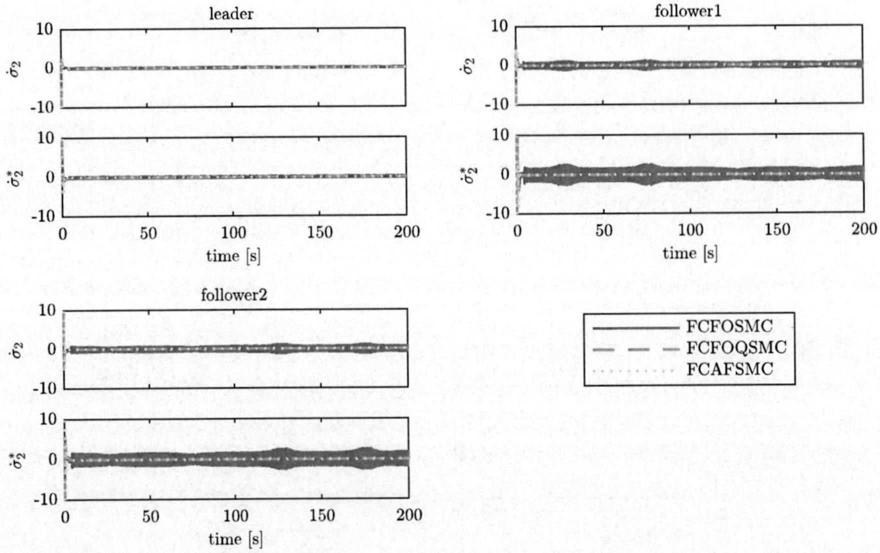

Fig. 9.11 Derivative of sliding surfaces $\dot{\sigma}_2$ and new sliding surfaces $\dot{\sigma}_2^*$ in the ideal scenario

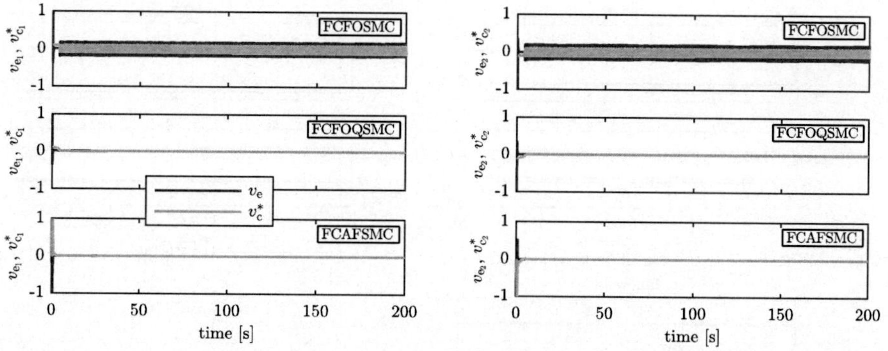

Fig. 9.12 Leader DWMR—Auxiliary velocity tracking errors and control velocities $\boldsymbol{v}_c^*$ in the ideal scenario

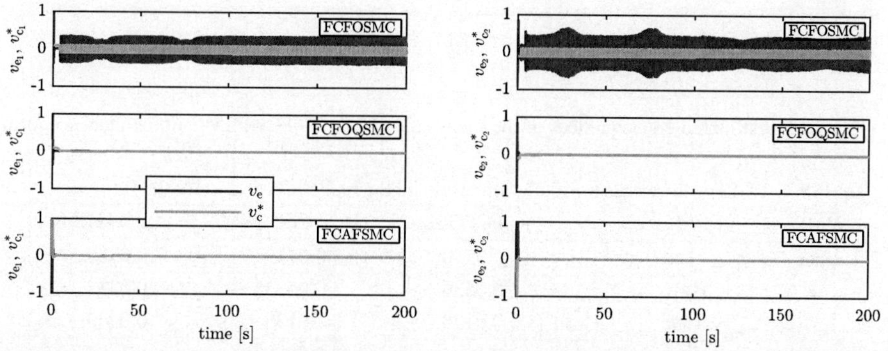

Fig. 9.13 Follower DWMR 1—Auxiliary velocity tracking errors and control velocities $\boldsymbol{v}_c^*$ in the ideal scenario

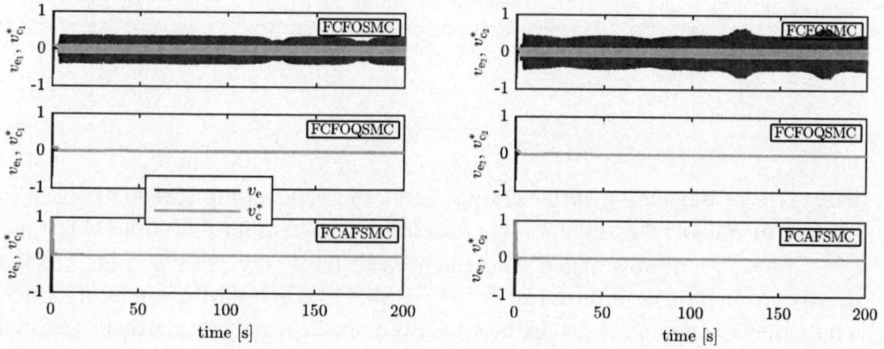

Fig. 9.14 Follower DWMR 2—Auxiliary velocity tracking errors and control velocities $\boldsymbol{v}_c^*$ in the ideal scenario

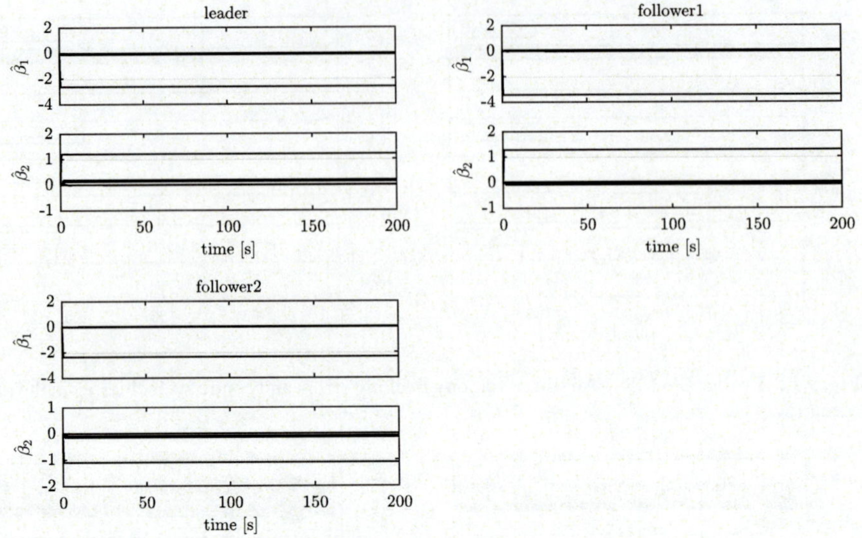

Fig. 9.15 Estimated consequences of the fuzzy rules $\hat{\boldsymbol{\beta}}$ for the FCAFSMC in the ideal scenario

Table 9.4 RMS of the errors—Simulation results in the realistic scenario for the formation control

DWMR	Error	FCFOSMC	FCFOQSMC	FCAFSMC
Leader	x_e	0.0750	0.0719	0.0580
	y_e	0.0536	0.0533	0.0532
	θ_e	0.0178	0.0135	0.0243
Follower 1	x_e	0.1005	0.0988	0.1135
	y_e	0.0327	0.0331	0.0340
	θ_e	0.0260	0.0221	0.0127
Follower 2	x_e	0.0549	0.0539	0.0661
	y_e	0.0481	0.0486	0.0447
	θ_e	0.0276	0.0153	0.0304

derivatives of the sliding surfaces (Fig. 9.23) and new sliding surfaces (Fig. 9.24), the control velocities $\boldsymbol{v}_c^*$ that provide the compensation control portions (Figs. 9.25, 9.26, and 9.27), the estimated consequences of the fuzzy rules $\hat{\boldsymbol{\beta}}$ (Fig. 9.28). As already ceaselessly described, also for FCFOSMC, FCFOQSMC, and FCAFSMC, it is important to emphasize that in these results there are oscillations around zero in the behavior of some variables caused by the influence of the chattering phenomenon. As can be seen, these oscillations are generally well attenuated to try the compensation of the uncertainties and disturbances considered in the MobileSim simulator.

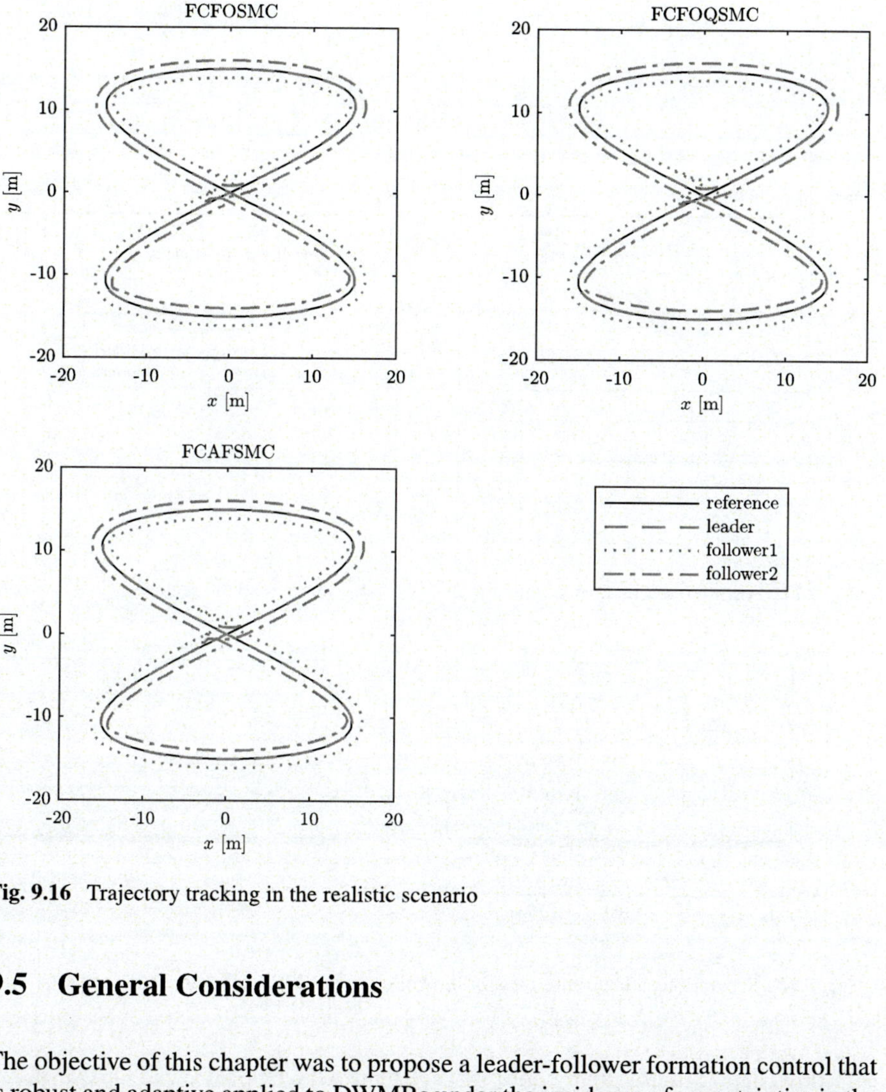

Fig. 9.16 Trajectory tracking in the realistic scenario

9.5 General Considerations

The objective of this chapter was to propose a leader-follower formation control that is robust and adaptive applied to DWMRs under the incidence of uncertainties in the resolution of the trajectory tracking control and desired formation problems. Such formation control is considered the FOSMC-B, however, this control has one of its disadvantages the chattering phenomenon due to the portion of discontinuous control. To mitigate the chattering phenomenon, the FOSQSMC-B was used, however, this control also requires knowledge of the lower and upper limits of uncertainties and disturbances in addition to the high values of the gain **G** to ensure robustness and satisfactory performance in the trajectory tracking and desired formation. To address the disadvantages of these two controls, the AFSMC-A was used, which online updates its outputs and contains a fuzzy logic inference system that is a feasible tool to approximate any real continuous nonlinear system to arbitrary accuracy [1–4].

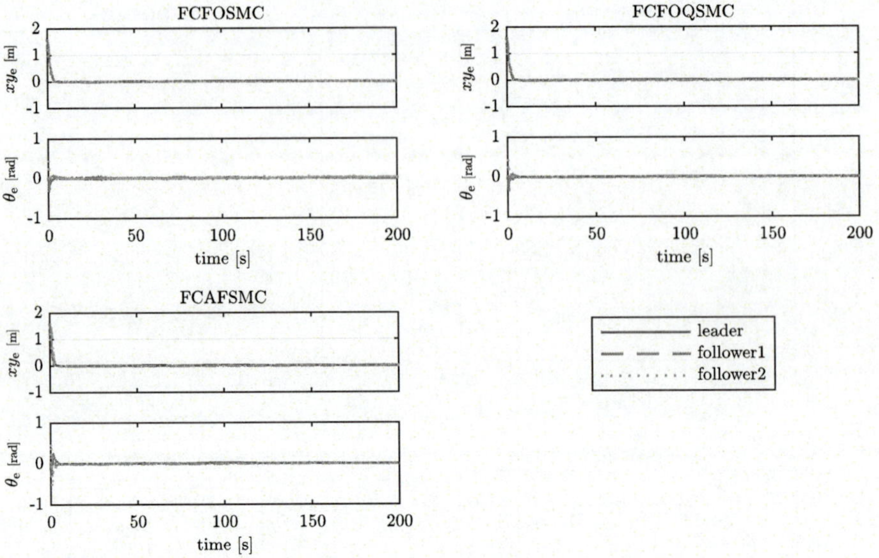

Fig. 9.17 Posture tracking errors, with $xy_e = \sqrt{x_e^2 + y_e^2}$, in the realistic scenario

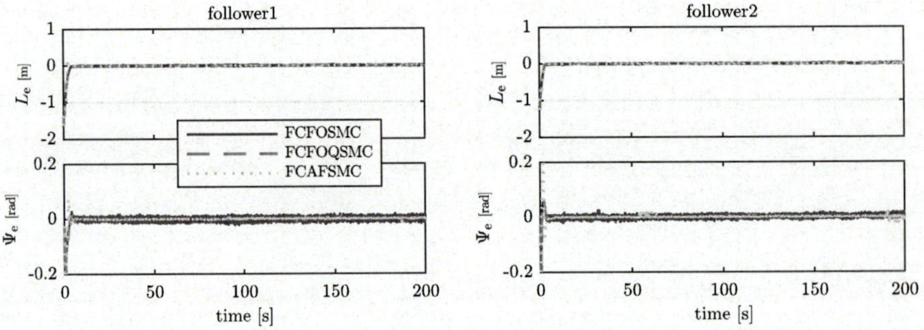

Fig. 9.18 Separation and barrier errors for the Follower DWMRs in the realistic scenario

In this chapter, these three controls were named FCFOSMC, FCFOSQSMC, and FCAFSMC respectively.

In the literature is emphasized that the meaning of losing the invariance principle can destroy the sliding mode [34, 35] and has little practical meaning [39], however, the FCFOSQSMC and FCAFSMC still ensure robustness acceptable to uncertainties and disturbances and generate smooth control efforts. Moreover, the simulations and real-time implementations have been verified that the invariance also has little practical meaning for both FCFOQSMC and FCAFSMC.

Research can be extended to carry out experimental tests of the FCAFSMC in real DWMRs; the generalization of the FCAFSMC for the other possible classes of DWMRs and other dynamic systems, for example, unmanned aerial vehicles

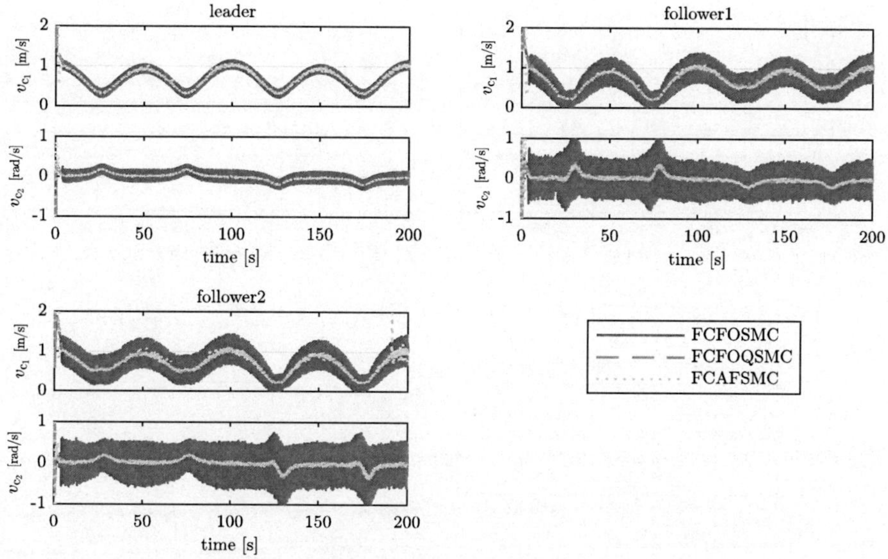

Fig. 9.19 Control velocities in the realistic scenario

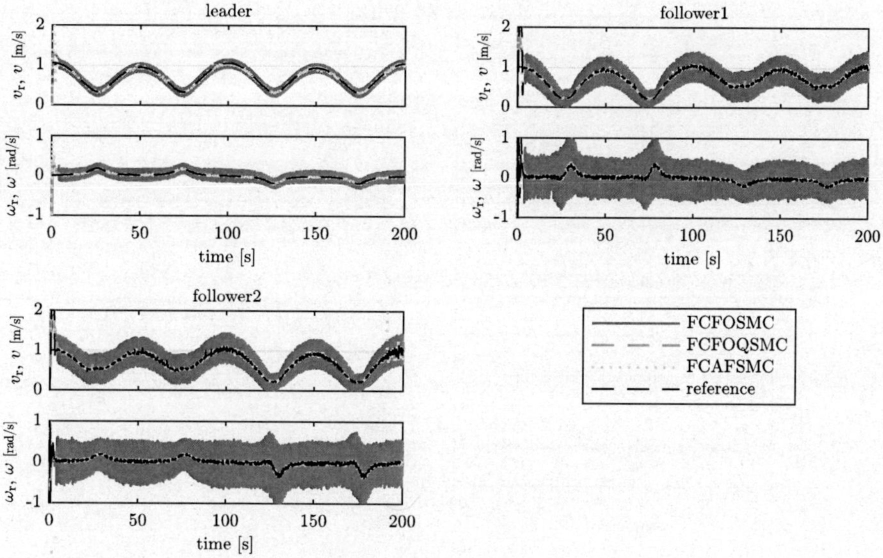

Fig. 9.20 Velocities of the DWMRs and reference velocities in the realistic scenario

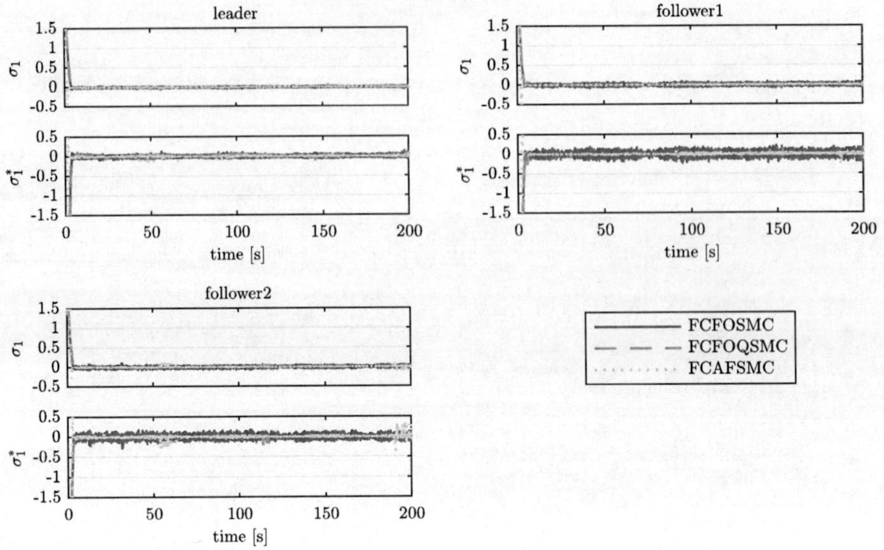

Fig. 9.21 Sliding surfaces σ_1 and new sliding surfaces σ_1^* in the realistic scenario

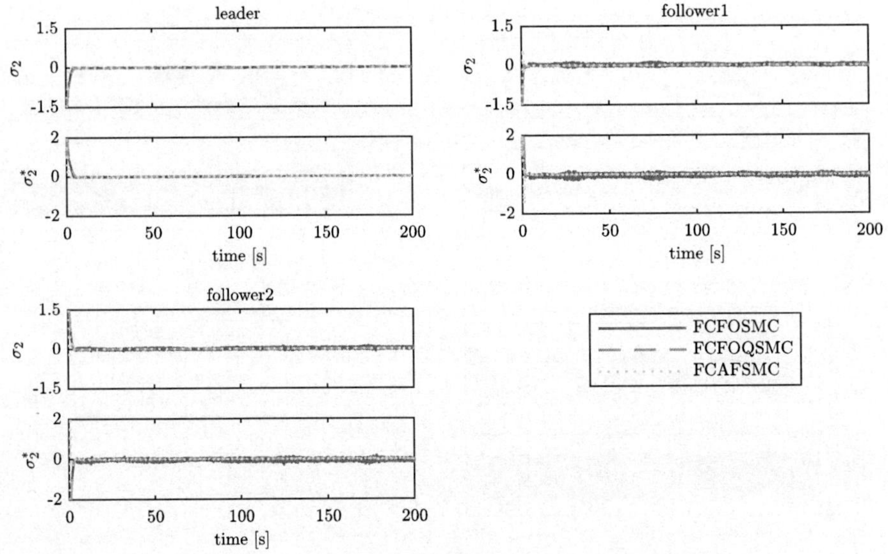

Fig. 9.22 Sliding surfaces σ_2 and new sliding surfaces σ_2^* in the realistic scenario

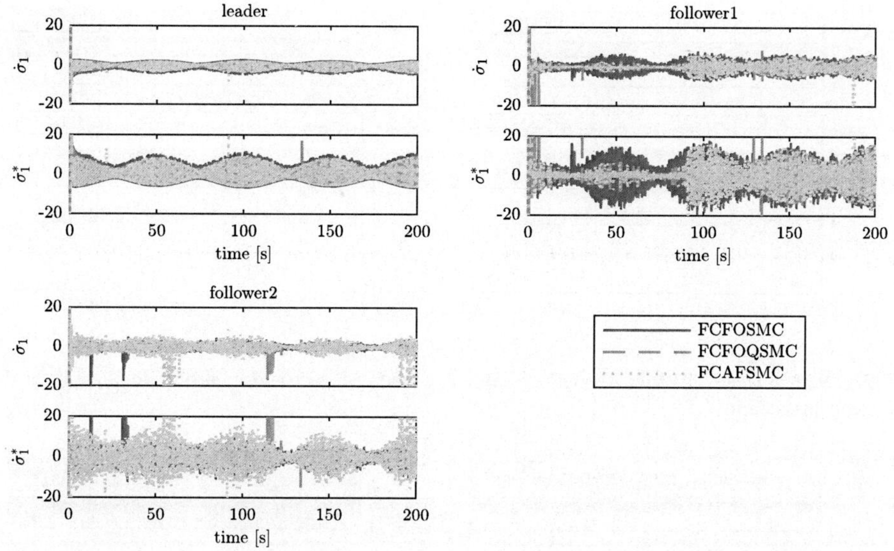

Fig. 9.23 Derivative of sliding surfaces $\dot{\sigma}_1$ and new sliding surfaces $\dot{\sigma}_1^*$ in the realistic scenario

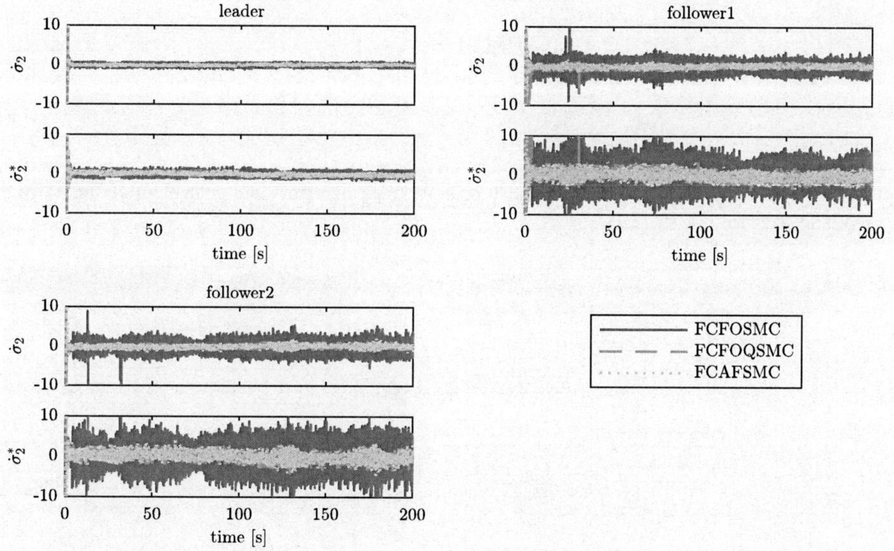

Fig. 9.24 Derivative of sliding surfaces $\dot{\sigma}_2$ and new sliding surfaces $\dot{\sigma}_2^*$ in the realistic scenario

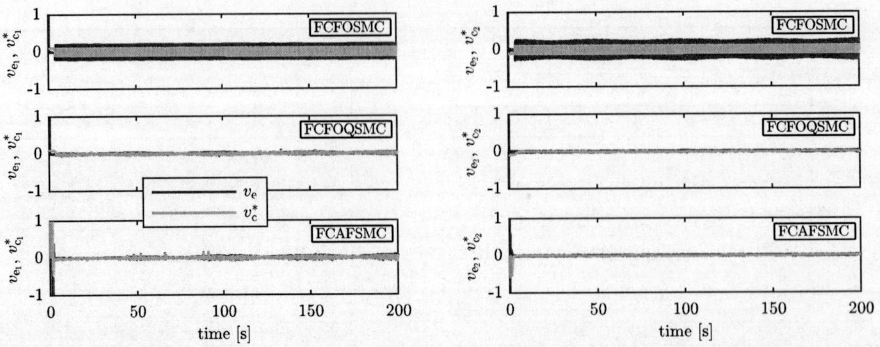

Fig. 9.25 Leader DWMR—Auxiliary velocity tracking errors and control velocities $\boldsymbol{v}_c^*$ in the realistic scenario

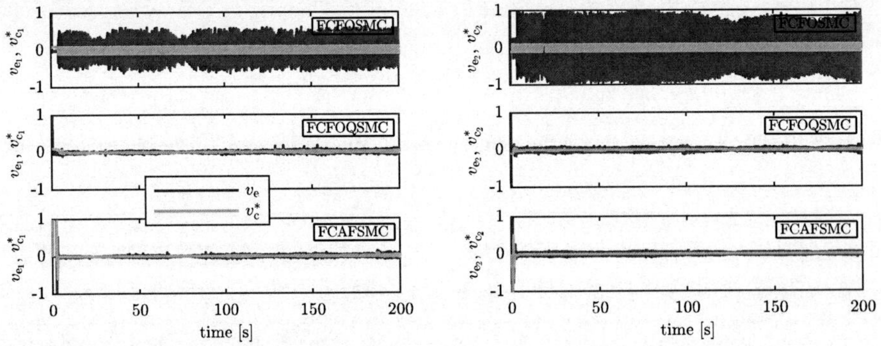

Fig. 9.26 Follower DWMR 1—Auxiliary velocity tracking errors and control velocities $\boldsymbol{v}_c^*$ in the realistic scenario

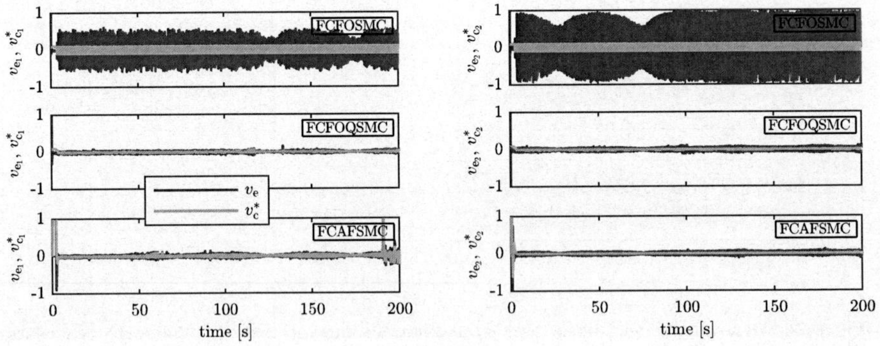

Fig. 9.27 Follower DWMR 2—Auxiliary velocity tracking errors and control velocities $\boldsymbol{v}_c^*$ in the realistic scenario

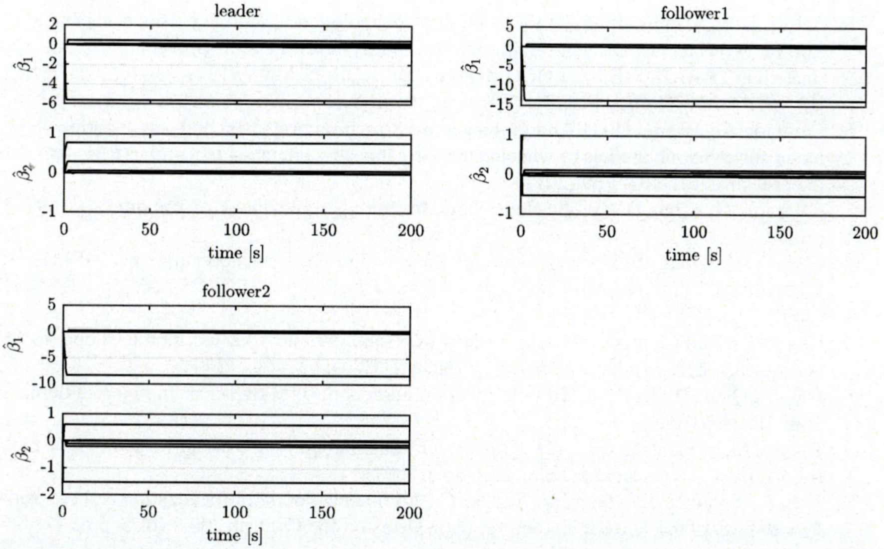

Fig. 9.28 Estimated consequences of the fuzzy rules $\hat{\boldsymbol{\beta}}$ for the FCAFSMC in the realistic scenario

(UAVs); determination of a methodology to adjust the parameters and gains of the FCAFSMC by using Design Optimization toolbox of the Matlab/Simulink software [23]; conducting a comparative study of FCFOSMC, FCFOSQSMC, and FCAFSMC with other kinematic leader-follower formation controllers in the literature, such as [10, 13, 24, 31, 41], taking into account mainly controllers that handle the approach to mitigate the chattering phenomenon using fuzzy logic; treatment of the static/dynamic obstacle avoidance problem in the formation control of DWMRs.

Finally, the design factors that can influence the achievement of different trajectory tracking and formation performances are the same considered in Sect. 7.7.

References

1. Begnini, M., Bertol, D., Martins, N.: Practical implementation of a simple and effective robust adaptive fuzzy variable structure trajectory tracking control for differential wheeled mobile robots. Int. J. Innov. Comput. Inf. Control: IJICIC **13**, 341–364 (2017)
2. Begnini, M., Bertol, D., Martins, N.: A robust adaptive fuzzy variable structure tracking control for the wheeled mobile robot: simulation and experimental results. Control Eng Practice **64**, 27–43 (2017)
3. Begnini, M., Bertol, D., Martins, N.: Design of an adaptive fuzzy variable structure compensator for the nonholonomic mobile robot in trajectory tracking task. Control Cybern. **47**, 239–275 (2018)
4. Begnini, M., Bertol, D., Martins, N.: Practical implementation of an effective robust adaptive fuzzy variable structure tracking control for a wheeled mobile robot. J. Intell. Fuzzy Syst. **35**, 1087–1101 (2018)

5. Bertol, D.W.: Contributions to locomotion of nonholonomic mobile robots using fuzzy control. Ph.D thesis. Department of Automation and Systems, Federal University of Santa Catarina, CONTRIBUIÇÕES À LOCOMOÇÃO DE ROBÔS MÓVEIS NÃO-HOLONÔMICOS USANDO CONTROLE FUZZY (2015) (in Portuguese)
6. Campion, G., Bastin, G., D'Andréa-Novel, B.: Structural properties and classification on kinematic and dynamic models of wheeled mobile robots. Nelineinaya Dinamika (Russian J. Nonlinear Dyn.) **7**(4), 733–769 (2011)
7. Campion, G., Chung, W.: Wheeled robots. In: Springer Handbook of Robotics, pp. 391–410 (2008)
8. Chen, Y.-Q., Wang, Z.: Formation control: a review and a new consideration. In: 2005 IEEE/RSJ International Conference on Intelligent Robots and Systems, 2005 (IROS 2005), pp. 3181–3186 (2005)
9. Chwa, D.: Sliding mode tracking control of nonholonomic wheeled mobile robots in polar coordinates. IEEE Trans. Control Syst. Technol. **12**(4), 637–644 (2004)
10. Dai, Y., Qian, D., Lee, S.: Multiple robots motion control to transport an object. Filomat **32**, 1547–1558 (2018)
11. Daigle, M.J., Koutsoukos, X.D., Biswas, G.: Distributed diagnosis in formations of mobile robots. IEEE Trans. Robot. **23**(2), 353–369 (2007)
12. Das, A., Spletzer, J., Kumar, V., Taylor, C.: Ad hoc networks for localization and control. In: Proceedings of the 41st IEEE Conference on Decision and Control, 2002, vol. 3, pp. 2978–2983 (2002)
13. Defoort, M., Floquet, T., Kokosy, A., Perruquetti, W.: Sliding-mode formation control for cooperative autonomous mobile robots. IEEE Trans. Ind. Electron. **55**(11), 3944–3953 (2008)
14. Desai, J.P., Ostrowski, J., Kumar, V.: Controlling formations of multiple mobile robots. In: Proceedings of 1998 IEEE International Conference on Robotics and Automation, 1998, vol. 4, pp. 2864–2869 (1998)
15. Desai, J.P., Ostrowski, J.P., Kumar, V.: Modeling and control of formations of nonholonomic mobile robots. IEEE Trans. Robot. Autom. **17**(6), 905–908 (2001)
16. Dierks, T., Jagannathan, S.: Control of nonholonomic mobile robot formations: backstepping kinematics into dynamics. IEEE International Conference on Control Applications, 2007. CCA 2007, pp. 94–99 (2007)
17. Dierks, T., Sarangapani, J.: Asymptotic adaptive neural network tracking control of nonholonomic mobile robot formations. J. Intell. Robotic Syst. **56**, 153–176 (2009)
18. Elyoussef, E.S., Martins, N.A., De Pieri, E.R., Moreno, U.F.: PD-super-twisting second order sliding mode tracking control for a nonholonomic wheeled mobile robot. In: Proceedings of the 19th World Congress of the International Federation of Automatic Control (IFAC World Congress), vol. 11, pp. 3827–3832 (2014)
19. Elyoussef, E.S., Martins, N.A., Bertol, D.W., de Pieri, E.R., Moreno, U.F.: Simulation results and practical implementation of a PD-super-twisting second order sliding mode tracking control for a differential wheeled mobile robot. Int. J. Comput. Appl. Technol. **63**(3), 213–227 (2020)
20. Ferrara, A., Rubagotti, M.: Second order sliding mode control of a mobile robot based on a harmonic potential field. IET Control Theory Appl. **2**(9), 807–818 (2008)
21. Fierro, R., Das, A.K., Kumar, V., Ostrowski, J.P.: Hybrid control of formations of robots. In: IEEE International Conference on Robotics and Automation, 2001. Proceedings 2001 ICRA, vol. 1, pp. 157–162 (2001)
22. Fierro, R., Lewis, F.L.: Control of a nonholonomic mobile robot using neural networks. IEEE Trans. Neural Netw. **9**(4), 589–600 (1998)
23. Freire, F., Martins, N., Splendor, F.: A simple optimization method for tuning the gains of PID controllers for the autopilot of Cessna 182 aircraft using model-in-the-loop platform. J. Control Autom. Electr. Syst. **29**, 441–450 (2018)
24. Gao, X., Huang, Q., Wan, M., Liu, C.: Sliding mode formation control of nonholonomic robots. In: 2010 International Conference on Artificial Intelligence and Computational Intelligence, vol. 1, pp. 67–71 (2010)

25. Kanjanawanishkul, K.: Formation control of mobile robots: survey. Eng. ubu. ac. th. 50–64 (2005)
26. Khanzadeh, A., Pourgholi, M.: Fixed-time leader–follower consensus tracking of second-order multi-agent systems with bounded input uncertainties using non-singular terminal sliding mode technique. IET Control Theory Appl. **12**, 679–686 (2018)
27. Martins, F.N., Sarcinelli-Filho, M., Bastos, T.F., Carelli, R.: Dynamic modeling and adaptive dynamic compensation for unicycle-like mobile robots. In: 2009 International Conference on Advanced Robotics, pp. 1–6, June 2009
28. Martins, N.A.., Alencar, M., Lombardi, W.C., Bertol, D.W., De Pieri, E.R., Ferasoli Filho, H.: Trajectory tracking of a wheeled mobile robot with uncertainties and disturbances: proposed adaptive neural control. Control and Cybern. **44**, 47–98 (2015)
29. Oriolo, G.: Wheeled robots. In: Encyclopedia of Systems and Control, pp. 1–9. Springer (2014)
30. Park, B.S., Yoo, S.J., Park, J.B., Choi, Y.H.: Adaptive neural sliding mode control of non-holonomic wheeled mobile robots with model uncertainty. IEEE Trans. Control Syst. Technol. **17**(1), 207–214 (2009)
31. Rashid, M., Yakub, F., Ali, M., Mamat, N., Putra, M., Roslan, S., Shah, H., Aras, M.: Comprehensive review on controller for leader-follower robotic system. Ind. J. Geo-Marine Sci. **48**, 985–1007 (2019)
32. Shao, J., Xie, G., Wang, L.: Leader-following formation control of multiple mobile vehicles. Control Theory Appl. IET **1**(2), 545–552 (2007)
33. Siciliano, B., Khatib, O.: In: Springer Handbook of Robotics. Springer (2008)
34. Solea, R., Cernega, D.: Modeling and performance evaluation for trajectory tracking control of a wheeled mobile robot. In: Proceedings of the 1st Workshop on Energy, Transport, and Environment Control Applications (ETECA'2009), pp. 1–7 (2009)
35. Solea, R., Cernega, D.: Sliding mode control for trajectory tracking problem—Performance evaluation. In: Proceedings of the 19th International Conference on Artificial Neural Networks (ICANN'2009), pp. 865–874. Springer (2009)
36. Spry, S., Hedrick, J.K.: Formation control using generalized coordinates. In: 43rd IEEE Conference on Decision and Control, 2004, CDC, vol. 3, pp. 2441–2446 (2004)
37. Sukhatme, G.S., Montgomery, J.F., Vaughan, R.T.: Experiments with cooperative aerial-ground robots. In: Robot Teams: From Diversity to Polymorphism, pp. 345–367. AK Peters (2001)
38. Toibero, JuanMarcos, Roberti, Flavio, Fiorini, Paolo, Carelli, Ricardo: Hybrid formation control for non-holonomic wheeled mobile robots. In: Lee, Sukhan, Suh, IlHong, Kim, MunSang (eds.) Recent Progress in Robotics: Viable Robotic Service to Human. Lecture Notes in Control and Information Sciences, vol. 370, pp. 21–34. Springer, Berlin, Heidelberg (2008)
39. Wang, S., Gao, W.: Robustness and invariance of variable structure systems with multiple inputs. In: Proceedings of the American Control Conference (ACC'1995), vol. 1, pp. 1035–1039. IEEE (1995)
40. Wang, Z.P., Ge, S.S., Lee, T.H.: Adaptive neural network control of a wheeled mobile robot violating the pure nonholonomic constraint. In: Proceedings of the 43rd IEEE Conference on Decision and Control (CDC'2004), vol. 5, pp. 5198–5203. IEEE (2004)
41. Zhao, Y., Park, D., Moon, J., Lee, J.: Leader-follower formation control for multiple mobile robots by a designed sliding mode controller based on kinematic control method. In: 2017 56th Annual Conference of the Society of Instrument and Control Engineers of Japan (SICE), pp. 186–189 (2017)